Advances in Plasma Processes for Polymers II

Advances in Plasma Processes for Polymers II

Editor

Choon-Sang Park

MDPI

Basel • Beijing • Wuhan • Barcelona • Belgrade • Novi Sad • Cluj • Manchester

Editor
Choon-Sang Park
Electrical Engineering
Milligan University
Johnson City
United States

Editorial Office
MDPI
St. Alban-Anlage 66
4052 Basel, Switzerland

This is a reprint of articles from the Special Issue published online in the open access journal *Polymers* (ISSN 2073-4360) (available at: www.mdpi.com/journal/polymers/special_issues/QP51AA2AB8).

For citation purposes, cite each article independently as indicated on the article page online and as indicated below:

Lastname, A.A.; Lastname, B.B. Article Title. *Journal Name* **Year**, *Volume Number*, Page Range.

ISBN 978-3-0365-8609-0 (Hbk)
ISBN 978-3-0365-8608-3 (PDF)
doi.org/10.3390/books978-3-0365-8608-3

Contents

Preface

Polymerized nanoparticles and nanofibers can be prepared using various processes, such as chemical synthesis, the electrochemical method, electrospinning, ultrasonic irradiation, hard and soft templates, seeding polymerization, interfacial polymerization, and plasma polymerization. Among these processes, plasma polymerization and aerosol-through-plasma (A-t-P) processes have versatile advantages, especially due to having a "dry" process, for the deposition of plasma polymer films and carbon-based materials with functional properties suitable for a wide range of applications, such as electronic and optical devices, protective coatings, and biomedical materials. Furthermore, it is well known that plasma polymers are highly cross-linked, pinhole-free, branched, insoluble, and adhere well to most substrates. In order to synthesize polymer films using plasma processes, therefore, it is very important to increase the density and electron temperature of plasma during plasma polymerization.

This Special Issue aimed to compile original and cutting-edge research works in the fields of plasma process, polymerization, synthesis, characterization, treatment, modification, manufacturing, and applications of functional plasma-processed polymers.

Choon-Sang Park
Editor

polymers

MDPI

Review

Strategies for Improved Wettability of Polyetheretherketone (PEEK) Polymers by Non-Equilibrium Plasma Treatment

Gregor Primc

Department of Surface Engineering, Jozef Stefan Institute, Jamova cesta 39, 1000 Ljubljana, Slovenia; gregor.primc@ijs.si

Abstract: Polyetheretherketone (PEEK) is the material of choice in several applications ranging from the automotive industry to medicine, but the surface properties are usually not adequate. A standard method for tailoring surface properties is the application of gaseous plasma. The surface finish depends enormously on the processing parameters. This article presents a review of strategies adapted for improved wettability and adhesion of PEEK. The kinetics of positively charged ions, neutral reactive plasma species, and vacuum ultraviolet radiation on the surface finish are analyzed, and synergies are stressed where appropriate. The reviewed articles are critically assessed regarding the plasma and surface kinetics, and the surface mechanisms are illustrated. The directions for obtaining optimal surface finish are provided together with the scientific explanation of the limitations of various approaches. Super-hydrophilic surface finish is achievable by treatment with a large dose of vacuum ultraviolet radiation in the presence of oxidizing gas. Bombardment with positively charged ions of kinetic energy between about 100 and 1000 eV also enable high wettability, but one should be aware of excessive heating when using the ions.

Keywords: non-equilibrium gaseous plasma; PEEK; functionalization; wettability

Citation: Primc, G. Strategies for Improved Wettability of Polyetheretherketone (PEEK) Polymers by Non-Equilibrium Plasma Treatment. *Polymers* **2022**, *14*, 5319. https://doi.org/10.3390/polym14235319

Academic Editor: Choonsang Park

Received: 2 November 2022
Accepted: 29 November 2022
Published: 5 December 2022

1. Introduction

Polyetheretherketone (PEEK) is a thermoplastic polymer with excellent mechanical and electrical properties and chemical stability even at temperatures up to 250 °C and above. These properties make PEEK an attractive material for application in extreme conditions and thus suitable as an insulation material in automotive and aerospace industries [1]. Furthermore, the properties also make PEEK promising for use in medicine, for example, as a material for dental implants [2,3]. As known for most polymers, the mechanical properties are further improved by enforcement with different fillers, often in the form of fibers. The excellent chemical stability, however, may be a drawback in numerous applications where PEEK should be applied as a coating or should be coated with another material.

The poor adhesion properties are due to the PEEK's chemical structure, which prevents interaction with most other materials. It is a moderately hydrophobic polymer with a static water contact angle (WCA) of about 90°. The interaction of all polymers with foreign material is improved by grafting various functional groups on its surface. The grafting may result in improved wettability, which is usually desired before coating with various materials, such as metallization, gluing, printing, or decorating with nanomaterials. Grafting with functional groups may not be suitable when PEEK in a liquid state is to adhere in the form of a thin film to another material. In such cases, the material coated with a thin PEEK film should be pre-treated to ensure improved adhesion, and thus, adequate mechanical properties of the product.

A review of the strategies for improving the adhesive properties of PEEK by the treatment with non-equilibrium gaseous plasma is the topic of this article. First, the literature on surface modification of solid PEEK by gaseous plasmas is reviewed and discussed, and second, the surface modifications using various approaches are illustrated. Lastly, the missing results are recognized, and future research is recommended.

2. Grafting of Functional Groups on the PEEK Surface and Morphological Changes

As early as 1998, Inagaki et al. [4] reported the surface functionalization of PEEK using oxygen plasma. Oxygen plasma is a natural choice for the surface functionalization of fluorine-free polymers [5]. The reactive species formed in oxygen upon plasma conditions interact chemically with the polymer surface, causing either a direct substitution of H atoms bonded to carbon with oxygen or bond breakage. The dangling bonds are occupied with oxygen atoms, thus forming a variety of oxygen-containing functional groups such as hydroxyl, epoxy, carbonyl, carboxyl, etc. [6]. Theoretically, these surface functional groups should ensure an increase in the polar component of the surface energy and, hence, an increase in the wettability. The increased wettability should ensure good coating adhesion on the PEEK's surface or adhesion between two PEEK samples. The exact mechanisms of interaction between oxygen plasma and polymer surface on the atomic scale are yet to be elaborated. Among recent spotless theories, Ventzek's group elaborated on the kinetics of surface functionalization of polystyrene [7], and the results were confirmed by a carefully designed experiment [8]. No suitable theory has been reported for PEEK, though. On the contrary, Inagaki et al. [4] reported a rather poor functionalization but significant modifications of the PEEK surface film. Namely, the oxygen plasma treatment caused the formation of low molecular weight fragments on the PEEK's surface. The fragments were easily washed away, and the resulting WCA observed after washing was not improved dramatically. Inagaki et al. used two configurations for plasma treatment, i.e., the glowing plasma, sustained by an electrodeless radiofrequency (RF) discharge in the capacitive mode, and the flowing afterglow. The difference between these two configurations is in the reactive species available for surface modification of a polymer. Afterglow consists of neutral oxygen atoms in the ground state and some metastable neutral molecules, whereas the glowing plasma also consists of charged particles and is a significant source of radiation [9]. The most important for polymer surface modification is radiation in the vacuum ultraviolet (VUV) part of the spectrum [10]. VUV photons have energies well above the binding energy of atoms in polymers and thus create dangling bonds in the surface film.

Inagaki et al. [4] presented plots of the WCA versus the treatment time between 5 and 120 s and discharge powers from 20 and 60 W. In all cases, the WCA dropped significantly after the shortest treatment time and remained unchanged thereafter. When treated in the glowing plasma, PEEK exhibited a WCA of about 17° at the lowest discharge power of 20 W and 9° at the largest power of 60 W. After washing, oxygen-plasma-treated PEEK exhibited a WCA of 62° and 70° for samples treated at the powers of 20 and 60 W, respectively. Therefore, the larger discharge power caused poorer wettability after washing away the molecular fragments. Even more contradicting results were obtained in the afterglow, where the WCA measured on washed samples was 44° and 77° for samples treated at the powers of 10 and 60 W, respectively. Some samples were mounted in a box with an MgF_2 window to test the influence of VUV radiation on the surface wettability. Even a minute of treatment with VUV radiation arising from oxygen plasma did not cause any detectable modification of PEEK's wettability.

The surface structure was determined from high-resolution X-ray photoelectron spectra (XPS) by Inagaki et al. [4]. The deconvolution of the C1s and O1s peaks led to the conclusion that the treatment in the glowing plasma did not change the surface structure much, but the afterglow treatment caused an increased concentration of both C–O and C=O groups, wherein the increase was twofold for the C–O group and less for the C=O group. The satellite peak characteristic for aromatic polymers vanished after treating PEEK both in the glowing plasma and afterglow, indicating the destruction of the aromatic ring in the surface film probed by XPS. The [O]/[C] concentration after treatments with glowing plasma or afterglow increased from 0.13 to about 0.20. The concentration of highly polar groups, such as O–C=O, was below the detection limit of XPS. Based on the results reported by Inagaki et al. [4], one can conclude that the weakly ionized, highly dissociated oxygen plasma causes preferential degradation and etching of the PEEK surface film rather than functionalization with highly polar functional groups. The effect is illustrated in Figure 1.

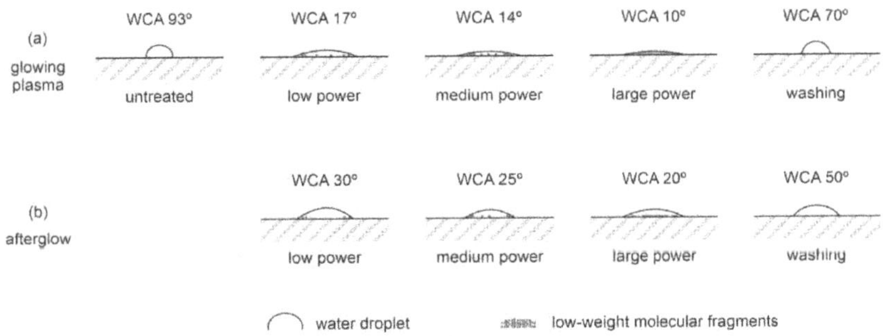

Figure 1. Schematic surface effects during PEEK treatment with weakly ionized oxygen plasma (**a**) and oxygen plasma afterglow (**b**).

Results similar to those of Inagaki et al. [4] were reported by Narushima and Ikeji [11], except that they used a temporal afterglow instead of the flowing afterglow. The temporal afterglow was obtained by pulsing the RF generator operating at the frequency of 13.56 MHz and the power of 50 W. The oxygen pressure was set to 13 Pa. The plasma pulses lasted 10 μs and were repeated every millisecond. The authors stated that the electrons began to decrease at about 100 μs whereas the oxygen radicals at about 100 ms. Taking into account these values, the density of neutral radicals was almost constant over the treatment time of 1 min, but the charged particles and VUV radiation were applied in pulses. The surface finish, as reported by Narushima and Ikeji [11], was independent of the treatment mode (continuous or pulsed). In both cases, the concentration of C−O functional groups as deduced from XPS spectra increased from 20 to 25%, and C=O from 5 to 6%. No other functional groups were found on the PEEK's surface after either of the plasma treatments. The [O]/[C] ratio, as deduced from XPS survey spectra, increased from 0.16 for untreated samples to 0.23 for plasma-treated samples. The difference between the continuous and pulsed modes was in the etching rate, which was found to be 5 and 1 μg/cm^{-2} min^{-1} for continuous and pulsed treatments, respectively. The WCA observed after the treatments were 66° in the case of the continuous mode and 62° for pulsed mode. The pulsed mode, therefore, somehow enabled better wettability than the continuous mode, but neither were sufficient to achieve a very low water contact angle. Interestingly enough, the adhesion force between deposited copper and the plasma-treated PEEK samples was better in the continuous mode, at 800 mN cm^{-2}, whereas in the pulsed mode, it was 360 mN cm^{-2}. The adhesion force was much lower, 40 mN cm^{-2} for the untreated PEEK samples. The oxygen plasma treatment, therefore, assured an order of magnitude larger adhesion force.

More recently, Han et al. [12] also probed oxygen plasma for the surface activation of PEEK samples for application in dentistry. The samples were prepared by 3D printing and exposed to oxygen plasma. Plasma was sustained by an RF discharge at 40 kHz, and the pressure and power were 100 Pa and 100 W, respectively. The treatment time was 15 min and the WCA 25 and 38° for the as-printed and polished samples, respectively. The authors did not mention any washing, but the hydrophobic recovery caused an increased WCA to about 60° after 3 weeks of aging. The surface finish was, therefore, similar to that shown in Figure 1.

The observation of Inagaki et al. [4] of PEEK's surface wettability is not in agreement with the recent paper by Arikan et al. [13]. Arikan and co-workers performed a systematic study of the surface kinetics versus the fluence of VUV radiation and found rich surface chemistry. The [O]/[C] concentration after treatment with VUV radiation increased to over 0.3 even after receiving a moderate fluence of VUV radiation of about 1 J/cm^2. Furthermore, Arikan et al. [13] also observed both COOH and COOR groups on the surface of VUV-treated PEEK. The optimal adhesive strength on VUV-treated PEEK was obtained already after 5 s irradiation with VUV radiation.

Arikan et al. [13] used an Xe excimer lamp that emits radiation in a broad wavelength range from about 150 to 190 nm, peaking at the wavelength of 172 nm (corresponding photon energy 7.2 eV). The flux of VUV radiation was adjusted by changing the treatment time from 3 to 720 s, corresponding to the photon doses between 5×10^{16} and 13×10^{18} cm^{-2}. The corresponding energy absorbed by the PEEK samples due to VUV photon absorption was between 57 and 14,500 mJ cm^{-2}. The samples were placed in the ambient air next to the VUV lamp. Upon these conditions, excellent hydrophilization was achieved. The authors did not report WCAs, but mentioned that the surface energy of the PEEK samples treated even at moderate VUV fluences was as high as 70 mJ m^{-2}, which was close to the detection limit using water droplets. The polar component was about 25 mJ m^{-2} for samples treated by VUV radiation, while the dispersive component remained practically unchanged at about 45 mJ m^{-2}.

The authors also estimated the etching rate due to VUV irradiation and found rather linear behavior at the rate of about 7 nm J^{-1} cm^2 [13]. The etching is explained by the formation of dangling bonds in the PEEK surface film. The dangling bonds interact with molecular oxygen (since the samples were kept in the air in close proximity to the VUV lamp) during the treatment. The etching was accompanied by nanostructuring of the surface. Atomic force microscopy (AFM) revealed the appearance of well-defined periodical structures of circular shape and a typical lateral dimension below 100 nm. The nanostructured surface already appeared at moderate VUV fluences (190 mJ cm^{-2}) and did not change much with prolonged treatment since the AFM image at 190 mJ cm^{-2} was practically identical.

XPS characterization revealed a gradual increase in the [O]/[C] ratio, which peaked at the VUV fluence of about 1000 mJ cm^{-2}. It slowly decreased with prolonged treatment time, but remained over 0.3 even at the largest fluence of 14,000 mJ cm^{-2}. The maximal [O]/[C] ratio was about 0.37 for amorphous and 0.33 for semi-crystalline PEEK. Unlike Inagaki et al. [4], Arikan et al. [13] also observed highly polar functional groups on the PEEK surface after VUV treatments. The concentration of various polar groups increased between 190 and 6840 mJ cm^{-2}, with the concentration of O−C=O groups deduced from the high-resolution XPS C1s peak as large as 10% at the largest VUV fluence.

The almost optimal surface finish of VUV-treated PEEK samples resulted in an excellent adhesive bond strength. While the strength of untreated samples was only 3 and 5 MPa for semi-crystalline and amorphous PEEK, respectively, it rose to about 20 MPa for both samples after being exposed to the VUV fluence of about 100 mJ cm^{-2}, and reached about 25 MPa at the fluence of about 1000 mJ cm^{-2}. The VUV irradiation of PEEK samples thus not only causes excellent wettability but adhesive bond strength as well. The surface effects upon treatment of PEEK by 172 nm VUV radiation is illustrated in Figure 2. The pristine samples are exposed to a rather intensive VUV radiation and simultaneously to air. Air at ambient conditions consists of stable molecules capable of chemical interaction with dangling bonds formed on the PEEK's surface upon VUV irradiation. No nitrogen was observed on the PEEK's surface, so the dangling bonds indeed interacted only with oxygen-containing molecules. The air moisture was not reported by Arikan et al. [13], so it would not be justified to exclude interaction with water vapor. Still, the oxygen content in the air, even at large humidity and room temperature, is much larger than the concentration of water molecules, so it is reasonable to assume the interaction of O_2 with dangling bonds. The interaction causes the formation of various oxygen-containing functional groups, while prolonged irradiation causes extensive etching and desorption of interaction products such as carbon oxides and water vapor. The net effect after the prolonged treatment is thinning the PEEK samples without a significant effect on surface morphology or chemical structure. The combination of highly polar surface functional groups and nanostructured polymer morphology always results in a super-hydrophilic surface finish [6]. The irradiation with

VUV photons from the Xe excimer lamp on the PEEK surface finish was also elaborated in [14].

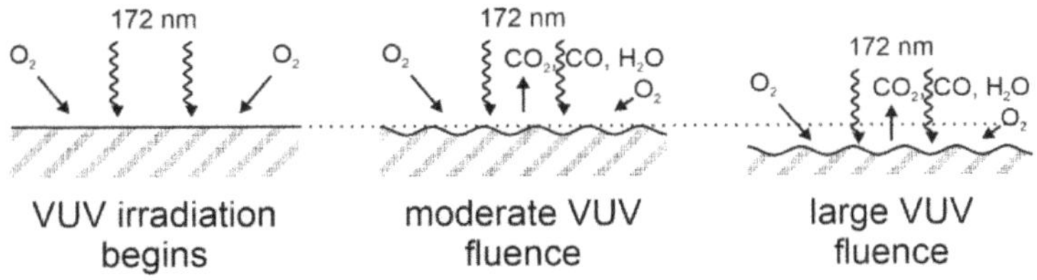

Figure 2. Schematic of the surface effects during PEEK treatment with VUV radiation from an Xe excimer lamp.

Modification of the PEEK surface properties by VUV radiation was also reported in a recent paper by Yoshida et al. [15]. They used a lamp emitting light at 185 or 254 nm wavelengths (probably a mercury lamp) operating at the power of 200 W. The PEEK samples were placed in an oxygen atmosphere at a pressure of 20 mbar and about 140 mm away from the lamp. The irradiation times were 1, 5, and 10 min, but the data on the photon fluxes are not available. At such conditions, Yoshida et al. [15] reported a roughly exponential decrease in the WCA with the treatment time, but the WCA after 5 min irradiation was still about 30°. The discrepancy with results reported by Arikan et al. [13] could be explained by the different wavelengths used by the authors. Another explanation could be different doses of photons, but the dose was only reported by Arikan et al. [13].

Extreme UV radiation was used for the surface modification of PEEK samples by Czwartos [16]. The EUV was supplied in pulses and the intensity was large enough to cause the formation of plasma above the polymer surface. The initial WCA was 74° and it dropped to 64° after 200 pulses of EUV irradiation. The XPS characterization revealed significant modifications of the surface composition and formation of various functional groups. The discrepancy between rich functionalization and a rather marginal increase in wettability may be explained by thermal effects.

An alternative approach to bond scission in the surface film of PEEK polymers by absorption of VUV radiation is the application of ion beams. High-energy ions cause radiation damage over a rather thick surface film and also significant heating, so they were found inappropriate by Kim et al. [17]. Instead, Kim et al. used relatively low-energy ion beams (below 1 keV). The penetration depth of such ions in PEEK was estimated to be about 8 nm using the TRIM96 code. The source of Ar^+ ions was a hollow cathode discharge, and the ion fluence varied between 1.6×10^{18} and 1.3×10^{20} cm^{-2}. The ion fluences used by Kim et al. [17] were therefore similar to the VUV fluence used by Arikan et al. [13]. Oxygen was introduced in the vacuum chamber at a low flow rate of a few sccm to ensure vacuum conditions upon treatment of PEEK samples. Such a low pressure suppressed the loss of Ar^+ ions' kinetic energy at elastic collisions with neutral gas in the processing chamber.

Kim et al. [17] monitored the surface chemistry by XPS. The irradiation with Ar^+ ions in the absence of oxygen enabled a gradual increase in the [O]/[C] ratio, from an initial 0.13 to 0.16 after the irradiation time of 30 s, and 0.22 after the irradiation time of 120 s. The surface functionalization in the irradiation chamber without introducing oxygen may be explained by the residual atmosphere, which usually consists of water vapor in hermetically tight vacuum systems. The ion irradiation causes bond scission in the surface film, and the dangling bonds interact with H_2O to form oxygen functional groups. The base pressure in the reaction chamber during treatment with Ar^+ ions was about 0.01 Pa, so the flux of molecules in the residual atmosphere was still as large as about 3×10^{20} $m^{-2}s^{-1}$. The deconvolution of the high-resolution C1s peak revealed the appearance of the O−C=O groups on the PEEK's surface. The resulting WCA after irradiation with Ar^+ ions was about 70°, so the hydrophilicity was slightly improved.

Kim et al. [17] reported even better hydrophilization upon treatment with ions and the presence of oxygen in the reaction chamber. In this case, the WCA of about 50° was obtained already at the treatment time of about a minute. Prolonged treatment did not cause any further decrease in the WCA. The [O]/[C] ratio of 0.28 was obtained at irradiation in the oxygen atmosphere at low pressure, and the concentration of the O−C=O groups on the PEEK's surface was about 4%. Copper was deposited on treated samples, and the shear strength increased from 10 to 60 mN cm^{-2} at the treatment time of about 2 min. Prolonged treatment resulted in somewhat lower shear strength, which may be explained by heating of the PEEK's surface due to Ar^+ bombardment. Etching was not mentioned by Kim et al., but it is expected that the treatment causes the removal of the oxidized surface layer due to the combined effects of surface oxidation and ion bombardment.

The method introduced by Kim et al. [17], therefore, enabled moderate hydrophilicity of PEEK surfaces. It is better than treatment with oxygen plasma, as reported by Inagaki et al. [4] and Narushima and Ikeji [11], but not as effective as irradiation with VUV radiation, as reported by Arikan et al. [13]. Both energetic Ar^+ ions and VUV radiation from the Xe excimer lamp are capable of breaking bonds in the surface film, but the energy of ions is much larger than the photon energy, so the thermal effects cannot be excluded when using Kim's method, which is illustrated in Figure 3.

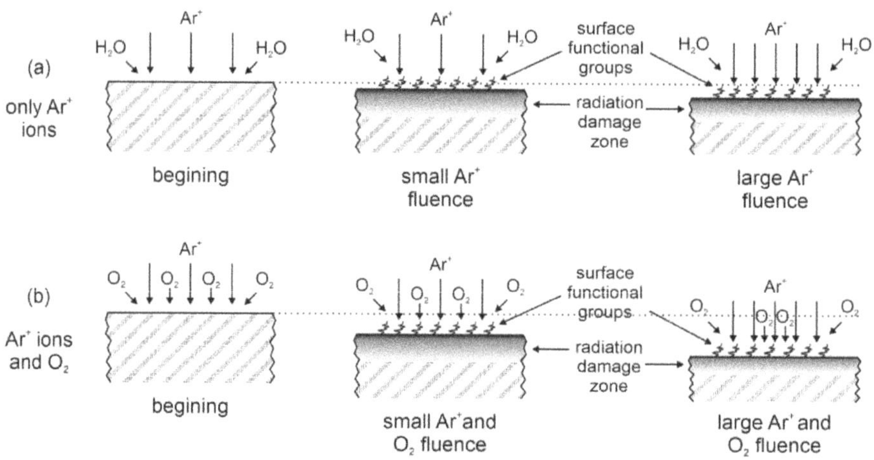

Figure 3. Illustration of the surface effects upon irradiation of PEEK with Ar^+ ions in the absence of oxygen (**a**) and oxygen presence (**b**).

Energetic ion beams were also used by Awaja et al. [18] to improve the adhesion of biological cells on the PEEK surface. They used ions created in plasma sustained in O_2 and CF_4 to bombard the polymer surface with ions having kinetic energy between 2 and 20 keV. Such energetic ions caused etching and functionalization, but the minimum WCA was about 40°. The cell adhesion increased with decreasing WCA. The rather inadequate wettability could be explained by thermal effects because other authors reported a super-hydrophilic surface finish upon irradiation with energetic ions [19]. Actually, cell adhesion was found to be optimal in the limited range of ion energies and treatment time. The best results were observed at the bias voltage of 6 kV and the treatment time of about 100 s.

More recently, Kruse et al. [20] also reported improved adhesion of biological cells on PEEK substrates using the plasma immersion ion implantation technique. Energetic nitrogen ions bombarded the polymer surface in pulses of duration 45 μs with a repetition frequency of 1500 Hz. The ion kinetic energy was 10 keV. A super-hydrophilic surface finish was reported for the treatment time of 20 min. The observations reported in [20] clearly illustrate the need to prevent overheating of the polymer surface upon ion bombardment. Treatment with energetic nitrogen ions was also reported by Zheng et al. [21], while Zhao et al. [22] used the plasma immersion ion-implantation technique with water vapor and ammonia plasmas. Biocompatible PEEK properties are also improved by the implantation of metallic ions [23].

A variety of gases were used for sustaining plasmas suitable for modification of the PEEK's surface [2]. Prolonged treatment of PEEK with oxygen plasma was reported by Botel et al. [24]. Plasma was sustained in oxygen or a mixture of oxygen and argon at the pressure of 30 Pa with a capacitively coupled discharge powered by a 100 kHz RF generator at the power of 200 W. The reported temperature of the samples during plasma treatment was 70 °C, and the treatment times were either 3 or 35 min. The authors reported the WCA after the treatment of 0.0° for all samples except those treated for 35 min in Ar/O_2 plasma, where the WCA was 2.8°. The authors reported a super-hydrophilic surface finish of the PEEK samples. No details about the plasma parameters are disclosed in [24], hence it is difficult to explain the reasons for the super-hydrophilic surface finish. One possible explanation would be bombardment with O_2^+ ions if the samples were placed on the powered electrode. Namely, the low-frequency capacitively coupled discharge operates at the voltage of several 100 V. Self-biasing occurs on the surface of samples placed on the powered electrode, so the samples are bombarded with O_2^+ (and perhaps also O^+) ions of several 100 eV kinetic energy. The rather large discharge power—as compared to [4] and [11]—of 200 W may also assure for rather extensive radiation in the VUV range. The synergy between ions and VUV radiation may lead to significant surface structural modifications. Another explanation may be in the appropriate roughness of the samples. Namely, the samples were first polished and then sand-blasted, resulting in a roughness R_a of about 0.7 µm. The nanoscale roughness was not reported, but bombardment with energetic oxygen ions should cause a rich morphology on the nanometer scale. It is known that the required conditions for the super-hydrophilic surface finish are both functionalization with highly polar functional groups and rich morphology on the submicrometer scale [6]. A feasible explanation of the surface effects using Botel's method [24] is illustrated in Figure 4.

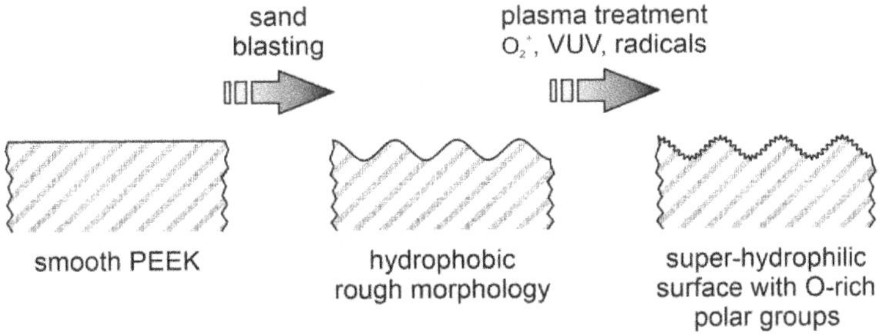

Figure 4. Schematic of the methods for super-hydrophilic surface finish of PEEK samples.

Despite the excellent wettability, the shear bond strength between PEEK samples and veneering composites did not change significantly because of plasma treatments. The feasible explanation for this rather unexpected observation is the application of special bonding agents purchased from Visiolink. A possible application of this technique is in dentistry. The same application was addressed by Akay et al. [25], but they used plasma-treated PEEK-PMMA (polymethyl methacrylate) composites.

Dental applications of plasma-treated PEEK were also addressed by Younis et al. [26]. They probed plasmas of various gases, including argon, air, oxygen, and nitrogen. They used a commercial plasma reactor intended for dentistry. Plasma was sustained in a cylindrical tube with an RF generator operating at the frequency of 40 kHz, and the output power was 100 W. No details of the discharge coupling or plasma parameters are reported in [26]. The treatment time was 10 min, and the authors mentioned that the PEEK samples were kept at 20 °C during the plasma treatment. The gas pressure was 30 Pa. The shear bond strength was measured after the plasma treatments. It increased by a factor of two as compared to the untreated PEEK samples and was independent of the type of gas used. The shear strength was almost the same as for glued samples.

Contrary to [26], Fedel et al. [27] also probed argon, oxygen, and nitrogen plasmas, and found nitrogen plasma particularly useful. A capacitively coupled RF discharge was powered with the generator operating at the frequency of 13.56 MHz and the output power of 100 W. The ultimate pressure in the discharge chamber was 1.6×10^{-2} Pa, and the working pressure was 0.67 Pa. The treatment time was 2 min. The samples were rather large, with dimensions 10 cm × 10 cm. Nitrogen plasma treatment significantly increased the strength of self-bonding in both crystalline and amorphous samples. The bonding strength was more than doubled when pressing for 4 h and tripled after pressing for 7.5 h at 200 °C. As expected, the plasma treatment did not have a measurable influence on the crystallinity, which was monitored by X-ray diffraction (XRD). The authors explained improved adhesion by surface functionalization and formation of dangling bonds, but did not use a surface-sensitive technique to elaborate the effect.

The surface chemistry upon treatment of PEEK with nitrogen plasma was discussed by Wang et al. [28]. They used a low-temperature plasma reactor powered by an RF generator operating at the frequency of 13.56 MHz and voltage of 500 V. No other details about the discharge system or the nitrogen pressure were disclosed. The treatment times were 15, 25, and 35 min. Scanning electron microscope (SEM) images revealed the evolution of a rich morphology on the submicrometer scale due to the plasma treatment. The XPS survey spectra showed an appearance of nitrogen on PEEK surfaces, but the results were not quantified. The wettability of the as-synthesized PEEK was rather good with the WCA of 73°. Specifically, this value differs from values reported by other authors (just above 90°). The WCA was found to decrease gradually with increasing treatment time: it was 59° after treating for 15 min and 35° after treating for 25 min. The shear bond strength did not follow the evolution of the wettability because it was the highest after 25 min of plasma treatment, when the WCA was 53°. The shear strength was about three times larger than for untreated samples. The authors explained the improved adhesion properties by introducing nitrogen-containing functional groups. They proposed a chemical bond of the surface nitrogen group with the dimethacrylate elements of the Variolink. The proposed interaction between nitrogen plasma and the PEEK surface is illustrated in Figure 5.

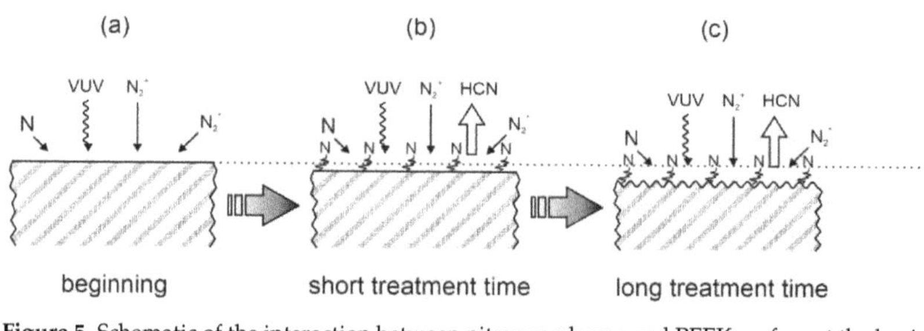

Figure 5. Schematic of the interaction between nitrogen plasma and PEEK surface at the beginning of treatment (**a**), after short (**b**) and long treatment time (**c**).

More recently, Bres et al. [29] used a powerful plasma torch for surface modification with either air or nitrogen plasma. The samples were placed in the afterglow to prevent melting, since the typical discharge power used for powering the atmospheric plasma torch was 1000 W. The surface tension increased from 53 to about 74 mN/m, and the resultant hydrophilicity depended on the distance between the glowing hot plasma and the sample. Large-impedance atmospheric-pressure discharges may be more suitable for deposition and tailoring surface properties of polymers [30,31], although the treatment uniformity over a large area is still a technological challenge [32].

3. Conclusions and Recommendations

This review on the non-equilibrium plasma techniques for surface modification of PEEK polymers explains the observations reported by various authors who probed different plasmas for achieving improved wettability. Plasma provides reactive species and radiation in the UV and VUV range, which are capable of causing both reversible and irreversible modifications. The surface finish depends enormously on the fluxes of reactive species and radiation, and the thermal effects should not be neglected in some cases. The major advance is providing readers with the influence of various species and radiation, which explains large variations in the surface wettability reported by authors who used different plasmas.

The literature survey indicates that the best conditions for improved wettability include treatment with VUV radiation in the presence of oxygen. Such a combination was also suitable for the activation of fluorine-containing polymers such as Teflon [33]. The VUV photons break even the strongest bonds in the polymer surface film [34]. The dangling bonds interact with gaseous radicals and even molecules in the ground electronic state, thus resulting in surface functionalization. The choice of the VUV source is not particularly limited since the radiation at a particular wavelength interacts with the polymer, irrespective of the source. The most common source of VUV radiation is gaseous plasma. Most plasmas irradiate in the VUV range, but the range of wavelengths and intensity depends enormously on the gases used in the plasma sources and the peculiarities of the discharges. Excimer and exciplex lamps operating at atmospheric pressure are sources of extensive VUV radiation. A VUV-transparent window should be placed between the plasma generated by excimer lamps and the PEEK samples.

Alternatively, the polymer samples are placed directly into the gaseous plasma. Most authors used this configuration for treating PEEK samples. The VUV radiation from low-pressure plasma sources useful for surface functionalization of PEEK samples will increase with increasing electron temperature and density, and thus with increasing discharge power (or, rather, power density). The VUV radiation from oxygen plasma arises from the resonant radiative transition of O-atoms [35]. The O-atoms, however, also interact chemically with the PEEK surface, which may lead to excessive chemical etching and the formation of low molecular weight fragments of poor adhesion on the PEEK's surface. Oxygen plasma is, therefore, useful for the desired surface finish of PEEK, providing the chemical interaction is suppressed by lowering the fluxes of reactive plasma species and favoring the flux of VUV radiation.

Apart from VUV radiation, the bond scission in the surface film of polymers is also achievable by treatment with energetic ions. In this case, one should be aware of surface charging and excessive heating. The surface charging will cause non-uniform treatment, so it is advisable to perform the treatment under the presence of low-energy charged particles, i.e., plasma conditions. The excessive heating can be suppressed by using moderately energetic ions, for example, in the range from 100 to 1000 eV. Such ions are found next to the polymer surface in cases where self-biasing occurs, for example, by placing the polymer samples on the powered electrode of a capacitively coupled RF discharge. Alternatively, it is advisable to apply energetic ions in pulses to prevent excessive heating and, thus, the loss of polar groups due to thermal effects.

Oxygen plasmas with a high density of reactive species (O-atoms in the ground and metastable states and metastable molecules), low VUV radiation, and no ability to accelerate positively charged ions onto the PEEK's surface are not recommended for functionalization of the PEEK's surface because of substantial etching. Such plasmas are sustained by electrodeless RF discharges in the continuous mode or in the afterglows (both temporal and flowing). The surface wettability is increased significantly using such plasmas rich in neutral radicals, but low molecular fragments are formed, which may suppress the adhesion properties.

Nitrogen plasma is an attractive alternative to oxygen because it is chemically less reactive than oxygen at a given discharge power. This means that the chemical etching will be lower than in the case of oxygen plasma. Still, chemical etching occurs and causes the formation of hydrogen cyanide, so some precautions are recommended when using plasmas containing nitrogen, including ammonia.

Finally, it should be stressed that the surface finish depends on the fluxes and fluences of chemically reactive species, VUV radiation, and positively charged ions. While some authors reported the fluxes of VUV radiation and ions, none reported the fluxes of radicals. The authors are therefore encouraged to measure the plasma parameters upon treatment of PEEK samples.

Funding: This research was funded by the Slovenian Research Agency, project No. L2-2616 (Selective functionalization of polymer components) and core funding P2-0082 (Thin-film structures and plasma surface engineering).

Data Availability Statement: Not applicable.

Conflicts of Interest: The author declares no conflict of interest.

References

1. Choudhury, S.S.; Pandey, M.; Bhattacharya, S. Recent Developments in Surface Modification of PEEK Polymer for Industrial Applications: A Critical Review. *Rev. Adhes. Adhes.* **2021**, *9*, 401–433. [CrossRef]
2. Wang, B.; Huang, M.; Dang, P.; Xie, J.; Zhang, X.; Yan, X. PEEK in Fixed Dental Prostheses: Application and Adhesion Improvement. *Polymers* **2022**, *14*, 2323. [CrossRef]
3. Soares Machado, P.; Cadore Rodrigues, A.C.; Chaves, E.T.; Susin, A.H.; Valandro, L.F.; Pereira, G.K.R.; Rippe, M.P. Surface Treatments and Adhesives Used to Increase the Bond Strength Between Polyetheretherketone and Resin-based Dental Materials: A Scoping Review. *J. Adhes. Dent.* **2022**, *24*, 233–245. [CrossRef] [PubMed]
4. Inagaki, N.; Tasaka, S.; Horiuchi, T.; Suyama, R. Surface Modification of Poly (aryl ether ether ketone) Film by Remote Oxygen Plasma. *J. Appl. Polym. Sci.* **1997**, *68*, 271–279. [CrossRef]
5. Vesel, A.; Mozetic, M. New developments in surface functionalization of polymers using controlled plasma treatments. *J. Phys. D Appl. Phys.* **2017**, *50*, 293001. [CrossRef]
6. Mozetic, M. Plasma-Stimulated Super-Hydrophilic Surface Finish of Polymers. *Polymers* **2020**, *12*, 2498. [CrossRef] [PubMed]
7. Fukunaga, Y.; Longo, R.C.; Ventzek, P.L.G.; Lane, B.; Ranjan, A.; Hwang, G.S.; Hartmann, G.; Tsutsumi, T.; Ishikawa, K.; Kondo, H.; et al. Interaction of oxygen with polystyrene and polyethylene polymer films: A mechanistic study. *J. Appl. Phys.* **2020**, *127*, 023303. [CrossRef]
8. Vesel, A.; Zaplotnik, R.; Mozetič, M.; Primc, G. Surface modification of PS polymer by oxygen-atom treatment from remote plasma: Initial kinetics of functional groups formation. *Appl. Surf. Sci.* **2021**, *561*, 150058. [CrossRef]
9. Zaplotnik, R.; Mozetic, M. Frontiers in the Interaction of Chemically Reactive Species From Gaseous Plasma With Hydrophobic Polymers. *Front. Phys.* **2022**, *10*, 896219. [CrossRef]
10. Popović, D.; Mozetič, M.; Vesel, A.; Primc, G.; Zaplotnik, R. Review on vacuum ultraviolet generation in low-pressure plasmas. *Plasma Process. Polym.* **2021**, *18*, 2100061. [CrossRef]
11. Narushima, K.; Ikeji, H. Plasma Surface Modification of Poly(aryl ether ether ketone) and Surface Metallization using Copper Metal. *Sen'i Gakkaishi* **2009**, *65*, 127–131. [CrossRef]
12. Han, X.; Sharma, N.; Spintzyk, S.; Zhou, Y.; Xu, Z.; Thieringer, F.M.; Rupp, F. Tailoring the biologic responses of 3D printed PEEK medical implants by plasma functionalization. *Dent. Mater.* **2022**, *38*, 1083–1098. [CrossRef] [PubMed]
13. Arikan, E.; Holtmannspötter, J.; Hofmann, T.; Gudladt, H.-J. Vacuum-UV of polyetheretherketone (PEEK) as a surface pre-treatment for structural adhesive bonding. *J. Adhes.* **2018**, *96*, 917–944. [CrossRef]
14. Arikan, E.; Holtmannspötter, J.; Zimmer, F.; Hofmann, T.; Gudladt, H.-J. The role of chemical surface modification for structural adhesive bonding on polymers—Washability of chemical functionalization without reducing adhesion. *Int. J. Adhes. Adhes.* **2019**, *95*, 102409. [CrossRef]
15. Yoshida, M.; Nakanishi, G.; Yamanaka, H.; Iwamori, S. Enhanced adhesion of copper plating to polyether ether ketone based on active oxygen species generated under ultraviolet irradiation. *Surf. Interface Anal.* **2022**, *54*, 759–766. [CrossRef]
16. Czwartos, J.; Budner, B.; Bartnik, A.; Wachulak, P.; Butruk-Raszeja, B.A.; Lech, A.; Ciach, T.; Fiedorowicz, H. Effect of Extreme Ultraviolet (EUV) Radiation and EUV Induced, N_2 and O_2 Based Plasmas on a PEEK Surface's Physico-Chemical Properties and MG63 Cell Adhesion. *Int. J. Mol. Sci.* **2021**, *22*, 8455. [CrossRef]
17. Kim, S.; Lee, K.J.; Seo, Y. Polyetheretherketone (PEEK) surface functionalization by low-energy ion-beam irradiation under a reactive O_2 environment and its effect on the PEEK/copper adhesives. *Langmuir* **2004**, *20*, 157–163. [CrossRef]
18. Awaja, F.; Bax, D.V.; Zhang, S.; James, N.; McKenzie, D.R. Cell Adhesion to PEEK Treated by Plasma Immersion Ion Implantation and Deposition for Active Medical Implants. *Plasma Process. Polym.* **2012**, *9*, 355–362. [CrossRef]
19. Tsougeni, K.; Vourdas, N.; Tserepi, A.; Gogolides, E.; Cardinaud, C. Mechanisms of oxygen plasma nanotexturing of organic polymer surfaces: From stable super hydrophilic to super hydrophobic surfaces. *Langmuir* **2009**, *25*, 11748–11759. [CrossRef]

20. Kruse, H.V.; McKenzie, D.R.; Clark, J.R.; Suchowerska, N. Plasma ion implantation of 3D-printed PEEK creates optimal host conditions for bone ongrowth and mineralisation. *Plasma Process. Polym.* **2021**, *18*, 2000219. [CrossRef]
21. Zheng, Y.; Gao, A.; Bai, J.; Liao, Q.; Wu, Y.; Zhang, W.; Guan, M.; Tong, L.; Geng, D.; Zhao, X.; et al. A programmed surface on polyetheretherketone for sequentially dictating osteoimmunomodulation and bone regeneration to achieve ameliorative osseointegration under osteoporotic conditions. *Bioact. Mater.* **2022**, *14*, 364–376. [CrossRef]
22. Zhao, Y.; Wong, H.M.; Lui, S.C.; Chong, E.Y.; Wu, G.; Zhao, X.; Wang, C.; Pan, H.; Cheung, K.M.; Wu, S.; et al. Plasma Surface Functionalized Polyetheretherketone for Enhanced Osseo-Integration at Bone-Implant Interface. *ACS Appl. Mater. Interfaces* **2016**, *8*, 3901–3911. [CrossRef]
23. Wan, R.; Wang, X.; Lei, L.; Hu, G.; Tang, H.; Gu, H. Enhanced anti-microbial activity and osseointegration of Ta/Cu co-implanted polyetheretherketone. *Colloids Surf. B Biointerfaces* **2022**, *218*, 112719. [CrossRef]
24. Botel, F.; Zimmermann, T.; Sutel, M.; Muller, W.D.; Schwitalla, A.D. Influence of different low-pressure plasma process parameters on shear bond strength between veneering composites and PEEK materials. *Dent. Mater.* **2018**, *34*, e246–e254. [CrossRef]
25. Akay, C.; Israfil, N.; Pat, S. Enhancement of Adhesive Bonding Properties of Polyetheretherketone based Materials using Plasma Surface Modifications. *J. Adhes. Dent.* **2022**, *24*, 117–124. [CrossRef]
26. Younis, M.; Unkovskiy, A.; ElAyouti, A.; Geis-Gerstorfer, J.; Spintzyk, S. The Effect of Various Plasma Gases on the Shear Bond Strength between Unfilled Polyetheretherketone (PEEK) and Veneering Composite Following Artificial Aging. *Materials* **2019**, *12*, 1447. [CrossRef]
27. Fedel, M.; Micheli, V.; Thaler, M.; Awaja, F. Effect of nitrogen plasma treatment on the crystallinity and self-bonding of polyetheretherketone (PEEK) for biomedical applications. *Polym. Adv. Technol.* **2020**, *31*, 240–247. [CrossRef]
28. Wang, Y.; Liu, C.; Chen, L.; Bai, J.; Wang, D.; Gan, K.; Liu, H. The effect of nitrogen plasma treatment on adhesive properties of PEEK. *J. Adhes. Sci. Technol.* **2020**, *34*, 2672–2684. [CrossRef]
29. Brès, L.; Gherardi, N.; Naudé, N.; Rives, B. Experimental investigations of a remote atmospheric pressure plasma by electrical diagnostics and related effects on polymer composite surfaces. *Eur. Phys. J. Appl. Phys.* **2021**, *95*, 30801. [CrossRef]
30. Jang, H.J.; Jung, E.Y.; Parsons, T.; Tae, H.S.; Park, C.S. A Review of Plasma Synthesis Methods for Polymer Films and Nanoparticles under Atmospheric Pressure Conditions. *Polymers* **2021**, *13*, 2267. [CrossRef]
31. Jung, E.Y.; Park, C.S.; Jang, H.J.; Iqbal, S.; Hong, T.E.; Shin, B.J.; Choi, M.; Tae, H.S. Optimization of Atmospheric Pressure Plasma Jet with Single-Pin Electrode Configuration and Its Application in Polyaniline Thin Film Growth. *Polymers* **2022**, *14*, 1535. [CrossRef] [PubMed]
32. Kim, D.H.; Park, C.-S.; Shin, B.J.; Seo, J.H.; Tae, H.-S. Uniform Area Treatment for Surface Modification by Simple Atmospheric Pressure Plasma Treatment Technique. *IEEE Access* **2019**, *7*, 103727–103737. [CrossRef]
33. Lojen, D.; Zaplotnik, R.; Primc, G.; Mozetič, M.; Vesel, A. Optimization of surface wettability of polytetrafluoroethylene (PTFE) by precise dosing of oxygen atoms. *Appl. Surf. Sci.* **2022**, *598*, 153817. [CrossRef]
34. Vesel, A.; Lojen, D.; Zaplotnik, R.; Primc, G.; Mozetic, M.; Ekar, J.; Kovac, J.; Gorjanc, M.; Kurecic, M.; Stana-Kleinschek, K. Defluorination of Polytetrafluoroethylene Surface by Hydrogen Plasma. *Polymers* **2020**, *12*, 2855. [CrossRef] [PubMed]
35. Primc, G. Generation of Neutral Chemically Reactive Species in Low-Pressure Plasma. *Front. Phys.* **2022**, *10*, 5264. [CrossRef]

polymers

MDPI

Review

Review of Plasma Processing for Polymers and Bio-Materials Using a Commercial Frequency (50/60 Hz)-Generated Discharge

Hong Tak Kim [1], Cheol Min Jung [2], Se Hyun Kim [2,*] and Sung-Youp Lee [1,*]

1 Department of Physics, Kyungpook National University, Daegu 41566, Republic of Korea; zam89blue@gmail.com
2 Division of Chemical Engineering, Konkuk University, Seoul 05029, Republic of Korea; daesil159@gmail.com
* Correspondence: shkim@konkuk.ac.kr (S.H.K.); physylee@knu.ac.kr (S.-Y.L.)

Abstract: This manuscript introduces the properties and diverse applications of plasma generated using commercial frequencies of 50/60 Hz. Commercial frequency (CF) derived plasma exhibits characteristics similar to DC discharge but with an electrical polarity and a non-continuous discharge. Due to the low-frequency nature, the reactor configurations usually are capacitively coupled plasma type. The advantages of this method include its simple power structure, low-reaction temperature, and low substrate damage. The electrical polarity can prevent charge buildup on the substrates and deposited films, thereby reducing substrate damage. The simple, low-cost, and easy-to-operate power structure makes it suitable for laboratory-scale usage. Additionally, the various applications, including plasma-enhanced vapor deposition, sputtering, dielectric barrier discharge, and surface modification, and their outcomes in the CF-derived plasma processes are summarized. The conclusion drawn is that the CF-derived plasma process is useful for laboratory-scale utilization due to its simplicity, and the results of the plasma process are also outstanding.

Keywords: dielectric barrier discharge (DBD); commercial frequency; 50/60 Hz; plasma; PECVD; polymer; sputtering; surface modification

Citation: Kim, H.T.; Jung, C.M.; Kim, S.H.; Lee, S.-Y. Review of Plasma Processing for Polymers and Bio-Materials Using a Commercial Frequency (50/60 Hz)-Generated Discharge. *Polymers* **2023**, *15*, 2850. https://doi.org/10.3390/polym15132850

Academic Editor: Marián Lehocký

Received: 27 April 2023
Revised: 22 May 2023
Accepted: 26 May 2023
Published: 28 June 2023

1. Introduction

Plasma is a partially ionized gas consisting of equal numbers of positively charged ions, negatively charged electrons, and neutral particles. There are two main types of plasma: thermal plasma and non-thermal plasma. Thermal plasma is in thermal equilibrium, meaning that the electron and ion temperatures are the same. Thermal plasma is commonly used in high-temperature industrial applications such as welding, cutting, and melting materials [1]. On the other hand, non-thermal plasma is not in thermal equilibrium, meaning that the electron temperature is much higher than the ion temperature. Thus, the gas temperature of non-thermal plasma is low and is usually used in material processing [2–4]. In addition, non-thermal plasma is appealing because it can activate chemical and physical reactions at low temperatures compared to non-plasma methods [2–4]. Non-thermal plasma processes are plasma-based material processing techniques. These can be used in deposition techniques such as plasma-enhanced vapor deposition (PECVD) [5,6] and sputtering [7–10], etching (RIE) [11,12], surface modification [13,14], and cleaning processes [15,16]. As a result, the plasma method has been extensively employed in the micro/nanofabrication of electronic devices, surface cleaning and modification of various substrates, trash disintegration, and food applications [17,18]. In addition, a plasma discharge is a strong ultraviolet (UV) source, and UV alone can be used in various applications, such as the photolithography process and surface modification [19,20]. From the point of view of the generation method, the plasma sources can be divided into capacitively coupled plasma (CCP), inductively coupled plasma (ICP), and microwave plasma (MWP), as shown in Figure 1. The most prevalent method for generating plasma is CCP, which involves the use of two metal electrodes separated by a discharge gap. The power is applied to the electrodes and the plasma is directly generated using an electric field, as depicted in Figure 1b. In the case of ICP, an alternating-current (AC) power, typically at a frequency of 13.56 MHz, is applied to an external coil or antenna. The time-varying current flowing through the coil generates a magnetic field around the external coil (or antenna). Then, the electric field is induced by the magnetic field, as shown in Figure 1a. Finally, the plasma is generated using the induced electric field formed inside the reactor.

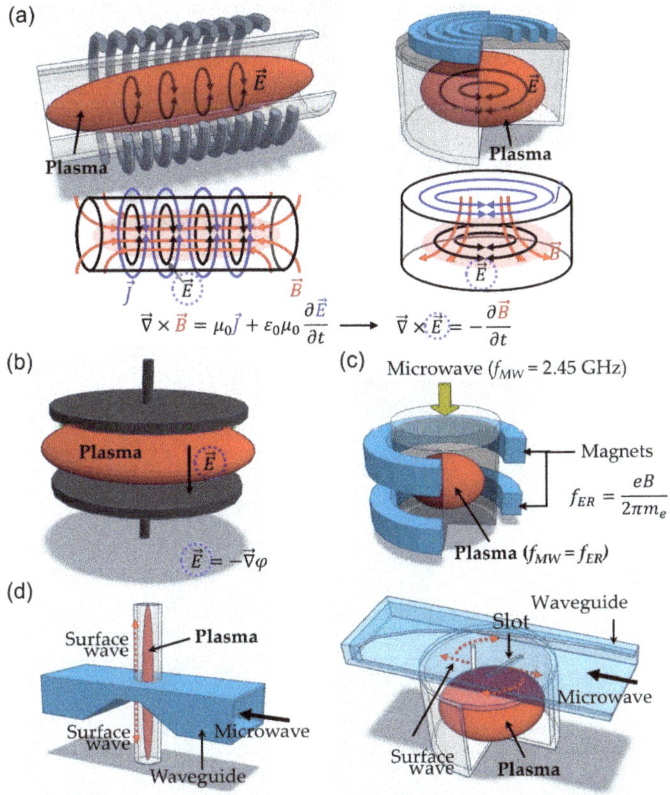

Figure 1. Schematic diagrams for plasma generation with different configurations: (**a**) inductively coupled plasma (ICP): Maxwell equations related to the generation of ICP and the schematic diagram for the applied current ($\vec{J}$: current density, solid blue line), induced magnetic field ($\vec{B}$, solid red line), and electric field (solid black line) in different ICP configurations; (**b**) capacitively coupled plasma ($\vec{E}$: electric field; φ: electric potential); (**c**) microwave plasma: electron resonant plasma (f_{MW}: microwave frequency; f_{ER}: electron resonance frequency); (**d**) microwave plasma: surface-wave sustained plasma.

The MWP can be classified as electron cyclotron plasma and surface-wave sustained plasma [2,3,21–24]. In electron cyclotron plasma, all free electrons rotate with the same frequency (electron cyclotron frequency: ECF) in a strong magnetic field. In this situation, the applied microwave frequency (MWF) matches the ECF, and the resonance occurs between the MWF and ECF. The resonating effect heats many electrons and generates a plasma discharge [21], as shown in Figure 1c. In surface-wave sustained plasma, the discharge is generated via the propagation of electromagnetic surface waves [22–24], as shown in Figure 1d. Table 1 shows a summary of plasma parameters at different plasma-generating methods. Similarly, in terms of power frequency, the plasma sources can be classified as direct current (DC), commercial frequency (CF, 50/60 Hz), mid-frequency (MF, ~kHz), radio frequency (RF, 13.56 MHz), and microwave frequency (MWF, 2.45 GHz). In most cases, a low-frequency, including 50/60 Hz, has been applied to the CCP method instead of the ICP method. The reason for this is that the slowly varying input current creates a magnetic field that makes it challenging to form an induced electric field sufficient to cause a discharge. Commercial-frequency (CF, 50/60 Hz) derived plasma is similar to DC discharge with polarity, and exhibits different properties compared to the high-frequency derived plasma. In addition, a CF power source has a very simple and easily made design, which is inexpensive to build. However, CF-derived plasma processes have not been widely used despite their many advantages, and studies have been performed by very few groups and focused on film depositions using plasma-enhanced chemical vapor deposition [25–40]. Recently, CF-derived discharge has been applied not only to low-pressure discharge but also to atmospheric discharge. This review focuses on the design of CF power sources, the properties of CF-derived plasma, and applications of CF-derived plasma processes.

Table 1. Summary of plasma parameters of different plasma generating methods.

Type	Frequency	Pressure (Torr)	T_e (eV)	n_e (cm^{-3})	Ref.
Positive Column	DC	10^{-2}–10	1–3	~10^9–10^{11}	[2–4]
CCP	DC, CF 13.56 MHz	10^{-2}–1	1–5	~10^8–10^{10}	[2,3,27–29,41]
ICP	13.56 MHz	10^{-4}–10^{-1}	1–10	10^{11}–10^{12}	[2–4,41]
Magnetron Sputtering	13.56 MHz	10^{-3}–1	1–5	10^{10}–10^{12}	[2–4,7,41]
ECR	2.45 GHz	10^{-4}–10^{-2}	2–7	10^{10}–10^{12}	[2–4,41]
Microwave plasma (Surfatron)	2.45 GHz	760	5	10^{12}–10^{15}	[42]
Microwave plasma (SLAN)	2.45 GHz	760	5	10^{11}	[42]
DBD	8 kHz 13.56 MHz	760	1–3	10^{14}	[43,44]
Thermal Arc	30 A-30 kA	76–76,000	1–10	10^{15}–10^{19}	[1–3]
Non-complete Thermal Arc	1-30 A	10^{-3}–100	0.2–2	10^{14}–10^{15}	[1–3]

2. Plasma Properties Generated Using a Commercial-Frequency (50/60 Hz) Power Source

2.1. Nonthermal Plasma

Typical classifications of plasma identify two main categories as equilibrium and non-equilibrium plasma [1–4]. In an equilibrium plasma, the temperatures of all species (electrons, ions, and neutral gases) in plasma are almost same. Thus, the gas temperature is high, which is called thermal plasma. The main heating mechanism is joule heating and thermal ionization. A representative example of thermal plasma is arc plasma. The main disadvantages of thermal plasma are high gas temperatures and low excitation selectivity. Thus, the applications of thermal plasma are limited to cutting, welding, and decomposition of various materials. In non-equilibrium plasma, the temperature of electrons is much higher than that of ions and neutral gases. Although the electron temperature is high, the gas temperature is low because neutral gases make up the majority of plasma. Thus, non-equilibrium plasma is called non-thermal plasma, and the electron impact process plays a crucial role in determining the plasma properties. Examples of non-thermal plasma are low-pressure glow discharge (CCP, ICP, and microwave plasma), atmospheric DBD plasma, and so on. Due to the low gas temperature and high excitation selectivity of non-thermal plasma, the plasma has been applied in many fields. The processing plasma can be classified into corona, the glow, and arc discharge according to the discharge current. The discharge current is in the range of 10^{-7}–10^{-5} A for corona discharge, 10^{-5}–1 A for glow discharge, and more than 1 A for arc discharge [45]. Thus, corona and glow discharge are non-thermal plasma, and arc plasma is thermal plasma. The properties of thermal and non-thermal plasma are summarized in Table 2. In addition, a plasma simulation is also a good tool for predicting discharge characteristics. There are a few plasma simulation studies that have used commercial frequencies; so, we briefly introduce plasma simulation here [46–48]. Plasma simulations can be divided into particle-in-cell (PIC), kinetic simulation and fluid simulation. PIC simulations do not require the assumption required in fluid simulation and has the advantage of high accuracy via directly calculating the motion of charged particles. However, if enough particles are not introduced, there is a statistical error, and the introduction of many particles requires a large number of calculations; so, there is a limit to the calculation speed. The Boltzmann equation solver, which uses the Eulerian scheme, while calculating the velocity space, has high accuracy. Still, it is difficult to calculate this using multi-dimensional calculations that deal with speeds and spaces in two or more directions. For this reason, the Boltzmann equation is solved in many cases due to collisions in 0 and 1 dimensions. Currently, the most used code uses a hybrid method that combines Monte Carlo collision calculation techniques and 0-dimensional Boltzmann solvers based on fluid techniques to easily handle various species [47]. These two codes improve the accuracy of ionization and the generation of excitation species caused by electrons and ions colliding with neutral gases. In the case of high pressure and various gas types, most researchers use fluid simulation.

Table 2. Comparisons of thermal and non-thermal plasma properties [41,49,50].

Properties	Thermal Plasma	Non-Thermal Plasma
Temperature	$T_e \gtrsim T_i \approx T_n$	$T_e \gg T_i \approx T_n$
Electron density (cm^{-3})	10^{15}–10^{20}	$<10^{13}$
Heating	Joule heating Thermal heating	Electron impact process
Characteristics	High gas temperature Low excitation selectivity	Low gas temperature High excitation selectivity
Examples	Arc discharge	Glow discharge

2.2. Design and Construction of CF power Source

A CF (50/60 Hz) power simply consists of a transformer and a voltage controller. Usually, the variable transformer, called VARIAC or power-stat, is used as the voltage controller, and the output voltage is boosted using a transformer with an appropriate step-up ratio. A secondary output voltage of the transformer is about 500–1000 V for low-pressure conditions, and several kV for high-pressure conditions, including atmospheric discharge. The voltmeter and ammeter are needed for the voltage and current measurements during plasma discharge. The CF power source does not necessitate a complicated power conversion system, such as a matching network and frequency generation components. As a result, the power structure is straightforward, cost-effective, and easy to use in a laboratory. At high-pressure and high-voltage discharge conditions, such as atmospheric conditions, the plasma tends to shift from glow discharge to arc spark due to a slow change in voltage polarity [25,26,51–53]. Thus, the power is often applied to the electrode through a suitable resistor to limit the discharge current. Moreover, it is recommended to use a suitable fuse to block the overcurrent for safety. Figure 2 shows the schematic diagram of a CF (50/60 Hz) power source.

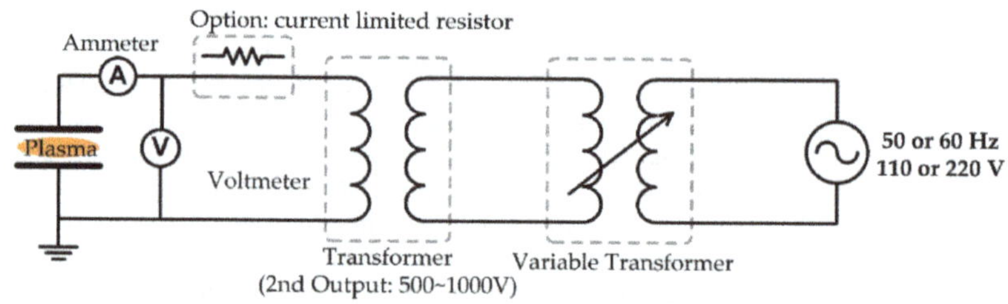

Figure 2. Schematic diagram of a low-frequency power source for plasma discharge generation.

2.3. Plasma Characterizations and Properties Generated Using CF power

The CF-derived plasma is a short-lived discrete discharge similar to the pulsed DC discharge with changing polarity. Figure 3a shows the oscillogram of current and voltage for a CF-derived H_2 discharge. The phase shift between applied voltage and discharge current was not observed in the CF-derived plasma, implying that the CF-derived plasma had resistive properties [27]. Kim et al. investigated plasma parameters of CF-derived discharge using hydrogen, argon, and hydrogen-methane gas [27,29,30,35]. The plasma was generated using the CCP type (electrode diameter: 10 cm, electrode gap: 3–5 cm, pressure: 0.4–1.2 Torr, Power: 10–50 W). The plasma parameters were acquired using a Langmuir probe method [2–4,35,45], and the electron energy distribution is assumed to follow the Maxwell distribution. In the plasma, the electron current (I_e) on the probe as a function of applied voltage (V) is given by:

$$I_e(V) = j_{es}A_p\exp\left(-\frac{e(V_p - V)}{kT_e}\right) \tag{1}$$

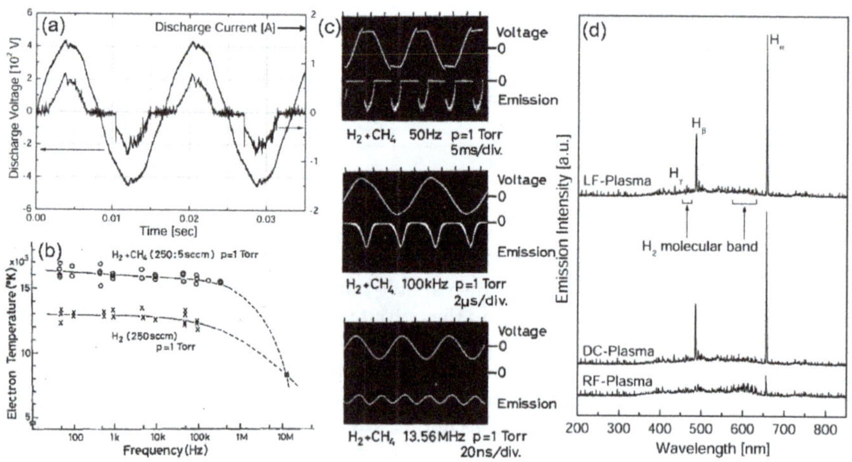

Figure 3. Properties of plasma discharge generated using a commercial frequency (CF, 50/60 Hz) power source and comparison of results for DC- and RF-derived plasma: (**a**) time-resolved applied voltage and discharge current for Ar plasma discharge [29], (**b**) change in the electron temperature in H_2 and H_2-CH_4 plasma as a function of plasma power frequency (○: H_2 + CH_4, ×: H_2) [28], (**c**) oscillograms of optical emission intensities and applied voltage at different power frequencies [28], and (**d**) optical emission spectra of H_2 plasma generated using DC, CF, and RF power sources [29]. ((**a**,**d**) reprinted with the permission from Ref. [29]. (2003, KPS). (**b**,**c**) reprinted from Ref. [28]. (1991, AIP Publishing).).

Where, j_{es} is the saturation current density, A_p is the surface area of the probe tip, V_p is the plasma potential, k is the Boltzmann constant, and T_e is the electron temperature (unit: eV). After taking the natural log in Equation (1) and plotting $\ln(I_e)$ with respect to V, the inverse of the slope for the graph is T_e. Using the values of T_e, A_p, and I_{es}, n_e is given by:

$$n_e = 3.78 \times 10^{11} \frac{I_{es}}{A_p T_e^{1/2}} \tag{2}$$

Where, the unit of T_e is the electron volt (eV). The measured values of T_e and n_e for the hydrogen, argon, and hydrogen-methane plasma were 1.5–3.3 eV and n_e = ~10^8 cm^{-3}. During Ar discharge, the increase in pressure from 400 to 1200 mTorr caused the decrease in T_e (3.3 eV $\rightarrow$ 2.6 eV), while T_e seldom changed at about 3.2 eV, according to the discharge current (pressure: 600 mTorr) [27,30]. In H_2-CH_4 plasma, T_e increased from 2.4 eV to 3.2 eV with the increase in power from 10 W to 45 W [30,35]. Lau et al. measured the discharge parameters of CF-derived Ar plasma using a Langmuir probe [54]. The plasma was generated using a CCP type (electrode diameter: 4.5 cm, electrode gap: 3 cm, power: 5–35 W, pressure: 0.9 Torr). The measured T_e and n_e were in the range of 2–3 eV and ~10^8 cm^{-3}, respectively. These parameters were similar values compared to those of the high-frequency plasma method. Shimozuma et al. obtained the T_e in H_2 and H_2-CH_4 plasma as a function of discharge power frequency, as shown in Figure 3b. The plasma was generated in a CCP type (electrode diameter: 12 cm, gap: 2 cm, pressure: 1.0–1.5 Torr), and T_e was evaluated using the relative intensity method, which is given by [27,28,45,55]:

$$\frac{I_{ij}}{I_{kl}} = \frac{\lambda_{kl} A_{ij} g_i}{\lambda_{ij} A_{kl} g_k} \exp\left(-\frac{E_i - E_k}{k_B T_e}\right) \tag{3}$$

where, I_{ij} and I_{kl} are the spectral intensity, λ_{ij} and λ_{kl} are the wavelengths, A_{ij} and A_{kl} are the transition probabilities, g_i and g_k are the statistical weights, and k_B is the Boltzmann constant. T_e values at 1 kHz and 13.56 MHz were 16,000 K (1.4 eV) and 8200 K (0.71 eV), respectively. The discharge mode also changed from discontinuous to continuous at a power frequency of about 200 kHz, as shown in Figure 3c. In addition, the intensity ratio of the excited molecular hydrogen (H_2*) to H_α and the ratio of H_2* to H_β in 13.56 MHz-generated plasma was much larger than that in CF-derived discharge [28]. The authors concluded that the electron temperature (T_e) in CF-derived plasma is higher than it is in RF-derived plasma [28]. However, the relationship between power frequency and electron temperature cannot be easily generalized without considering the specific system conditions, including the geometric configuration of electrodes, discharge gas type, discharge pressure, discharge voltage, and so on. Kim et al. studied the molecular excitations in H_2 plasma generated at different power frequencies, as shown in Figure 3d [29]. The intensity of molecular excitations in DC- and CF-generated plasma was less than that in RF-generated plasma. This meant that the energy consumption for molecular excitations in RF-generated discharge was much higher than that at DC- and CF-generated plasma. Thus, the main portion of input energy of the molecular gas discharge is consumed for the excitation of molecular vibration, and the energy loss is larger than that of high-frequency discharge [30,56,57]. This difference can be explained as a plasma heating mechanism. In non-thermal plasma, the electron impact processes play a crucial role in the energy transfer, and there are two modes to heat plasma: ohmic heating and stochastic heating. Electrons absorb energy from the applied electric field, and electron-particle collisions produce ohmic heating in bulk plasma. On the other hand, the momentum transfer due to sheath oscillation leads to stochastic heating at the sheath edge [2,3]. Generally, electron heating at a low frequency is dominated by the direct absorption of energy from the electric field. In contrast, the heating mechanism shifts towards stochastic heating processes at a high frequency.

In summary, CF-derived plasma was a discontinuously discharged, while DC- and RF-derived plasma were continuously discharged. The electron temperature and density were in the range of a few eV and 10^8–10^9 cm^{-3}, respectively, and these values are similar to those of other CCP plasma. The direct comparison of plasma parameters for discontinuous and continuous discharges required some attention. In the case of continuous discharge, the mean value could express plasma characteristics because the plasma maintained the glow discharge after discharge build-up. In periodically repeating plasma, the build-up from Townsend to glow discharge was repeated according to power frequency, and this implied the plasma characteristics periodically changed in one discharge period. Ohmic heating was the main heating mechanism in DC- and CF-derived plasma, and the heating mechanism shifted to stochastic heating in RF-derived plasma. Considering plasma processing, CF- and RF-derived plasmas could be applied to the processing of most materials, such as metals and dielectrics. In contrast, they could be applied only to metal deposition and surface treatments for DC plasma. CF-derived plasma has a slower processing speed than DC- and RF-derived discharges do because of the short plasma duration time. Table 3 summarizes the comparisons of DC-, CF-, and RF-derived plasma characteristics in CCP-type discharges. In addition, it is also important to understand the difference between CF- and kHz-derived plasma. Both are discrete discharges, and the difference between CF- and kHz-derived plasmas is the degree of pre-ionization in the off-state of discharges. Pre-ionization refers to the charges that remain in the off-state. As the power frequency in the discontinuous discharge mode increased, the effect of the pre-ionized charge also played an important role in generating plasma, which mainly affected the discharge firing voltage and discharge lag time [58–60]. Thus, CF plasma has a higher discharge firing voltage than kHz-derived plasma does.

Table 3. Comparisons of direct current (DC)-, commercial frequency (50/60 Hz)-, and radio frequency (RF, 13.56 MHz)-derived plasma characteristics in CCP-type discharge [2,3,27–29,41].

Properties	DC	CF	RF
Power frequency	0	50/60 Hz	13.56 MHz
Discharge type	Continuous	Discrete	Continuous
Main heating mechanism	Ohmic	Ohmic	Ohmic, Stochastic
Deposition/treatment material	Metal	Metal, dielectric	Metal, dielectric
Deposition/treatment rate	Lower in DC and RF discharge		
Plasma parameters	Similar in DC and RF discharge (T_e: 1–5 eV, n_e: ~10^8–10^{10})		

3. Plasma Processes Using CF-Derived Discharge

3.1. Film Depositions on Polymer and Polymeric Materials Using CF-Derived Discharge

Chemical vapor deposition (CVD) is a common thin film growth method based on a chemical reaction on a substrate surface [61]. Typical CVD uses thermal energy to drive the chemical reaction and usually requires high temperatures. PECVD is a kind of CVD that combines thermal and plasma-derived chemical reactions [2–4]. The plasma is used to dissociate reactant gases that then recombine on the sample surface into the desired material. Chemical reactions can activate the usage of plasma, and the reaction temperature can be drastically lowered. Thin film formation using the commercial frequency (CF) PECVD technique was first developed by Shimozuma et al. [31]. They grew Si_3N_4, SiO_2, and a-Si:H films on GaAs and Si substrates using CF-PECVD [31–34]. The as-deposited films showed good properties, and the deposition temperature was relatively low, below 200 °C compared to another plasma CVD method. Kim et al. also reported the growth of a-C:H films on glass substrates at room temperature and acquired crack-free films [35]. In addition, wear-resistant depositions, including TiN and TiC films, were performed at relatively low temperatures (350–500 °C), and the resulting films exhibited good properties [36–40]. However, the film depositions using CF-PECVD were not applied to polymer substrates. The film depositions on polymer substrates were performed using the CF sputtering technique [62–65].

Magnetron sputtering deposition is a kind of physical vapor deposition. This technique uses a confined plasma to sputter vaporized atoms from the precursor target, and the vaporized atoms form thin films on the surface of the samples [2–4,7–10]. The samples can be held at RT or heated at the desired temperature during film depositions. Moreover, oxide or nitride materials can be obtained using a reactive sputtering technique. This technique involves the sputtered atoms reacting with gases, such as oxygen and nitrogen, upon condensation on the sample. The film deposition on polymer substrates has many limitations due to low heat-deflection temperature and the weak surface of polymers. Thus, polymer substrates can be easily damaged under the deposition process, and this can affect the characteristics of as-deposited films on polymer and polymer-like materials. In this respect, CF-generated magnetron sputtering can be considered one of the growth techniques on polymer and polymer-like substrates. Jung et al. and Kim et al. reported the deposition of transparent conductive oxides on various polymers using the same CF-sputtering system [62,63], which was a conventional type sputtering gun, as shown in Figure 4a. Jung et al. studied the effects of film thickness and growth temperature for ITO film depositions on polyether-sulfone (PES) substrate [62]. A distance of 10 cm was kept between the ITO target (In_2O_3:SnO_2 = 9:1 wt%, 3 inches) and the substrate, and the applied voltage was maintained at 280 V. Ar was used as the sputtering gas (pressure: 2.2 mTorr, flow rate: 30 sccm), and the growth rate of the film was about 7 nm/min. As the growth temperature increased from RT to 140 °C, the sheet resistance changed from 69 to 151 Ω/sq., the roughness decreased from 1.1 to 0.8 nm, and the optical band gap increased from 3.61 to 3.83 eV. Due to the film thickness change from 134 to 237 nm, the surface roughness increased from 1.2 to 2.2 nm, and the sheet resistance decreased from 316 to 56 Ω/sq., whereas other properties showed a little variation. All films showed good transparency (~85%) in the visible wavelength region and an amorphous phase. Kim et al. reported the growth of indium zinc oxide (IZO) film on various polymers such as PES, PET (polyethylene terephthalate), and PC (polycarbonate) [63]. The sputtering system was the same as mentioned above [62], and the difference was the target material, which was composed of In_2O_3 and ZnO (9:1 wt%). The Ar gas flow rate was set to 30 sccm, and the working pressure was maintained at 1.9 mTorr to generate plasma. The plasma was generated at different voltages between 300 V and 320 V, and the polymer substrates were kept at RT during IZO growth. With an increase in the applied voltage, the growth rate of IZO films increased from 4.5 nm/min to 6.5 nm/min. The resulting IZO films were amorphous with an average transmittance of approximately 84% and had a very smooth surface with a surface roughness (R_{rms}) of about 2 nm. These excellent properties demonstrate that the CF technique is suitable for polymer substrates. Another application of CF-derived plasma is film deposition on a paper substrate. Kim et al. reported an ITO film deposition on a paper sheet using the CF DC-pulsed magnetron sputtering technique [65]. They used a different concept to design the sputtering gun, as shown in Figure 4a; Figure 4c represents the schematic diagram of the CF DC-pulsed sputtering for ITO film deposition on paper sheets. They modified the magnetic array inside a magnetron gun to acquire a magnetic field parallel to the substrate. The size of the ITO target (In_2O_3:SnO_2 = 9:1 wt%) was 416 mm × 298 mm, and its distance from the substrate was 60 mm. Ar-O_2 mixing gas (Ar:O_2 = 98:2, 1 sccm) and pure Ar gas (185 sccm) were utilized as sputtering gases (working pressure: 3 mTorr).

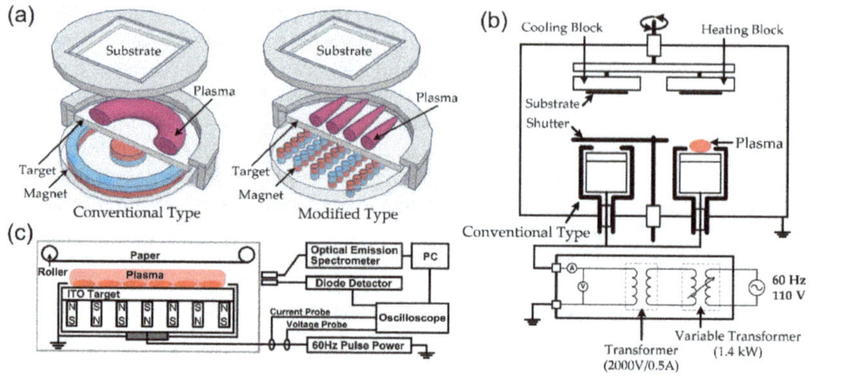

Figure 4. Magnetron sputtering system for transparent conductive thin film deposition using a commercial-frequency (CF) power source: (**a**) comparison between conventional sputtering gun (**left**) and modified sputtering gun with parallel magnet array (**right**), (**b**) ITO and IZO film depositions on glass and polymer substrates using CF (60 Hz) sputtering [63,64], and (**c**) ITO film depositions on paper substrates using CF (60 Hz) DC-pulsed sputtering [65]. ((**c**) reprinted with permission from Ref. [65]. 2021, Elsevier).

The plasma was generated using DC-pulsed power with a negative square pulse type and a frequency of 60 Hz. The film growth was carried out on paper sheets at a power of 93 W (292 V, 32 mA), with a growth rate of 40 nm/min. The ITO films that were deposited exhibited desirable characteristics, including a high level of transmittance (>85%), a good sheet resistance (40 Ω/sq.), a hydrophobic surface with a water droplet contact angle of 116°, and a cubic crystal structure with a grain size of 20.5 nm. Most of all, the fabric structure of the paper after ITO film depositions seldom changed compared to the original paper, and the ITO-deposited sheet did not bend due to the residual stress of ITO films. Compared to DC and RF methods, CF-derived plasma sputtering could reduce the rapid heating of samples due to the incidence of high-energy particles. And this method could prevent sample damage from occurring due to the accumulation of charged particles. Because CF-derived plasma continuously changes the polarity of the discharge [65,66]. On the other hand, the plasma processing time was longer than that of continuous discharges. These results [62–65] indicate that the CF-derived plasma sputtering method is effective for film depositions on soft substrates, including papers, polymers, and fibers.

3.2. Surface Modifications of Polymer and Polymer-like Materials Using CF-Derived Discharge

CF-derived plasma techniques have been widely explored for surface modification of various materials. Surface modification of polymers is especially crucial in achieving desirable surface properties such as adhesion, wettability, and biocompatibility [13,14,41,66–69]. To this end, Lau and colleagues conducted a study on the surface modification of polytetrafluoroethylene (PTFE) films using Ar plasma generated using CF power at a frequency of 50 Hz [54]. The discharge reactor was a traditional diode configuration (see Figure 5a), and Ar plasma (applied voltage: 240 V, pressure: 120 Pa/0.9 Torr) was used to modify the surface of PTFE films. The measured electron temperature and density were 2–3 eV and ~10^8 cm^{-3}, respectively. After plasma exposure, the contact angle for the water was changed from 114° to 91° with varying applied power. They claimed that wettability increased from incorporating oxygen-containing functional groups on the treated surface and the concomitant reduction in fluorine elements. Bhak et al. performed the surface treatment of PES using a magnetized plasma and a CF (60 Hz) power source [70]. The traditional diode-type configuration was used to generate plasma discharge. Each electrode had an array of Nd-Fe-B permanent magnets, similar to a magnetic mirror, as shown in Figure 5b. After subjecting the water droplets to Ar plasma treatment (pressure: 800 mTorr, power: 20 W, treatment time: 0–20 min), the contact angle decreased from 80° to 30°, while the surface roughness of the films remained unchanged (~3.0 nm). These results indicate that the magnetized plasma did not cause any damage to the surface of the films, and it was attributed to the magnetic field's ability to trap energetic particles.

The application of plasma surface treatment has proven to be advantageous in the field of biomaterials, specifically for gelatin. Gelatin is a transparent and colorless protein commonly derived from the partial hydrolysis of native collagen. Gelatin is a biocompatible, biodegradable, non-immunogenic, and non-antigenic material [71,72]. Gelatin has been widely used as a gelling substance in food, medications, drug capsules, photographic films, and cosmetics. Prasertsung et al. reported the surface modification of cross-linked gelatin films using oxygen, nitrogen, and air glow discharge generated using a 50 Hz power source [71,73,74], and the configuration of the discharge cell was a traditional diode structure, as shown in Figure 5c. After the O_2, N_2, and air plasma treatment, in this order (pressure: 1 mbar, treatment time: 1–13 s), the contact angle for a water droplet dramatically decreased to about 23° compared to untreated films (water contact angle: 87°). For all cases, the surface energy of the gelatin films rose progressively up to a plasma exposure duration of 15 s, after which the energy values stabilized. Notably, the polar components of the surface energy increased, while the dispersive components decreased, indicating the emergence of polar functional groups on the surface of the gelatin films as a result of the plasma treatments. The plasma, which was oxygen-containing, facilitated the formation of polar functional groups, including C=O and O-C=O groups, on the polymer substrates' surface, resulting in the surface's hydrophilic properties [75] and N_2 plasma treatment played a similar role on the surface of gelatin films. Prasertsung et al. also studied the attachment and growth behavior of mouse fibroblasts (L239) and rat bone marrow-derived mesenchymal stem cells (MSCs) on nitrogen plasma-treated gelatin films using the same plasma treatment system. The suitable water contact angle and O/N ratio of nitrogen plasma-treated gelatin film for best L929 and MSC attachment were 27°–32° and 1.4, respectively [71,74].

Recently, atmospheric plasma was generated using a CF (50/60 Hz) power source, and the plasma was used as the surface treatment method. Atmospheric plasma can be widely used for surface modification because of the eco-friendly properties of the plasma and inexpensive methods compared to vacuum plasma treatments. Usually, nonthermal atmospheric plasma is generated using high-frequency power sources of several kHz or more [76–79]. Choi et al. developed the nonthermal atmospheric pressure plasma torch, operated by a 60 Hz power source with a neon transformer (output voltage: 4 kV, output current: 120 mA) [80]. The cold plasma torch (N_2 flow rate: 40 slm, applied voltage: ~2 kV), as shown in Figure 5d, was applied to polypropylene polymer substrates to enhance the bonding strength, which is evaluated by the lap shear strength. The plasma torch showed a low gas temperature below 60 °C, and the mainly observed excitation emissions were N_2^* ($C^3\Pi_u \rightarrow B^3\Pi_g$) at a wavelength of 337.1 nm, N_2^* ($B^3\Pi_u^+ \rightarrow A^3\Sigma_u^+$) at a wavelength of 715.3 nm, and N_2^+ ($B^2\Sigma_u^+ \rightarrow X^2\Sigma_g^+$) at a wavelength of 391.4 nm. As a result, the maximum bond strength of about 10.5 MPa was obtained at the optimal condition, and the value is about 60 times higher than that of plasma-untreated samples. According to the authors, the use of excited species such as N_2^* and N_2^+ can create polymer-excited states, which increases the population of oxygen and nitrogen atoms on the polymer surface and subsequently improves bonding strength.

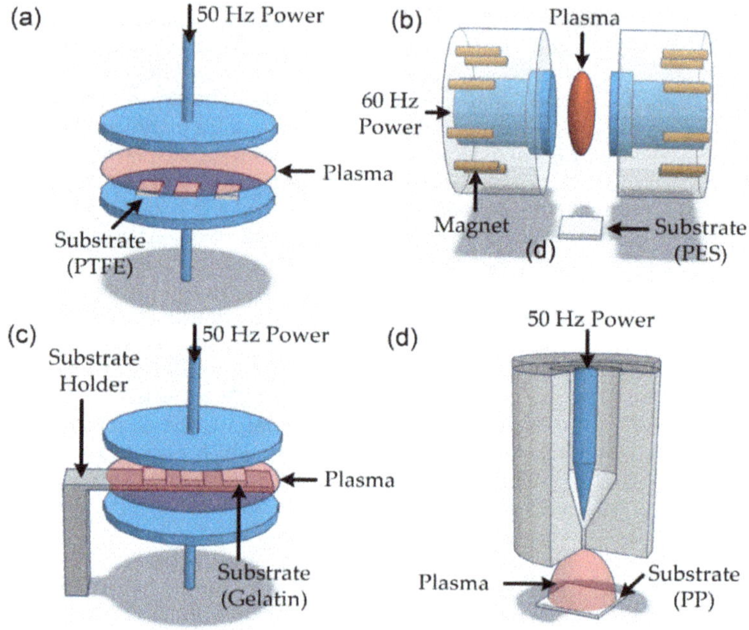

Figure 5. Polymer surface treatments using commercial-frequency (50/60 Hz) plasma source: (**a**) simple diode configuration for surface modification of polytetrafluoroethylene [54], (**b**) magnetic enhanced diode-electrode arrangement for surface treatment of polyether-sulfone film [70], (**c**) simple diode configuration for plasma generation (the substrate holder was positioned at the center of two electrodes) [71,73,74], (**d**) plasma torch configuration [80].

Joshi et al. employed a surface treatment technique using a DBD generated using a 50 Hz power source [81]. Two rectangular copper electrodes (size: 5 cm × 3.5 cm × 1 cm, gap distance: 3.5 mm) and the lower electrode were covered by a polycarbonate (PC, size: 10 cm × 8 cm × 0.2 cm) plate, resulting in the type of DBD (see Type 2 in Figure 6). The plasma discharge was generated via applying a voltage of 13 kV, and a mixture of air-argon gas (Ar flow rate: 2slm) was used as the treatment gas. Following plasma treatment lasting between 5 and 60 s, the water contact angle on untreated PP films was 93.7° (with a surface energy of 36.7 mJ/m^2). However, after a treatment time of just 5 s, the contact angle decreased significantly to 67.3° (with a surface energy of 41.6 mJ/m^2), and no further change was observed with increased treatment time.

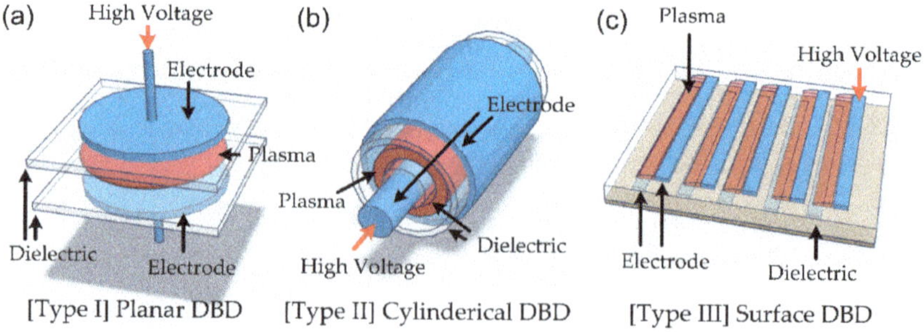

Figure 6. Schematic diagram of typical dielectric barrier discharge (DBD) configurations: (**a**) planar-type DBD source, (**b**) cylindrical type DBD source, and (**c**) surface-discharge type DBD sources.

3.3. In-Package Cold Plasma Treatment on Organic Materials Using CF-Derived Plasma

In-package plasma is a type of plasma technology that involves using a plasma source integrated into a packaging material or product [82–88]. This allows for the inactivation of microorganisms and other pathogens on the product's surface without additional processing steps. The plasma source used in in-package plasma systems is typically a non-thermal plasma, which means that it operates at relatively low temperatures compared to thermal plasmas. Non-thermal plasma sources can generate a variety of reactive species, such as ozone, hydrogen peroxide, and other radicals, which can effectively inactivate microorganisms on surfaces. The generation method for in-package plasma is usually a dielectric barrier discharge (DBD). The DBD is a form of CCP, and the plasma is generated between two electrodes separated by an insulating (or dielectric) barrier, as shown in Figure 6. The dielectric barrier can stop excessive currents and prevent arc (or spark) formation. The DBD has been widely applied in various fields: ozone generation, ultraviolet (UV) and excimer light sources, polymer surface modification, bio/medical applications, air pollutant cleaning, etc. [3,41,67–69,89–96]. Usually, mid-frequency power (typically 10–30 kHz) is used to generate DBD, and DC discharge cannot be applied due to the existence of the dielectric barrier [3,41,67,93]. CF power is also applied to generate DBD plasma, which does not require an impedance-matching network. This simplicity and affordability make DBD systems more practical for industrial applications. DBD devices can be constructed in various configurations, and planar, cylindrical, and surface-discharge type configurations (Figure 6) are widely used.

To create an in-package plasma system, the plasma source is typically integrated into the packaging material, as shown in Figure 7. The in-package plasma can be generated using planar-type DBD and surface-discharge type DBD. The plastic packaging materials for the in-package plasma are usually polymer-based materials such as LDPE (low density polyethylene), a PE pouch, a rigid PP/Cryovac pouch, and PET films. Without additional processing steps, this generates the plasma directly on the product's surface. One of the advantages of in-package plasma is that it can be used to inactivate micro-organisms on products sensitive to heat or other processing methods. Rana et al. studied microbial reduction in strawberries using the in-package plasma [82]. The DBD system consisted of Al plate electrodes (gap: 30 mm), and the in-package plasma was generated inside the LDPE packet at a voltage of 60 kV (power frequency: 50 Hz). The LDPE was filled with air and contained the strawberry sample. Thus, the LDPE acted as both the container and the dielectric barrier. After plasma exposure, the reduction in bacteria, yeast, and mold was 2.1-$\log_{10}$ CFU/g. Here, CFU (colony-forming unit) is a unit, used to estimate the number of microbial cells. In addition, an n $\log_{10}$ reduction means that the concentration of remaining contaminants is only 10^{-n} times that of the original. For example, a 1-$\log_{10}$, 2-$\log_{10}$, 3-$\log_{10}$, and 4-$\log_{10}$ corresponds to 90%, 99%, 99.9%, and 99.99% reduction, respectively, from the initial concentration. Mahonot et al. reported the reduction in the natural microflora of carrots after in-package plasma [85]. The plasma was generated using planar-type DBD (60 Hz, 80–100 kV), as shown in Figure 7a. The DBD system comprised two Al plate electrodes (diameter: 158 mm) and a polypropylene dielectric layer (thickness: 2 mm). A PP box containing carrots was sealed with polymeric film (Cryovac BB3050, Sealed Air Corp. Charlotte, NC, USA) and positioned between two electrodes. After the plasma treatment, a 2.1-$\log_{10}$ CFU/g reduction was observed for aerobic mesophiles and yeast. Zhao et al. investigated the reduction in yeast and micro-organisms on fermented vegetables (radish paocai) using the in-package plasma treatment. The DBD system comprised two Al circular electrodes and two PP barrier layers. The packed paocai was positioned between two barrier layers (gap: 40 mm) and was plasma treated three times at a voltage of 60 kV (power frequency: 60 Hz) for 60 s. After the treatment, the reduction rate of micro-organisms ranged from 4.54 to 5.61 log CFU/g. Los et al. and Ziuzina et al. also reported the reduction in bacteria, fungi, and pest insects using in-package plasma [97,98]. The DBD system consisted of two round Al electrodes (diameter: 15 mm) and a PP container with a sealed PP bag (Cryovac B2630, Sealed Air Corp., Charlotte, NC, USA). The wheat and pest insect (red flour beetle) were plasma-treated at 80 kV (power frequency: 50 Hz). The reduction rate for bacteria and fungi on wheat was 1.5 and 2.5 log10 CFU/g, respectively. A mortality of 95.0–100% for preadult stages can be achieved within seconds of treatment, but longer plasma exposure (5 min) is required to kill adult insects. In-package plasma can be used to sterilize fresh produce or other food products without damaging the product or altering its quality. In addition, in-package plasma can be used in various other applications, such as medical devices, electronics, and other products requiring surface sterilization or decontamination. Overall, in-package plasma is a promising technology that has the potential to improve the safety and quality of a wide range of products while also reducing the need for additional processing steps and increasing efficiency in various industries.

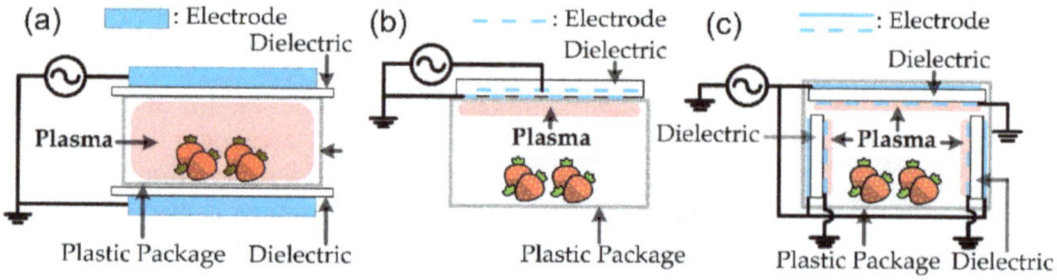

Figure 7. Schematic diagram of typical configurations for the generation of in-package cold plasma: (**a**) in-package plasma using planar-type DBD source located outside the package, (**b**) in-package plasma using surface-discharge type DBD source positioned outside the package, and (**c**) in-package plasma using surface-discharge type DBD source positioned inside the package. (Adapted with permission from Ref. [88]. 2019, Elsevier).

3.4. Comparison of CF-Derived Plasma and Cold Atmospheric Pressure Plasma Jet

Cold atmospheric-pressure plasma jets (APPJs) are a point of interest in many fields, and their applications range from lab-scale applications to industrial production [99–103]. Especially, compact, affordable, easy-to-use, and flexible plasma tools have been in great demand in lab-scale research. In fact, the cost is one of the important factors in lab-scale applications. There are two major cost factors in plasma generation equipment, which are the vacuum equipment and power source. Atmospheric plasmas do not require vacuum equipment, and APPJ satisfies these needs. In the CF-derived plasma, the power source is inexpensive, and it is possible to make it yourself because the structure of the power source is very simple. In addition, atmospheric pressure DBD discharge is possible, as mentioned in the previous session. In conclusion, APPJ and CF-derived plasma are cost-effective and easy to use, allowing a variety of applications. Figure 8 shows the schematic diagram of typical configurations for APPJ. The kINPen is a representation of a cold APPJ, which was developed at the Leibniz Institute for Plasma Science and Technology and commercialized by neoplas GmbH [104,105]. The schematic structure of kINPen is shown in Figure 8b. The kINPen is typically driven at a sinusoidal frequency of 1 MHz and can be operated with an inert gas and molecular gas mixture (up to 2%). The gas temperature is ~40 °C, the plasma temperature is up to 4 eV, and the estimated electron density is ~10^{14} cm^{-3} [104]. These properties are suitable for the processing of biomaterials and are applied to medicine, biology, dental treatment, genetics, etc. Also, this is used in the surface modification of many materials, surface cleaning, and decomposition of contaminants. The piezoelectric cold plasma generator (PCPG) is also another type of APPJ pencil [106], which is based on the resonant piezoelectric transformer [107]. Piezoelectric transformer, which is a simple PZT ($PbZr_{1-x}Ti_xO_3$) ceramic rod, has a step-up ratio of about 1000 times and an output voltage of 10 kV or more is possible with a low input power of 25 W. PCPG was commercialized by relyon plasma GmbH, and the schematic diagram is shown in Figure 8d. PCPG can use various discharge modes, including DBD mode. This can operate with very low power input (typically ~10 W), and the advantage is the wireless plasma pencil. The kINPen and PCPG have the advantages of affordable equipment, a low gas temperature, the use of additive gas, and excellent portability. For CF-derived plasma, the power construction cost is very low. Especially in the case of atmospheric DBD discharge, plasma generation equipment can be inexpensively built since vacuum equipment is not required. In addition, the plasma volume can be easily increased to change the electrode configuration and reduce sample damages due to the production of slow-changing, discontinuous discharge. The comparison between kINPEN-, PCPG-, and CF-derived plasma is summarized in Table 4.

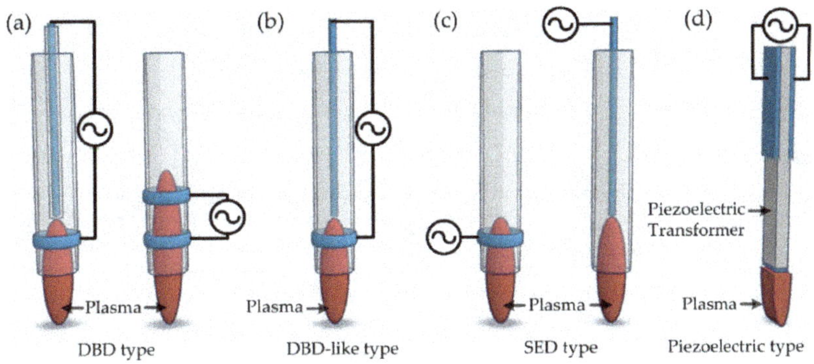

Figure 8. Schematic diagram of typical configurations for the generation of cold atmospheric pressure plasma jet: (**a**) dielectric barrier discharge (DBD) jets, (**b**) DBD-like discharge jet, (**c**) single electrode discharge (SED) jets, and (**d**) piezoelectric direct plasma jet.

Table 4. Comparisons between kINPen-, PCPG- (piezoelectric cold plasma generator), and CF-derived plasma as lab-scale equipment.

Type	Frequency	Pressure (Torr)	Characteristics	Ref.
kINPen	Typically, 1 MHz	~760	Low gas temperature: ~40 °C T_e = 4 eV, n_e = ~10^{14} cm^{-3} Wired portable Affordable cost	[95]
PCPG	40-90 kHz	~760	Low gas temperature: ~40 °C Low input power: typically, 10 W Wireless portable Affordable cost	[97]
CF-derived Plasma	50/60 Hz	~1	T_e = 1.4–3.2 eV, n_e = ~10^8 cm^{-3} Inexpensive power source	[27–30,49]
		~760	Inexpensive power source High voltage: 50–80 kV	[89,90]

4. Conclusions

This review describes the characteristics and applications of plasma using a frequency of 50/60 Hz as the plasma power source. Due to low-frequency, the plasma reactor configurations in most plasma processes are of the CCP type. This method can be utilized in a wide range of applications, such as thin film deposition, surface oxidation/nitridation of thin films, and surface treatment of polymers. This method exhibits properties similar to those of DC-generated plasma but with electrically bipolar features, which prevent damage to the substrate due to charge accumulation. Furthermore, unlike DC power, this process can be applied to almost all materials, including dielectrics and polymers. The simple structure and low cost of CF power are particularly advantageous, making it a viable option for laboratory-scale manufacturing and use. Consequently, the CF-derived plasma process is helpful at a laboratory scale due to its simplicity, and the results for the plasma processes are also outstanding.

Author Contributions: H.T.K., C.M.J., S.H.K. and S.-Y.L. contributed to concept. H.T.K., C.M.J., S.H.K. and S.-Y.L. wrote the original manuscript draft and reviewed references. All authors equally contributed to the critical review and editing of the manuscript. All authors have read and agreed to the published version of the manuscript.

Funding: This research was supported by Basic Science Research Program through the National Research Foundation of Korea (NRF) funded by the Ministry of Education (NRF-2018R1D1A1B07049180).

Institutional Review Board Statement: Not applicable.

Data Availability Statement: Not applicable.

Conflicts of Interest: The authors declare no conflict of interest.

References

1. Sahoo, A.; Tripathy, S. Development in plasma arc welding process: A review. *Mater. Today Proc.* **2021**, *41*, 363–368. [CrossRef]
2. Lieberman, M.A.; Lichtenberg, A.J. *Principles of Plasma Discharges and Materials Processing*, 2nd ed.; Wiley & Sons, Inc.: Hoboken, NJ, USA, 2005.
3. Fridman, A. *Plasma Chemistry*; Cambridge University Press: Cambridge, UK, 2008.
4. Hitchon, W.N.G. *Plasma Processes for Semiconductor Fabrication*; Cambridge University Press: Cambridge, UK, 1999.
5. Kabongo, G.L.; Mothudi, B.M.; Dhlamini, M.S. Advanced Development of Sustainable PECVD semitransparent Photovoltaics: A Review. *Front. Mater.* **2021**, *8*, 762030. [CrossRef]
6. Yi, K.; Liu, D.; Chen, X.; Yang, J.; Wei, D.; Liu, Y.; Wei, D. Plasma-Enhanced Chemical Vapor Deposition of Two-Dimensional Materials for Applications. *Acc. Chem. Res.* **2021**, *54*, 1011–1022. [CrossRef] [PubMed]
7. Gudmundsson, J.T. Physics of technology of magnetron sputtering discharge. *Plasma Sources Sci. Technol.* **2020**, *29*, 113001. [CrossRef]
8. Liang, J.; Liu, Q.; Li, T.; Luo, Y.; Lu, S.; Shi, X.; Zhang, F.; Asiri, A.M.; Sun, X. Magnetron sputtering enabled sustainable synthesis of nanomaterials for energy electrocatalyst. *Green Chem.* **2021**, *23*, 2834–2867. [CrossRef]
9. Larhlimi, H.; Ghailane, A.; Makha, M.; Alami, J. Magnetron sputtered titanium carbide-based coatings: A review of science and technology. *Vacuum* **2022**, *197*, 110853. [CrossRef]
10. Deng, Y.; Chen, W.; Li, B.; Wang, C.; Kuang, T.; Li, Y. Physical vapor deposition technology for coated cutting tools: A review. *Ceram. Int.* **2020**, *46 Pt B*, 18373–18390. [CrossRef]
11. Huff, M. Recent Advances in Reactive Ion Etching and Applications of High-Aspect Ratio Microfabrication. *Micromachines* **2021**, *12*, 991. [CrossRef]
12. He, T.; Wang, Z.; Zhong, F.; Fang, H.; Wang, P.; Hu, W. Etching Techniques in 2D Materials. *Adv. Mater. Technol.* **2019**, *4*, 1900064. [CrossRef]
13. Vajpayee, M.; Singh, M.; Ledwani, L. Non-thermal plasma treatment of cellulosic biopolymer to enhance its surface property for various applications: A review. *Mater. Today Proc.* **2021**, *43*, 3250–3255. [CrossRef]
14. Laroussi, M. Cold Plasma in Medicine and Healthcare: The New Frontier in Low Temperature Plasma Applications. *Front. Phys.* **2020**, *8*, 74. [CrossRef]
15. Ma, C.; Nikiforov, A.; Hegemann, D.; Geyter, N.D.; Morent, R.; Ostrikov, K. Plasma-controlled surface wettability: Recent advances and future applications. *Int. Mater. Rev.* **2023**, *68*, 82–119. [CrossRef]
16. Sanito, R.C.; You, S.-J.; Wang, Y.-F. Degradation of contaminants in plasma technology: An overview. *J. Hazard. Mater.* **2022**, *424*, 127390. [CrossRef] [PubMed]
17. Friedrich, J. *The Plasma Chemistry of Polymer Surfaces: Advanced Techniques for Surface Design*; Wiley-VCH: Weinheim, Germany, 2012.
18. Laroque, D.A.; Seó, S.T.; Valencia, G.A.; Laurindo, J.B.; Carciofi, B.A.M. Cold plasma in food processing: Design, mechanisms, and application. *J. Food Eng.* **2022**, *312*, 110748. [CrossRef]
19. Yang, D.-K.; Wang, D.; Huang, Q.-S.; Song, Y.; Wu, J.; Li, W.-X.; Wang, Z.-S.; Tang, X.-H.; Xu, H.-X.; Liu, S.; et al. The development of laser-produced plasma EUV light source. *Chip* **2022**, *1*, 100019. [CrossRef]
20. Jaritz, M.; Behm, H.; Hopmann, C.; Kirchheim, D.; Mitschker, F.; Awakowicz, P.; Dahlmann, R. The effect of UV radiation from oxygen and argon plasma on the adhesion of organosilicon coatings on polypropylene. *J. Phys. D-Appl. Phys.* **2017**, *50*, 015201. [CrossRef]
21. Jiang, Y.; Peng, S.; Wu, W.; Ma, T.; Zhang, J.; Ren, H.; Li, K.; Zhang, T.; Wen, J.; Xu, Y.; et al. A miniaturized ECR plasma flood gun for wafer charge neutralization. *Rev. Sci. Instrum.* **2020**, *91*, 033319. [CrossRef]
22. Zamri, A.A.; Ong, M.Y.; Nomanbhay, S.; Show, P.L. Microwave plasma technology for sustainable energy production and the electromagnetic interaction within the plasma system: A review. *Environ. Res.* **2021**, *197*, 111204. [CrossRef]
23. Deng, P.; Xiao, W.; Wang, F.; Zhang, Z. Design of a Novel Microwave Plasma Source Based on Ridged Waveguide. *Prog. Electromagn. Res. Lett.* **2021**, *101*, 19–27. [CrossRef]
24. Zalieckas, J.; Pobedinskas, P.; Greve, M.M.; Eikehaug, K.; Haenen, K.; Holst, B. Large area microwave plasma CVD of diamond using composite right/left-handed materials. *Diam. Relat. Mat.* **2021**, *116*, 108394. [CrossRef]

25. Fujiwara, T.; Sato, T.; Sekikawa, J.; Yamada, H. Transient glow discharge in nitrogen after the breakdown. *J. Phys. D-Appl. Phys.* **1994**, *27*, 826–829. [CrossRef]
26. Okazaki, S.; Kogoma, M.; Uehara, M.; Kimura, Y. Appearance of stable glow discharge in air, argon, oxygen and nitrogen at atmospheric pressure using a 50 Hz source. *J. Phys. D-Appl. Phys.* **1993**, *26*, 889–892. [CrossRef]
27. Kim, H.T.; Kim, C.-D.; Pyo, M.S.; Park, C. Electrical and optical characteristics of Ar plasma generated by low-frequency (60 Hz) power source. *Korean J. Chem. Eng.* **2014**, *31*, 1892–1897. [CrossRef]
28. Shimozuma, M.; Tochitani, G.; Tagashira, H. Optical emission diagnostics of H_2+CH_4 50 Hz–13.56 MHz plasmas for chemical vapor deposition. *J. Appl. Phys.* **1991**, *70*, 645–648. [CrossRef]
29. Kim, H.T.; Park, D.K.; Choi, W.S. Measurements of plasma parameters in low-frequency (60 Hz) hydrogen discharge. *J. Korean Phys. Soc.* **2003**, *42*, S916–S919.
30. Kim, H.T. Diagnostics and Some Applications of Low-Frequency (60 Hz) Plasma. Ph.D. Thesis, Kyungpook National University, Daegu, Republic of Korea, 2002.
31. Shimozuma, M.; Kitamori, K.; Ohno, H.; Hasegawa, H.; Tagashira, H. Room temperature Deposition of Silicon Nitride Films Using Very Low Frequency (50 Hz) Plasma CVD. *J. Electron. Mater.* **1985**, *14*, 573–586. [CrossRef]
32. Tochitani, G.; Shimozuma, M.; Tagashira, H. Deposition of silicon oxide films from TEOS by low-frequency plasma chemical vapor deposition. *J. Vac. Sci. Technol. A* **1993**, *11*, 400–405. [CrossRef]
33. Hashizume, T.; Hasegawa, H.; Tochitani, G.; Shimozuma, M. Annealing behavior of HF-treated GaAs capped with SiO_2 films prepared by 50-Hz plasma-assisted chemical vapor deposition. *Jpn. J. Appl. Phys.* **1992**, *31*, 3794–3800. [CrossRef]
34. Tochitani, G.; Shimozuma, M.; Tagashira, H. Properties of hydrogenated amorphous silicon films prepared by low-frequency (50 Hz) plasma-enhanced chemical-vapor deposition. *J. Appl. Phys.* **1992**, *72*, 234–238. [CrossRef]
35. Kim, H.T.; Sohn, S.H. Effects of applied power on hydrogenated amorphous carbon (a-C:H) film deposition by low frequency (60 Hz) plasma-enhanced chemical vapor deposition. *Vacuum* **2012**, *86*, 2148–2151. [CrossRef]
36. Shimozuma, M.; Date, H.; Iwasaki, T.; Tagashira, H.; Yoshino, M.; Yoshida, K. Three-dimensional deposition of TiN film using low frequency (50 Hz) plasma chemical vapor deposition. *J. Vac. Sci. Technol. A* **1997**, *15*, 1897–1901. [CrossRef]
37. Yoshino, M.; Shimozuma, M.; Date, H.; Rodrigo, A.; Tagashira, H. Properties of TiN Films on Heated Substrate Below 550 °C by 50 Hz Plasma-Enhanced Chemical Vapor Deposition. *Jpn. J. Appl. Phys.* **2000**, *39*, 1299–1302. [CrossRef]
38. Kim, H.T.; Kim, M.J.; Sohn, S.H. Characterization of TiN thin films grown by low-frequency (60 Hz) plasma enhanced chemical vapor deposition. *J. Phys. Chem. Solids* **2012**, *73*, 931–935. [CrossRef]
39. Kim, H.T.; Chae, C.S.; Han, D.H.; Park, D.K. Effect of substrate temperature and input power on TiN film deposition by low-frequency (60 Hz) PECVD. *J. Korean Phys. Soc.* **2000**, *37*, 319–323.
40. Kim, H.T.; Lee, S.-Y.; Lee, H.R.; Park, C. Effects of growth temperature on titanium carbide (TiC) film formation using low-frequency (60 Hz) plasma-enhanced chemical vapor deposition. *Korean J. Chem. Eng.* **2018**, *35*, 246–250. [CrossRef]
41. Roth, J.R. *Industrial Plasma Engineering Vol. 2: Applications to Nonthermal Plasma Processing*; Institute of Physics Publishing: Bristol, UK, 2001.
42. Meichester, J.; Schmidt, M.; Schneider, R.; Wagner, H.-E. *Nonthermal Plasma Chemistry and Physics*; CRC Press: Boca Raton, FL, USA, 2012.
43. Hong, Y.; Niu, J.; Pan, J.; Bi, Z.; Ni, W.; Liu, D.; Li, J.; Wu, Y. Electron temperature and density measurement of a dielectric barrier discharge argon plasma generated with tube-to-plate electrodes in water. *Vacuum* **2016**, *130*, 130–136. [CrossRef]
44. Park, S.; Choe, W.; Moon, S.Y.; Yoo, S.J. Spatio-temporally resolved electron temperature in argon radio-frequency capacitive discharge at atmospheric pressure. *Plasma Sources Sci. Technol.* **2015**, *24*, 032006. [CrossRef]
45. Wong, C.S.; Mongkolnavin, R. *Elements of Plasma Technology*; Springer: Singapore, 2016.
46. Brieda, L. *Plasma Simulation by Example*; CRC Press: Boca Raton, FL, USA, 2019.
47. Kushner, M.J. Hybrid modelling of low temperature plasma for fundamental investigations and equipment design. *J. Phys. D Appl. Phys.* **2009**, *42*, 194013. [CrossRef]
48. Cohen, B.I. Perspectives on research in computational plasma physics with applications to experiments. *IEEE Trans. Plasma Sci.* **2020**, *48*, 757–767. [CrossRef]
49. Tendero, C.; Tixier, C.; Tristant, P.; Desmaison, J.; Leprince, P. Atmospheric pressure plasmas: A review. *Spectroc. Acta Part B-Atom. Spectr.* **2006**, *61*, 2–30. [CrossRef]
50. Mumtaz, S.; Khan, R.; Rana, J.N.; Javed, R.; Iqbal, M.; Choi, E.H.; Han, I. Review on the Biomedical and Environmental Applications of Nonthermal Plasma. *Catalysis* **2023**, *13*, 685. [CrossRef]
51. Kekez, M.M.; Barrault, M.R.; Craggs, J.D. Spark channel formation. *J. Phys. D-Appl. Phys.* **1970**, *3*, 1886–1896. [CrossRef]
52. Baeva, M.; Lofhagen, D.; Becker, M.M.; Uhrlandt, D. Fluid modelling of DC argon micro-plasmas: Efects of the electron transport description. *Plasma Chem. Plasma Process.* **2019**, *39*, 949–968. [CrossRef]
53. Saifutdinov, A.I. Numerical study of various scenarios for the formation of atmospheric pressure DC discharge characteristics in argon: From glow to arc discharge. *Plasma Sources Sci. Technol.* **2022**, *31*, 094008. [CrossRef]
54. Lau, Y.T.; Chiu, W.S.; Lee, H.C.; Woo, H.J.; Chin, O.H.; Tou, T.Y. A low power 50 Hz argon plasma for surface modification of polytetrafluoroethylene. *Plasma Sci. Technol.* **2022**, *24*, 114001. [CrossRef]
55. Griem, H.R. *Principle of Plasma Spectroscopy*; Cambridge University Press: Cambridge, UK, 1992.
56. Eletskiĭ, A.V.; Smirnov, B.M. Nonuniform gas discharge plasma. *Phys. Usp.* **1996**, *39*, 1137. [CrossRef]

57. Baranov, V.Y.; Ul'yanov, K.N. Contraction of a positive column. I, II(Positive column contraction in Ar and Ar-Cs mixtures, developing theory for nonisothermal local collision arc). *Sov. Phys. Tech. Phys.* **1969**, *14*, 176.
58. Zhao, Z.; Li, J. Repetitively pulsed gas discharges: Memory effect and discharge mode transition. *High Volt.* **2020**, *5*, 569–582. [CrossRef]
59. Huang, B.-D.; Takashima, K.; Zhu, X.-M.; Pu, Y.-K. The influence of the repetition rate on the nanosecond pulsed pin-to-pin microdischarges. *J. Phys. D App. Phys.* **2014**, *47*, 422003. [CrossRef]
60. Huang, B.-D.; Carbone, E.; Takashima, K.; Zhu, X.-M.; Czarnetzki, U.; Pu, Y.-K. The effect of the pulse repetition rate on the fast ionization wave discharge. *J. Phys. D App. Phys.* **2018**, *51*, 225202. [CrossRef]
61. Xu, Y.; Yan, X.-T. *Chemical Vapour Deposition: An Integrated Engineering Design of Advanced Materials*; Springer: London, UK, 2010.
62. Jung, S.K.; Kim, M.C.; Sohn, S.H.; Park, D.K.; Lee, S.H.; Park, L.S. Properties of indium tin oxide on polymer films deposited by low-frequency magnetron sputtering method. *Mol. Cryst. Liquid Cryst.* **2006**, *459*, 167–177. [CrossRef]
63. Kim, E.Y.; Jung, S.K.; Sohn, S.H.; Park, D.K.; Lee, S.M.; Park, L.S.; Lee, S.H.; Song, K.D. Properties of indium zinc oxide films on various polymer substrates deposited by low-frequency 60 Hz magnetron sputtering. *Mol. Cryst. Liquid Cryst.* **2007**, *470*, 241–250. [CrossRef]
64. Lee, S.H.; Lee, D.K.; Seo, S.H.; Oh, J.H.; Jung, S.K.; Sohn, S.H.; Park, D.K. Influence of oxygen flow rate on the properties of ITO films prepared by low-frequency (60 Hz) magnetron sputtering. *Mol. Cryst. Liquid Cryst.* **2006**, *459*, 221–229. [CrossRef]
65. Kim, H.T.; Jung, S.K.; Lee, S.-Y. Properties of ITO films deposited on paper sheets using a low-frequency (60 Hz) DC-pulsed magnetron sputtering. *Vacuum* **2021**, *187*, 110056. [CrossRef]
66. Kim, H.T.; Jung, S.K.; Kim, D.-E.; Park, C.Y.; Lee, S.-Y. Wettability control of paper through substitution between the hydroxyl group and carbon elements using argon-carbon plasma treatment. *Vacuum* **2022**, *205*, 111398. [CrossRef]
67. Wolf, R.A. *Atmospheric Pressure Plasma for Surface Modification*; Scrivener Publishing LLC: Salem, MA, USA, 2013.
68. Kusano, Y.; Norrman, K.; Singh, S.V.; Leipold, F.; Morgen, P.; Bardenshtein, A.; Krebs, N. Ultrasound enhanced 50 Hz plasma treatment of glass-fiber-reinforced polyester at atmospheric pressure. *J. Adhes. Sci. Technol.* **2013**, *27*, 825–833. [CrossRef]
69. Mei, D.; Tu, X. Atmospheric pressure non-thermal plasma activation of CO_2 in a packed-bed dielectric barrier discharge reactor. *ChemPhysChem* **2017**, *18*, 3253–3259. [CrossRef]
70. Bhak, J.-G.; Kim, H.T.; Kim, C.; Rhee, I. Damage-free surface treatment of polyethersulfone films using magnetized plasma generated by low-frequency (60 Hz) power. *Mol. Cryst. Liquid Cryst.* **2017**, *645*, 239–245. [CrossRef]
71. Prasertsung, I.; Kanokpanont, S.; Mongkolnavin, R.; Wong, C.S.; Panpranot, J.; Damrongsakkul, S. Plasma enhancement of in vitro attachment of rat bone-marrow-derived stem cells on cross-linked gelatin films. *J. Biomater. Sci.-Polym. Ed.* **2012**, *23*, 1485–1504. [CrossRef]
72. Takahashi, Y.; Yamamoto, M.; Tabata, Y. Osteogenic differentiation of mesenchymal stem cells in biodegradable sponges composed of gelatin and beta-tricalcium phosphate. *Biomaterials* **2005**, *26*, 3587–3596. [CrossRef]
73. Prasertsung, I.; Mongkolnavin, R.; Damrongsakkul, S.; Wong, C.S. Surface modification of dehydrothermal cross-linked gelatin film using a 50 Hz oxygen glow discharge. *Surf. Coat. Technol.* **2010**, *205*, S133–S138. [CrossRef]
74. Prasertsung, I.; Kanokpanont, S.; Mongkolnavin, R.; Wong, C.S.; Panpranot, J.; Damrongsakkul, S. Comparison of the behavior of fibroblast and bone marrow-derived mesenchymal stem cell on nitrogen plasma-treated gelatin films. *Mater. Sci. Eng. C-Mater. Biol. Appl.* **2013**, *33*, 4475–4479. [CrossRef]
75. Morent, R.; De Geyter, N.; Leys, C. Effects of operating parameters on plasma-induced PET surface treatment. *Nucl. Instrum. Methods Phys. Res. Sect. B-Beam Interact. Mater. Atoms* **2008**, *266*, 3081–3085. [CrossRef]
76. Jang, H.J.; Jung, E.Y.; Parsons, T.; Tae, H.-S.; Park, C.-S. A review of plasma synthesis methods for polymer films and nanoparticles under atmospheric pressure conditions. *Polymers* **2021**, *13*, 2267. [CrossRef] [PubMed]
77. Zabidi, N.Z.A.; Zaaba, S.K.; Sut, K.D.E.; Mohamed, C.W.S.R.; Masiman, R.I. A brief review on atmospheric air plasma. *J. Phys. Conf. Ser.* **2021**, *2071*, 012004. [CrossRef]
78. Reema; Khanikar, R.R.; Bailung, H.; Sankaranarayanan, K. Review of the cold atmospheric plasma technology application in food, disinfection, and textiles: A way forward for achieving circular economy. *Front. Physics* **2022**, *10*, 942952. [CrossRef]
79. Kumar, A.; Škoro, N.; Gernjak, W.; Pauč, N. Cold atmospheric plasma technology for removal of organic micropollutants from wastewater-a review. *Eur. Phys. J. D* **2021**, *75*, 283. [CrossRef]
80. Choi, Y.-H.; Kim, J.-H.; Paek, W.-T.; Ju, W.-T.; Hwang, Y.S. Characteristics of atmospheric pressure N_2 cold plasma torch using 60-Hz AC power and its application to polymer surface modification. *Surf. Coat. Technol.* **2005**, *193*, 319–324. [CrossRef]
81. Joshi, U.M.; Subedi, D.P. Surface treatment of polypropylene (PP) film by 50 Hz dielectric barrier discharge produced in air and argon/air mixture at atmospheric pressure. *AIP Conf. Proc.* **2015**, *1670*, 020003.
82. Rana, S.; Mehta, D.; Bansal, V.; Shivhare, U.S.; Yadav, S.K. Atmospheric cold plasma (ACP) treatment improved in-package shelf-life of strawberry fruit. *J. Food Sci. Technol.* **2020**, *57*, 102–112. [CrossRef]
83. María, C.; Pérez-Pizá, M.C.; Prevosto, L.; Grijalba, P.E.; Zilli, C.G.; Cejas, E.; Mancinelli, B.; Balestrasse, K.B. Improvement of growth and yield of soybean plants through the application of non-thermal plasma to seeds with different health status. *Heliyon* **2019**, *5*, e01495.
84. Mahnot, N.K.; Siyu, L.-P.; Wan, Z.; Keener, K.M.; Misra, N.N. In-package cold plasma decontamination of fresh-cut carrots: Microbial and quality aspects. *J. Phys. D App. Phys.* **2020**, *53*, 154002. [CrossRef]

85. Ganesan, A.R.; Tiwari, U.; Ezhilarasi, P.N.; Rajauria, G. Application of cold plasma on food matrices: A review on current and future prospects. *J. Food Process. Preserv.* **2021**, *45*, e15070. [CrossRef]
86. Zhou, R.; Rezaeimotlagh, A.; Zhou, R.; Zhang, T.; Wang, P.; Hong, J.; Soltani, B.; Mai-Prochnow, A.; Liao, X.; Ding, T.; et al. In-package plasma: From reactive chemistry to innovative food preservation technologies. *Trends Food Sci. Technol.* **2022**, *120*, 59–74. [CrossRef]
87. Perera, K.Y.; Prendeville, J.; Jaiswal, A.K.; Jai-wal, S. Cold Plasma Technology in Food Packaging. *Coatings* **2022**, *12*, 1986. [CrossRef]
88. Misra, N.N.; Yepez, X.; Xu, L.; Keener, K. In-package cold plasma technologies. *J. Food Eng.* **2019**, *244*, 21–31. [CrossRef]
89. Li, S.; Dang, X.; Yu, X.; Abbas, G.; Zhang, Q.; Cao, L. The application of dielectric barrier dis-charge non-thermal plasma in VOCs abatement: A review. *Chem. Eng. J.* **2020**, *388*, 124275. [CrossRef]
90. He, J.; Wen, X.; Wu, L.; Chen, H.; Hu, J.; Hou, X. Dielectric barrier discharge plasma for nanomaterials: Fabrication, modification and analytical applications. *Trac-Trends Anal. Chem.* **2022**, *156*, 116715. [CrossRef]
91. Niu, G.; Knodel, A.; Burhenn, S.; Brandt, S.; Franzke, J. Review: Miniature dielectric barrier discharge (DBD) in analytical atomic spectrometry. *Anal. Chim. Acta* **2021**, *1147*, 211–239. [CrossRef]
92. Ibra-him, J.; Al-Bataineh, S.A.; Michelmore, A.; Whittle, J.D. Atmospheric Pressure Dielectric Barrier Discharges for the Deposition of Organic Plasma Polymer Coatings for Biomedical Application. *Plasma Chem. Plasma Process.* **2021**, *41*, 47–83. [CrossRef]
93. Rosocha, L.A.; Anderson, G.K.; Bechtold, L.A.; Coogan, J.J.; Heck, H.G.; Kang, M.; McCulla, W.H.; Tennant, R.A.; Wantuck, P.J. *Nonthermal Plasma Techniques for Pollution Control: Part B—Electron Beam and Electrical Discharge Processing*; Penetrante, B.M., Schultheis, S.E., Eds.; Springer: Berlin, Germany, 1993.
94. Osawa, N.; Yoshioka, Y. Generation of low-frequency homogeneous dielectric barrier discharge at atmospheric pressure. *IEEE Trans. Plasma Sci.* **2012**, *40*, 2–8. [CrossRef]
95. Tay, W.H.; Kausik, S.S.; Yap, S.L.; Wong, C.S. Role of secondary emission on discharge dynamics in an atmospheric pressure dielectric barrier discharge. *Phys. Plasmas* **2014**, *21*, 044502. [CrossRef]
96. El-Zein, A.; Talaat, M.; El-Aragi, G.; El-Amawy, A. The characteristics of dielectric barrier discharge plasma under the effect of parallel magnetic field. *IEEE Trans. Plasma Sci.* **2020**, *48*, 1022–1029. [CrossRef]
97. Los, A.; Ziuzina, D.; Akkermans, S.; Boehm, D.; Cullen, P.J.; Van Impe, J.; Bourke, P. Improving microbiological safety and quality characteristics of wheat and barley by high voltage atmospheric cold plasma closed processing. *Food Res. Int.* **2018**, *106*, 509–521. [CrossRef] [PubMed]
98. Ziuzina, D.; van Cleynenbreugel, R.; Tersaruolo, C.; Bourke, P. Cold plasma for insect pest control: Tribolium castaneum mortality and defense mechanisms in response to treatment. *Plasma Process. Polym.* **2021**, *18*, e200017. [CrossRef]
99. Lu, X.P.; Liu, D.W.; Xian, Y.B.; Nie, L.L.; Cao, Y.G.; He, G.Y. Cold atmospheric-pressure air plasma jet: Physics and opportunities. *Phys. Plasmas* **2021**, *28*, 100501. [CrossRef]
100. Braný, D.; Dvorská, D.; Halašová, E.; Škovierová, H. Cold Atmospheric Plasma: A Powerful Tool for Modern Medicine. *Int. J. Mol. Sci.* **2020**, *21*, 2932. [CrossRef]
101. Bae, G.T.; Jang, H.J.; Jung, E.Y.; Lee, Y.R.; Park, C.-S.; Kim, J.Y.; Tae, H.-S. Development of an Atmospheric Pressure Plasma Jet Device Using Four-Bore Tubing and Its Applications of In-Liquid Material Decomposition and Solution Plasma Polymerization. *Polymers* **2022**, *14*, 4917. [CrossRef]
102. Winter, J.; Brandenburg, R.; Weltmann, K.-D. Atmospheric pressure plasma jets: An overview of devices and new directions. *Plasma Sources Sci. Technol.* **2015**, *24*, 064001. [CrossRef]
103. Adamovich, I.; Agarwal, S.; Ahedo, E.; Alves, L.L.; Baalrud, S.; Babaeva, N.; Bogaerts, A.; Bourdon, A.; Bruggeman, P.J.; Canal, C.; et al. The 2022 Plasma Roadmap: Low temperature plasma science and technology. *J. Phys. D-Appl. Phys.* **2022**, *55*, 373001. [CrossRef]
104. Reuter, S.; Von Woedtke, T.; Weltmann, K.-D. The kINPen-a review on physics and chemistry of the atmospheric pressure plasma jet and its applications. *J. Phys. D-Appl. Phys.* **2018**, *51*, 233001. [CrossRef]
105. Weltmann, K.-D.; Kindel, E.; Brandenburg, R.; Meyer, C.; Bussiahn, R.; Wilke, C.; von Woedtke, T. Atmospheric Pressure Plasma Jet for Medical Therapy: Plasma Parameters and Risk Estimation. *Contrib. Plasma Phys.* **2009**, *49*, 631–640. [CrossRef]
106. Korzec, D.; Hoppenthaler, F.; Nettesheim, S. Piezoelectric Direct Discharge: Devices and Applications. *Plasma* **2021**, *4*, 1–41. [CrossRef]
107. Carazo, A.V. Piezoelectrsic Transformers: An Historical Review. *Actuators* **2016**, *5*, 12. [CrossRef]

polymers

MDPI

Article

Atmospheric Pressure Plasma Polymerisation of D-Limonene and Its Antimicrobial Activity

Asad Masood [1], Naeem Ahmed [1], M. F. Mohd Razip Wee [1], Anuttam Patra [2], Ebrahim Mahmoudi [3] and Kim S. Siow [1,*]

[1] Institute of Microengineering and Nanoelectronics, University Kebangsaan Malaysia, Bangi 43600, Selangor, Malaysia
[2] Chemistry of Interfaces Group, Luleå University of Technology, SE-97187 Luleå, Sweden
[3] Department of Chemical and Process Engineering, Faculty of Engineering and Built Environment, Universiti Kebangsaan Malaysia, Bangi 43600, Selangor, Malaysia
* Correspondence: kimsiow@ukm.edu.my

Abstract: Antibacterial coating is necessary to prevent biofilm-forming bacteria from colonising medical tools causing infection and sepsis in patients. The recent coating strategies such as immobilisation of antimicrobial materials and low-pressure plasma polymerisation may require multiple processing steps involving a high-vacuum system and time-consuming process. Some of those have limited efficacy and durability. Here, we report a rapid and one-step atmospheric pressure plasma polymerisation (APPP) of D-limonene to produce nano-thin films with hydrophobic-like properties for antibacterial applications. The influence of plasma polymerisation time on the thickness, surface characteristic, and chemical composition of the plasma-polymerised films was systematically investigated. Results showed that the nano-thin films deposited at 1 min on glass substrate are optically transparent and homogenous, with a thickness of 44.3 ± 4.8 nm, a smooth surface with an average roughness of 0.23 ± 0.02 nm. For its antimicrobial activity, the biofilm assay evaluation revealed a significant 94% decrease in the number of Escherichia coli (*E. coli*) compared to the control sample. More importantly, the resultant nano-thin films exhibited a potent bactericidal effect that can distort and rupture the membrane of the treated bacteria. These findings provide important insights into the development of bacteria-resistant and biocompatible coatings on the arbitrary substrate in a straightforward and cost-effective route at atmospheric pressure.

Keywords: atmospheric pressure; plasma polymerisation; D-limonene; ASTM E2149; antimicrobial coating; *E. coli* bacteria

Citation: Masood, A.; Ahmed, N.; Razip Wee, M.F.M.; Patra, A.; Mahmoudi, E.; Siow, K.S. Atmospheric Pressure Plasma Polymerisation of D-Limonene and Its Antimicrobial Activity. *Polymers* **2023**, *15*, 307. https://doi.org/10.3390/polym15020307

Academic Editor: Choon-Sang Park

Received: 4 October 2022
Revised: 8 December 2022
Accepted: 13 December 2022
Published: 6 January 2023

1. Introduction

Infectious illness treatment is becoming more complex for physicians due to an increasing rate of antibiotic resistance around the world, causing major morbidity and death. Pharmaceutical companies are insufficiently responding to the rising demand for new antibiotics. Alternative medicinal approaches based on age-old herbal expertise utilising secondary plant compounds as promising antibacterial agents may aid in the development of new drugs [1]. Alkaloids, phenolics, terpenes, and saponins are examples of secondary metabolites known to assist and keep homeostasis in their surroundings [2]. Unlike the present synthetic antibiotics, these phytochemicals are recognised to have a weak antagonistic response while still possessing influential antibacterial properties.

Terpenes are a type of natural substance made up of cycloaliphatic and/or aromatic components. Limonene is one of the members of this family. It is thought to be a promising bioactive chemical since it has antioxidant, anticancer, anti-inflammatory, antiviral, and insecticidal activities. The virucidal activity of SARS-CoV-2 was reduced by about 6 logs in mouthwash that contained both D-limonene and cetylpyridinium chloride (CPC) [3]. Limonene can be used as an antibacterial agent because of its vast variety of medicinal applications. The science of antibacterial properties and mechanisms involved should be constructed to uncover the plausible function of the compounds against diverse germs to investigate natural compounds as medication in a much more practical route. Previous research on the antimicrobial activity of terpenes and their compounds has revealed that sub-lethal disruption to the cytoplasmic membranes is a possible mechanism for *E. coli* death [4–6]. Recent findings by Gupta et al. described the different sequential changes that occurred in *E. coli* cells after exposure to limonene, which caused cell rupture and eventually led to cell death [7]. Due to a growing interest in recent implantable medical devices, the issues arising from bacterial infection at the site of the devices have also become a major health challenge [8–11]. Therefore, there is a great need to produce an effective antimicrobial coating with various surface modifications for contact-killing bacteria on the device surfaces [12,13]. Lately, contact-active antimicrobial coatings have received increasing consideration [14,15].

Thin film technology advancements are critical for a range of research sectors, including microelectronics, biomedical, and anticorrosion applications [16–20]. Nanometre-thick films (<100 nm), for example, can be useful as bioactive layers for implant materials [21] and biomaterial films for anticorrosion purposes [22]. There has been a surge of interest in developing ecologically friendly films from renewable organic precursors [23]. This quest for alternative materials is motivated by a desire to reduce reliance on petroleum-derived resources and to generate goods with higher added value derived from low-cost sources [24]. These are induced via degradative chain transfer processes, which obstruct polymer synthesis [25]. Unlike traditional polymerisation techniques, plasma polymerisation allows thin film deposition from most of the organic precursors [26]. Terpenes originating from plants and fruits, such as terpinen-4-ol [27], geranium [28], and carvone [29] have been shown to create solid films using plasma polymerisation. During the plasma polymerisation reaction, ionic species are bombarded, resulting in radical sites that stimulate cross-linking of the deposited nano-thin films and grafting of new species [30].

Even though limonene has a chemical structure comparable to those terpenes-related precursors, no feasible experiments have been conducted yet using this precursor to make nano-thin films via APPP for antibacterial activity. For instance, a plasma coating with antibacterial surface modifications by an atmospheric pressure plasma jet (APPJ) process is a relatively new field due to its simplicity and convenience of usage [31,32]. In most APPJs and dielectric barrier discharges (DBDs) [33], one electrode is grounded while the other is energised with a high-voltage source to create an electric field between them. According to this method, stable plasma is formed when a flow of gas travels between the coaxial electrodes and is ionised by the electric field. As precursors for the deposited polymer coating, monomers must be vaporised before the plasma ignition. The contact between the precursor and the plasma in the air causes the monomers to fragment and induces the oxygen- and nitrogen-containing groups that dictate the coating characteristics [31]. The products generated from the precursor-plasma reactions, such as reactive plasma species, as well as non-reactive species, deposit on the substrate, where the adsorption and surface reactions take place simultaneously. Some of the published results have been reviewed here [34–37].

Developing a simple and green route for imparting antibacterial and hydrophobic properties is highly desirable because of economic and environmental reasons. In this work, we reported a rapid and one-step plasma polymerisation technique, utilising D-limonene as a precursor, to create nano-thin films at atmospheric pressure (AP) for antibacterial applications. To the best of our knowledge, the use of APPP of D-limonene as a potential antibacterial agent with a bactericidal effect has not yet been demonstrated. Previous work on the deposition of the D-limonene coating used the low-pressure plasma polymerisation technique approach, which necessitated the use of vacuum [38]. In our APPP approach, the impact of plasma polymerisation time on the physicochemical properties of the resultant nano-thin films was investigated here to demonstrate its feasibility. As an initial proof of concept, the results were incorporated to produce an effective antibacterial coating against

Gram-negative (*E. coli*) bacteria, in which *E. coli* exposed to the D-limonene coated surface were evaluated in vitro via field-emission scanning electron microscopy and fluorescence microscopy to access their survivability after incubation.

2. Materials and Methods

2.1. Materials

The essential oil (D-limonene, $C_{10}H_{16}$) was acquired from Sigma Aldrich (St. Louis, MO, USA) with a purity of >97%. It was utilised as received without any additional purification. Microscopic glass slides were selected as a model substrate because of their established record as a positive control. Each glass slide was ultrasonically cleaned for 30 min in acetone, ethanol, deionised (DI) water, and lastly in isopropyl alcohol (IPA). For plasma discharge, 99.99% pure Ar gas was employed in this work.

2.2. Experimental Setup of the AP Plasma System

Figure 1 shows a schematic diagram of the AP plasma polymerisation system where the plasma discharge was created in a plasma jet consisting of a quartz tube. A copper rod was placed into the quartz tube connected to a high voltage (HV) supply and acted as an HV electrode. The outside electrode, fastened around the quartz tube, serving as a grounded electrode, was likewise made of copper. A neon power supply with a 3 kV output voltage was used to power up the plasma jet. The inner and outer diameters of quartz tubes were about 3.0 and 5.0 mm, respectively. The distance between the HV electrode tip and the nozzle was maintained at 20 mm throughout the experiment, whereas the nozzle was 15 mm away from the target glass substrate. The deposition of AP plasma-polymerised D-limonene (AP-PP-lim) was carried out on the glass substrates of 5 cm × 5 cm size. Mass flow controllers (MFCs) were used to feed the argon into the system. To vaporise the liquid D-limonene monomer, the Ar gas was supplied into a glass bubbler at a constant flow rate of 130 sccm.

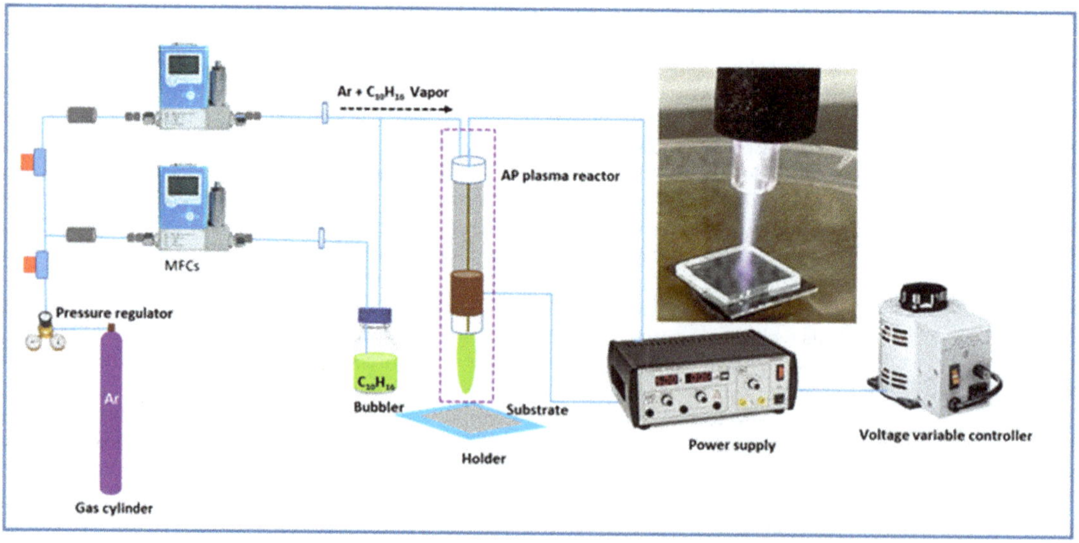

Figure 1. Schematic diagram of the APPP setup. The inset shows the photograph of plasma generated from the APPJ during the deposition process.

2.3. Characterisation of AP-PP-lim Nano-Thin Films

The static water contact angle (WCA) measurements were carried out on an automated Contact Angle Goniometer (Rame−Hart, Inc. model 100, Succasunna, NJ, USA) using a static sessile drop technique at room temperature. Two microlitre droplets of DI water were carefully placed on the samples and WCAs were recorded. The surface masking approach was utilised to determine the thickness of the AP-PP-lim nano-thin films on glass substrate [29]. The step height was measured with a surface profiler (model: Bruker Dektak XT). An atomic force microscopy (AFM) system (NT-MDT, Automated AFM NEXT) operating in non-contact mode was used to assess the roughness of the AP-PP-lim. A non-contact silicon gold cantilever beam (NSG-10), with a resonance frequency of 390 kHz and a force constant of 37 Nm^{-1}, was used, and the average roughness (R_a) for the scan regions of 10 μm × 10 μm was also measured.

The chemical composition of the pure D-limonene and AP-PP-lim samples was assessed using a Perkin Elmer Spectrum 400 FT-IR Spectrometer (1250 to 4000 cm^{-1} range). Based on 32 accumulated scans with a resolution of 4 cm^{-1}, the spectra were acquired. To analyse the spectra, adjust the baseline, and locate the relevant peaks, Spectrum™ 10 software was employed.

Furthermore, the surface chemistry of the samples was analysed using an XPS system (ULVAC-PHI, Quantera II) with an Al-K source that had an energy of 1486 eV and functioned at 50 W and 15 kV. All spectra were acquired with an emission angle of 45°. Each sample was subjected to five scan cycles. Based on the survey scan spectra, the PHI Multipak programme was used to estimate the atomic content of carbon (C) and oxygen (O). For the high-resolution spectra, the peak deconvolutions of C 1 s spectra were fitted using Gaussian curve fitting. During component fitting, the Shirley background was used to deduce the C 1 s spectra, and the full width at half maximum was set between 1.46 and 1.55 eV [39,40].

2.4. Microbiological Activity

The antimicrobial activity of AP-PP-lim coatings was evaluated based on the standard test method (ASTM E2149-20). A quantifiable antimicrobial testing technique was used to determine the antibacterial activity of non-leaching antibacterial agents [41]. The antibacterial properties of the coatings were tested using the Gram-negative (*E. coli*) bacteria (ATTC 25927). To grow these bacteria, Luria-Bertani (LB) broth was grown overnight at 37 °C. The inoculum (1×10^8 CFU mL^{-1}) and LB media were placed together at a ratio of 1:9 to make an inoculum (1×10^7 CFU mL^{-1}) and further diluted to 1×10^5 CFU mL^{-1}. *E. coli* bacteria were streaked onto agar (nutrient) plates and left overnight at 37 °C. The samples were placed into 12-well plates and incubated with inoculum (1×10^5 CFU mL^{-1}) for 24 h for biofilm formation studies. The Gram staining protocol was used to identify the Gram-negative bacteria by colouring the cells [42]. In a nutshell, the samples were steeped in crystal violet solution for 30 s before being rinsed in saline and then submerged in Gram iodine solution for another 30 s. The weak colours were subsequently decoloured using a decolouriser (GMO-CO2). The average surface area of bacteria was estimated using ImageJ® software (NIH and LOCI, Wisconsin, USA) after the sample images were optically captured at different spots using a fluorescence microscopy system (Olympus, BX53M). Each experiment was carried out three times. The morphology of the bacteria [43] on the control and treated samples was studied using a field-emission electron scanning microscopy (FESEM) system (Zeiss, Supra 55 VP). The sample preparation method was similar to those used for biofilm formation, with the exception that after 24 h of incubation, the samples were rinsed with saline solution to eliminate any loose biofilms. Lastly, the samples were dried at ≃35 °C for 2 h. Prior to FESEM imaging, a 99.99% platinum ultra-thin coating was applied to reduce the charging effect on the non-conducting sample. A live-dead fluorescence experiment was also used to examine the antibacterial activity of the plasma-polymerised film and 12-well plates were used to hold the prepared samples (AP-PP-lim and clean glass substrate). These samples were treated with an inoculum (1×10^7 CFUmL^{-1}) for 24 h at 37 °C. The samples were then preserved with 10% *v*/*v* neutral buffered formalin saline after being rinsed with phosphate-buffered saline (PBS). To determine the bacteria's viability, they were then stained using the LIVE/DEAD® BaclightTM bacterial viability kit (L7007, Invitrogen, ThermoFisher Scientific, Waltham, MA, USA). Before examining the stained samples under a fluorescence microscope, the samples were submerged in 0.9 *w*/*v* saline water to eliminate any extra staining. An Olympus BX51 fluorescence microscope was used to image the viability of bacteria on the AP-PP-lim and glass substrate at a magnification of 60×. The red fluorescent substance, propidium iodide, was seen at an excitation wavelength of 490 nm and an emission wavelength of 635 nm, whereas the green fluorescent substance, SYTO9, was seen at 480 nm and 500 nm. Each sample received five images from experiments performed in triplicate.

3. Results and Discussion

3.1. Characteristics of AP-PP-lim Nano-Thin Films

The sessile drops technique was used to test the wettability of plasma-polymerised films. The angle at the triple-phase contact line between the water droplet and the film determined the wettability. The adhesive forces between such a liquid drop and a substrate are theoretically a local response driven by interactions between the actual drop and the nearby vapour with the substrate, demanding drop-volume independence [44]. Droplet symmetry may be affected by the distribution of chemical composition and roughness of the surface [45]. The topography is an important factor to consider since roughness can increase contact angles in some circumstances [46]. Contact angle measurements, on the other hand, offer information regarding the nature of a topmost layer of the surface in the 0.5–1.0 nm range [47].

Figure 2a exhibits the average static WCA and thickness of AP-PP-lim nano-thin films deposited on the glass substrate at different plasma polymerisation times. The results were compared with the control sample, a clean glass substrate, which is completely wettable (i.e., WCA = 0°). After the plasma polymerisation at a minimum period of 1 min, a much higher static WCA was obtained at about 90.7 ± 1.1°. No significant changes in WCA were further observed regardless of plasma polymerisation time. The corresponding photographs of water drops on the control and AP-PP-lim samples are shown in Figure 2b.

From the profilometer measurements, as shown in Figure 2c–g, the thickness of AP-PP-lim deposited at 1, 3, 5, 7, and 9 min was estimated to be about 44.3 ± 4.8, 190.7 ± 7.3, and 281.6 ± 10.5, 322.8 nm ± 11.6 and 363.3 nm ± 10.3 nm, respectively. Meanwhile, no significant deviations of the film thickness were observed across the deposited films, for each deposition time, after plasma polymerisation. Furthermore, the surface morphologies of AP-PP-lim nano-thin films were also characterised based on the topographical AFM measurements (2-D and 3-D view), as shown in Figure 2h,i and Figure S1a–d. The results showed a relationship between the film thickness and plasma polymerisation time with a deposition rate of ~0.8 nm s^{-1}, indicating the reproducibility of the plasma polymerisation process when carried out in a controlled manner. By adjusting the plasma polymerisation time with the suggested atmospheric plasma method, one may control the resultant film thickness.

The AFM profiles showed a smooth, and complete coverage of AP-PP-lim (1 min) on the substrate, which correlates well with low average roughness (R_a) value (0.23 ± 0.02 nm) and root mean square roughness (R_q) value (0.27 ± 0.02 nm). The average roughness (R_a) and root mean square roughness (R_q) of AP-PP-lim nano-thin films (3, 5, 7 and 9 min) and control clean glass substrates are shown in Table S1 and Figure S1e. The results showed a linear relationship between the film roughness and plasma polymerisation time. Figures S2 and S3 showed that the nano-thin films deposited at various times on glass substrate are optically transparent. In this work, the R_a value of AP-PP-lim was comparable to that of low-pressure plasma-polymerised D-limonene reported previously [38].

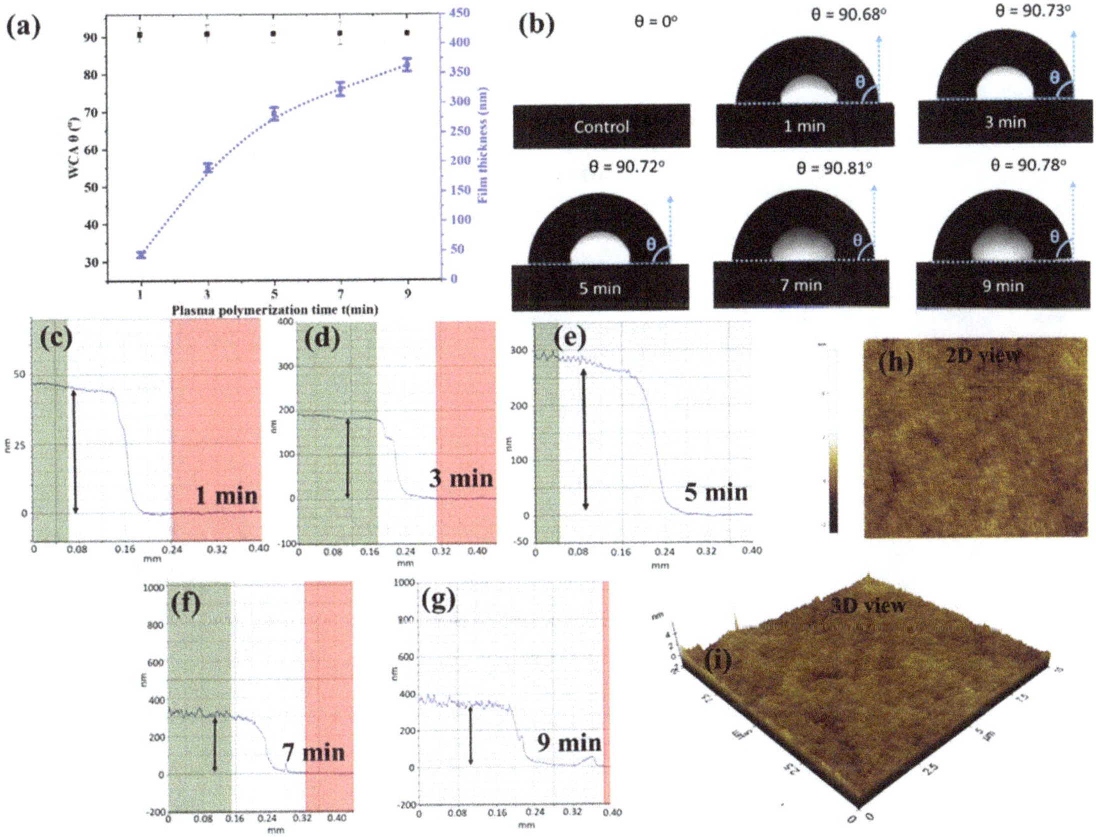

Figure 2. (**a**) Average static WCA and film thickness at different plasma polymerisation times. (**b**) Photographs of WCA of control and AP−PP−lim samples at different plasma polymerisation times. (**c**–**g**) Step height measurements of AP−PP−lim films at different plasma polymerisation times with their corresponding height profile plots. (**h**,**i**) 2D and 3D AFM profiles of the smooth AP−PP−lim nano-thin films deposited at 1 min.

In comparison, the thickness control in this work was almost comparable to the plasma polymerisation materials reported in the literature, in which the thickness and plasma polymerisation time were likewise linearly related due to a constant flow of precursors [48]. Meanwhile, much thicker films were found to be less effective in antibiofouling activity, possibly due to increased roughness that enhanced the attachment of bacteria on the surfaces [49]. Furthermore, plasma-polymerised coatings with hydrophobic or superhydrophobic surfaces could also greatly reduce the adhesion of bacteria on the surfaces [50]. Hence, in this work, the AP-PP-lim nano-thin films with adequate surface characteristics suitable for antimicrobial activity, deposited at a short period of 1 min, were considered for further characterisations.

The ATR-FTIR spectra of AP-PP-lim nano-thin films and D-limonene monomer samples are shown in Figure 3 and Figure S4. The molecules of D-limonene consist of C−H, C−C, and C=C bonds [38,51]. Bands include symmetric stretching of C−H bonds (2834 and 2856 cm^{-1}), asymmetric stretching of C−H (2920 and 2964 cm^{-1}), and unsaturated C−H bonds (3010, 3046, 3072, and 3083 cm^{-1}). The stretching band of the C=C bonds was seen at around 1644 cm^{-1}. Asymmetric C−H bending was presented at 1436 and

1451 cm^{-1}, while symmetric C−H bending was at 1375 cm^{-1}. Furthermore, out-of-plane bending of C−H bond was presented at 1310 cm^{-1} as different band with low intensities.

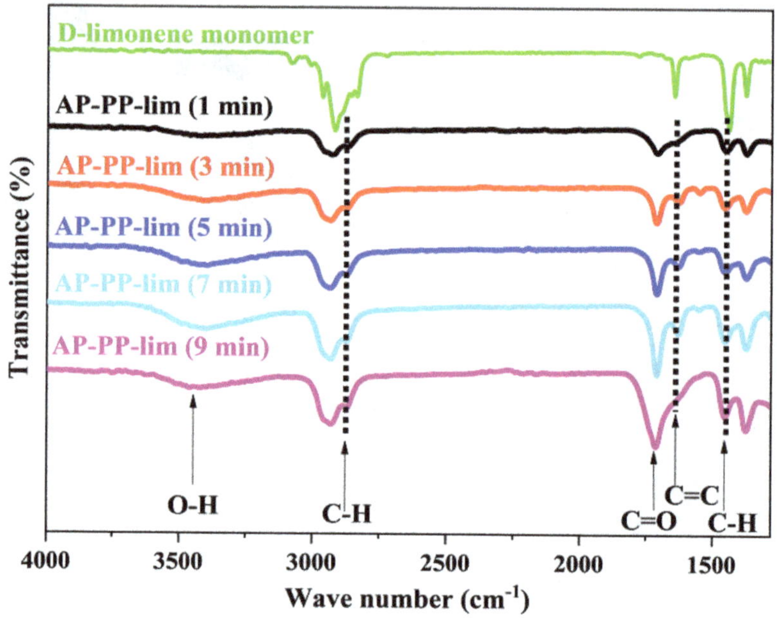

Figure 3. ATR-FTIR spectra of D-limonene monomer and AP−PP−lim deposited on a glass substrate at plasma polymerisation times of 1, 3, 5, 7 and 9 min.

Compared to the monomer D-limonene, the AP-PP-lim nano-thin films, in the surface phase, have broader and fewer noticeable bands because of the cross-linking nature of plasma polymers [52]. Based on Figure 3, AP-PP-lim nano-thin films show the following bands: (1647 cm^{-1}) represents unsaturated C=C stretching, (2926 and 2956 cm^{-1}) represent asymmetric and symmetric C−H stretching, (1374 and 1450 cm^{-1}) represent symmetric and asymmetric bending of C−H. Hence, the cross-linking properties of AP-PP-lim nano-thin films, with random bonding and cross-linking, might be associated with a variety of bonding environments [53,54].

Although the precursor molecule lacks oxygenated chemical groups, polar functionality on plasma-polymerised films arises when they come into contact with ambient air [55]. In atmospheric plasma, the free radicals in the plasma-polymerised films are often produced by plasma active species ablation or imperfect fragment bonding [56]. During the plasma polymerisation, these radicals rapidly interact with oxygen in the air to produce oxygenated species [57]. It is well-known that limonene easily oxidises to carveol, carvone and limonene oxide (Figure S5).

The present authors had already provided plausible mechanisms of plasma polymerisation followed by deposition of carvone on the surface [6]. Similar mechanisms would likely also hold for carveol and limonene oxide to polymerise and deposit on the surface. For instance, a band that arose with a weak intensity at about 1050 cm^{-1}, might have originated from the C−O bonds, while a band at about 1709 cm^{-1} was ascribed to the stretching of the C=O bonds that emerge due to similar causes. Another weak broadband at about 3340 cm^{-1} was also presented which corresponds to O−H stretching. Those chemical fingerprints often resemble the existing plasma-polymerised materials produced from essential oils and their constituents [48].

Importantly, the band presented in the AP-PP-lim nano-thin films at about 1647 cm^{-1} corresponds to the C=C bond. The unconjugated C=O bond appeared at 1709 cm^{-1}. Both

C=O and C=C peaks were similar to the carvone-based system, as reported earlier. Though the C=C peak was often absent in the films prepared with various natural precursors as well as precursors that formally containing double bonds [57–59], these authors had reported that for carvone moiety this peak appears almost at the same position for both pure compound and plasma polymerised compound [6]. This result suggested that under the deposition circumstances utilised in this work, chemical structures are partially retained during plasma polymerisation [60]. Furthermore, the same FTIR spectra were also found for the samples deposited at 3, 5, 7, and 9 min, suggesting the same characteristics of chemical bonding regardless of plasma polymerisation time (see Figures 3 and S4). As the outermost surface is always the same for all five deposited surfaces, the WCA is also the same, as shown in Figure 2. Moreover, it might be interesting to note that, as the WCA of the films are ~90°, the variation in the surface roughness does not influence WCA according to Wenzel equation [61].

$$\theta m = \cos^{-1} (r.\cos\theta Y)$$

where θ_m is the measured contact angle, θ_Y is the Young contact angle as defined for an ideal surface. r is the roughness ratio. It is defined as the ratio of true area of the solid surface to the apparent area. r = 1 for a smooth surface and >1 for a rough one. As θ_Y is ~90°, θ_m also become the same ~90° irrespective of r.

XPS analysis was used to evaluate the surface chemical composition of the AP-PP-lim films and confirm the possible chemical bonding on the surfaces. Figure 4a shows an XPS survey spectrum of AP-PP-lim films without any Si signal, thus suggesting that the plasma coating has completely masked the underlying substrate. The results of the XPS survey scan showed typical two C 1 s and O 1 s peaks at around 285 and 533 eV binding energies, respectively. It was observed that the intensity of the C 1 s peak is much higher than that of the O 1 s peak, indicating the highly preserved structure of C=C and C−H bonds after the plasma polymerisation. Considering the atomic concentrations of the AP-PP-lim films, carbon emerged as the dominant content with 84.7 atomic percent, whereas the remaining contribution to the oxygen content is 15.3 atomic percent.

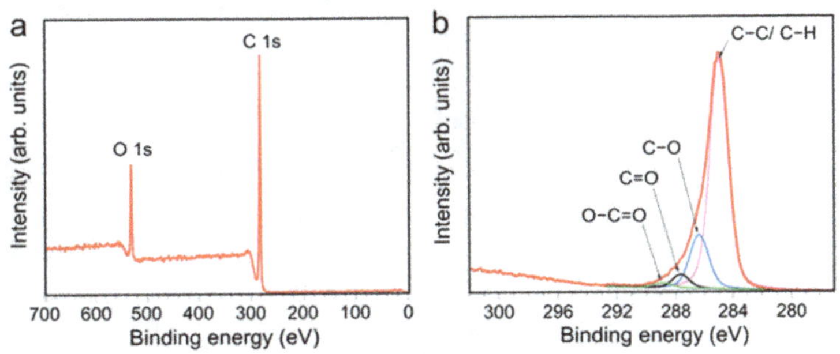

Figure 4. XPS of AP-PP-lim nano-thin films. (**a**) XPS survey spectrum and (**b**) core-level of C 1 s spectra.

The analysis of the C 1 s peak provided more detailed information about the chemical composition of AP-PP-lim (Figure 4b). The C 1 s peak could be deconvoluted into four components: hydrocarbons bonds (C−C/C−H) with a binding energy of 284.8 eV form the polymer backbone; and chemical groups such as hydroxyl (C−O–H)/ether (C–O–C), carbonyl (C=O), and carboxyl (O−C=O) with binding energies of 286.4 eV, 287.7 eV, and 289.0 eV, respectively, are incorporated in the polymer matrix [58]. The existence of C−O, C=O, and O−C=O bonds indicates that AP-PP-lim has been partially oxidised, and the findings correlated well with the FTIR spectra.

These findings were consistent with previous research on plasma-polymerised natural oils [58,62]. The relative concentrations of distinct carbon bonding states in relation to the overall carbon concentration in the sample was estimated using Gaussian curve fitting. Hydrocarbon species (79.1% of total carbon) formed the backbone of the coatings and reflected the chemical bonds of D-limonene. Compared to prior studies employing AP plasma [63], the quantity of hydrocarbons (C−C/C−H) dominated this coating composition. The second largest contribution came from C−O groups, which accounted for 16.5% of the total, with the remaining 3.0% and 1.4% contributed by C=O and O−C=O groups, respectively.

A proposed mechanism of AP-PP-lim chemical structure based on the ATR-FTIR and XPS analysis is shown in Figure 5. The plasma produced reactive species by ionising a flow of argon gas and D-limonene ($C_{10}H_{16}$) vapour that subsequently reacted with ambient air to form gas-phase reactive hydrocarbon (C_xH_y), oxygen (O), and nitrogen (N) species, respectively. During the dissociation in plasma, the $C_{10}H_{16}$ monomers partially broke into fragments that consisted of ions, radicals, and molecules. Due to its unique reaction process with ambient air, the fragments reacted with the oxygen reactive species and recombined onto the substrate to create a highly cross-linked polymer with new oxygen chemical groups (e.g., C−O, C=O, and O−C=O). In this work, the new chemical groups indicated the typical characteristics of plasma-polymerised nano-thin films that behaved similarly to the previous work conducted at low-pressure [38].

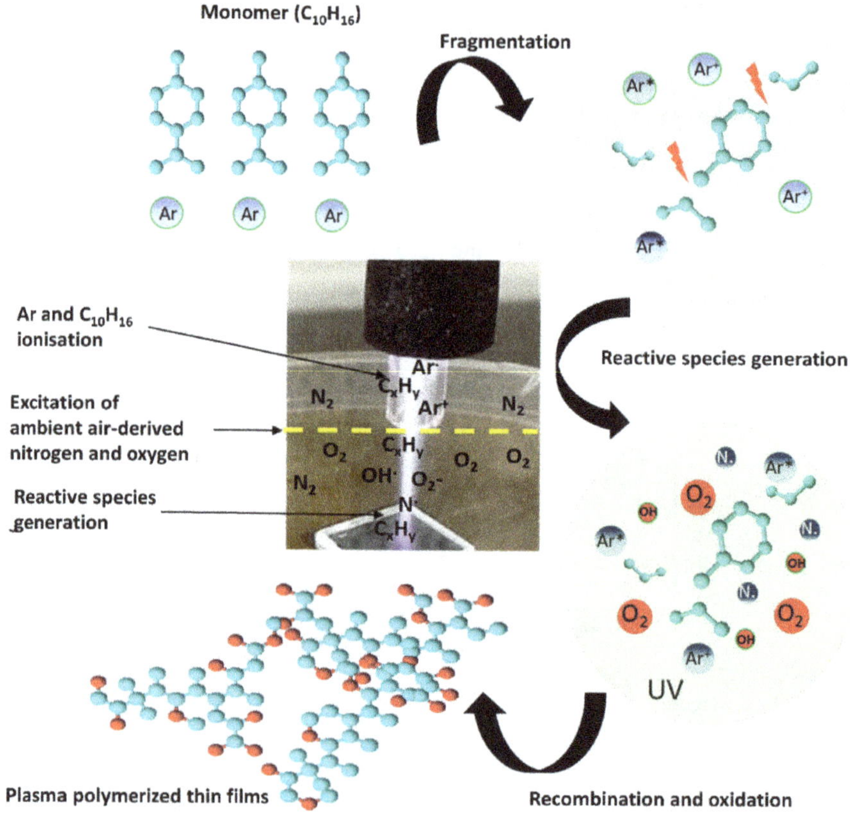

Figure 5. Schematic representation of plasma polymer derived from D-limonene monomer using the APPJ system.

3.2. *Antimicrobial Performance of AP-PP-lim Nano-Thin Films*

Antibacterial properties of D-limonene have attracted great interest in bio-fouling studies since it is a bio-safe essential oil that can act as an excellent antimicrobial agent [64]. It was also well known that D-limonene has superb antimicrobial properties to different species of bacteria strains such as *E. coli* and *Staphylococcus aureus* (*S. aureus*). Recent studies by Gupta et al. and Han et al. also reported that D-limonene possesses excellent antibacterial properties to prevent different kinds of bacterial propagation and their growth [7,65]. In this work, Gram-negative (*E. coli*) bacteria (which cause biofouling) were used to examine the antibacterial performance of the AP-PP-lim nano-thin films. Gram-negative bacteria, unlike Gram-positive bacteria, tend to be antibiotic resistance due to their thick cell walls [66]. As a result, if a coating is resistant to Gram-negative bacteria, it is likely to be resistant to Gram-positive bacterial growth as well.

The mechanisms by which contact-active antibacterial coatings affect cell death are often associated with disruption of the cell surface [67–69]. To examine this hypothesis, the morphologies of *E. coli* coated onto the control and treated samples with AP-PP-lim nano-thin films were imaged by FESEM. Prior to the evaluation, both control and AP-PP-lim samples were separately incubated with 10^5 CFU mL^{-1} of *E. coli* cell suspension for 5 h, then the cells underwent fixation, and substrates were treated with glutaraldehyde (2.5%), followed by ethanol (20–100%) dehydration and air drying. Figure 6a−d shows the FESEM images of the morphological changes in the treated *E. coli* between the control and AP-PP-lim sample. It was observed that the typical rod-shaped *E. coli* bacteria (about 2.00 μm long and 0.25−1.00 μm wide) on the control sample incubated with the cell suspension have remained almost unchanged. The *E. coli* on the control sample grew well with intact cytoplasmic membranes and formed abundant biofilms (Figure 6a,c).

In contrast, the treated *E. coli* on the AP-PP-lim were found to be isolated with a significant decrease in the bacterial colonies, as shown in Figure 6b. The shape of *E. coli* was shrivelled, which indicates that the AP-PP-lim could distort the bacteria upon their initial attachment on the surface. Interestingly, further rupture to the structure of dead *E. coli* was apparently seen in the yellow box of Figure 6d, indicating that the AP-PP-lim could disrupt and collapse the cytoplasmic membranes through the affected pores, leading to death. It was noted that the bacterial outer cover (cell membrane or cell wall) is the most probable cellular goal for D-limonene due to the creation of pores. Furthermore, lipids in the cell wall composition from numerous bacterial classes explained their various vulnerabilities to D-limonene. In this case, the findings suggested that the AP-PP-lim nano-thin films not only inhibit the propagation and growth of bacteria, but also possess a direct bactericidal effect.

To analyse the bacteria count, the fluorescence effects of biofilm assay and early attachment assay were quantitatively evaluated. Figure 6e,f shows the fluorescence microscopy images of treated *E. coli* bacteria adhered onto the control and AP-PP-lim sample after 24 h of incubation. It was revealed that the number of bacteria adhered onto the AP-PP-lim sample was much lower (with a count of $2.37 \times 10^4 \pm 0.9 \times 10^3$ cm^{-2}) than that of the control sample (with a count of $4.27 \times 10^5 \pm 1.3 \times 10^4$ cm^{-2}) (see Figure 6g). The findings suggest that the AP-PP-lim nano-thin films significantly contribute to the reduction in bacterial adhesion by up to 94% as compared to the control sample. The initial adhesion assay and the biofilm assay findings for fluorescence were qualitatively evaluated. Propidium iodide stains dead bacteria red by penetrating their ruptured membrane, while SYTO9 turned live bacteria green. After being incubated for 24 h, the fluorescence pictures of *E. coli*'s initial attachment to a clean glass substrate and an AP-PP-lim are shown in Figure 6h and 6i, respectively. AP-PP-lim demonstrated a much lower amount of *E. coli* adhered in comparison to a clean glass substrate and dead bacteria in red colour are visible, as shown in Figure 6i. It should be noted that the biofilm adhesion could be influenced by more factors than just surface reactivity. Variations in biofilm surface adherence may be due to the synergetic effects of multiple factors (e.g., reduced adhesion or destruction of intercellular signal conduction and communication), which might affect the contact of bacteria onto the surface, according to the mechanism reported previously [70].

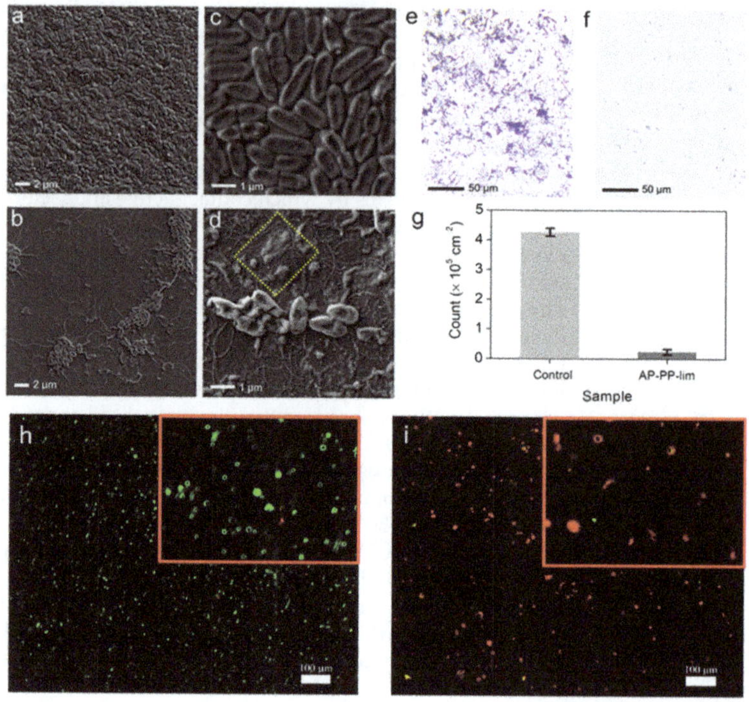

Figure 6. Antimicrobial performance of AP–PP–lim nano-thin films against Gram-negative (*E. coli*) bacteria. FESEM images of treated *E. coli* on the (**a**,**c**) control and (**b**,**d**) AP–PP–lim sample. The yellow square box indicates the rupture of *E. coli* cells. Fluorescent microscopic imaging of treated crystal violet-stained E. coli on the (**e**) control and (**f**) AP–PP–lim sample. (**g**) Count of treated *E. coli* per unit area (cm^{-2}) on the control and AP–PP–lim sample. Illustrative fluorescence views in live-dead fluorescence assay of (**h**) *E. coli* attached on clean glass substrate and (**i**) *E. coli* attached on AP–PP–lim. Samples were incubated for 24 h. Live bacteria are represented by green, whereas dead bacteria are represented by red.

4. Conclusions

In summary, optically transparent and smooth AP-PP-lim nano-thin films as an antibacterial coating were successfully deposited on glass substrates using the precursor D-limonene via a rapid and single-step APPP method. The physicochemical properties of the AP-PP-lim nano-thin films were systematically investigated in relation to the plasma polymerisation time. Under the circumstances utilised in this work, a high deposition rate of ~0.8 nm s^{-1} can be achieved to control the thickness of AP-PP-lim nano-thin films. Surface chemical analysis of the resultant nano-thin films indicated that the hydrocarbon molecules (e.g., C–C/ C–H) are partly retained from the fragmented D-limonene, along with new oxygen chemical groups (e.g., C–O, C=O, and O–C=O), resulting from the free radical reactions in ambient air under the plasma condition. For its antimicrobial behaviour, the biofilm assay evaluation exhibited a 94% decrease in the number of *E. coli*, compared to the control sample, suggesting the effectiveness of the AP-PP-lim nano-thin films as an antibacterial coating against Gram-negative bacteria. Overall, this APPP method can provide a new insight into the development of bacteria-resistant and biocompatible coatings based on precursors.

Supplementary Materials: The following supporting information can be downloaded at: https://www.mdpi.com/article/10.3390/polym15020307/s1. Figure S1: 2D and 3D AFM profiles of the smooth AP-PP-lim nano-thin films deposited were at 3-, 5-, 7- and 9-min; Figure S2: Transmittance and absorption spectra of AP-PP-lim thin films with different deposition time (1-, 3-, and 5-min) deposited on a glass substrate; Figure S3: Photograph of a glass substrate with optically transparent AP-PP-lim nano-thin films; Figure S4: ATR-FTIR spectra of pure D-limonene and the AP-PP-lim films at 1-, 3-, 5-, 7- and 9-min plasma polymerisation; Figure S5: Chemical structure of D-limonene and its oxidation products; Table S1: Average roughness (R_a) and root mean square roughness (R_q) of glass substrate and AP-PP-lim nano thin films were deposited at 3-, 5-. 7- and 9-min.

Author Contributions: A.M. wrote the main manuscript text. The design of the setup, deposition and characterisation of the plasma polymer materials were conducted by A.M. N.A., M.F.M.R.W., A.P., E.M. and K.S.S. provided support in the monitoring as well as verification of the results. All authors have read and agreed to the published version of the manuscript.

Funding: This work was supported by Universiti Kebangsaan Malaysia (UKM), Malaysia under Geran Universiti Penyelidikan (GUP-2022-071). A.P. acknowledges the financial support from PRECISE LTU.

Institutional Review Board Statement: Not applicable.

Data Availability Statement: The data used to support the findings of this study are available from the corresponding author upon request.

Acknowledgments: The authors acknowledge the financial support from Universiti Kebangsaan Malaysia grant GUP-2022-071 for this work. Special thanks to the research officers of Centre for Research and Instrumentation Management (CRIM) Centralised Lab, UKM, Malaysia for their technical assistance. A.P. is grateful for the financial support from PRECISE LTU.

Conflicts of Interest: The authors declare no conflict of interest.

References

1. Barros, C.H.; Casey, E. A Review of Nanomaterials and Technologies for Enhancing the Antibiofilm Activity of Natural Products and Phytochemicals. *ACS Appl. Nano Mater.* **2020**, *3*, 8537–8556. [CrossRef]
2. Kosová, K.; Vítámvás, P.; Urban, M.O.; Prášil, I.T.; Renaut, J. Plant Abiotic Stress Proteomics: The Major Factors Determining Alterations in Cellular Proteome. *Front. Plant Sci.* **2018**, *9*, 122. [CrossRef] [PubMed]
3. Rodríguez-Casanovas, H.J.; la Rosa, M.D.; Bello-Lemus, Y.; Rasperini, G.; Acosta-Hoyos, A.J. Virucidal Activity of Different Mouthwashes Using a Novel Biochemical Assay. *Healthcare* **2022**, *10*, 63. [CrossRef] [PubMed]
4. Prabuseenivasan, S.; Jayakumar, M.; Ignacimuthu, S. In Vitro Antibacterial Activity of Some Plant Essential Oils. *BMC Complement. Altern. Med.* **2006**, *6*, 39. [CrossRef] [PubMed]
5. Burt, S. Essential Oils: Their Antibacterial Properties and Potential Applications in Foods—A Review. *Int. J. Food Microbiol.* **2004**, *94*, 223–253. [CrossRef]
6. Masood, A.; Ahmed, N.; Mohd Razip Wee, M.F.; Haniff, M.A.S.M.; Mahmoudi, E.; Patra, A.; Siow, K.S. Pulsed Plasma Polymerisation of Carvone: Characterisations and Antibacterial Properties. *Surf. Innov.* **2022**, *40*, 1–13. [CrossRef]
7. Gupta, A.; Jeyakumar, E.; Lawrence, R. Strategic Approach of Multifaceted Antibacterial Mechanism of Limonene Traced in Escherichia Coli. *Sci. Rep.* **2021**, *11*, 13816. [CrossRef]
8. Hetrick, E.M.; Schoenfisch, M.H. Reducing Implant-Related Infections: Active Release Strategies. *Chem. Soc. Rev.* **2006**, *35*, 780–789. [CrossRef]
9. Hickok, N.J.; Shapiro, I.M. Immobilized Antibiotics to Prevent Orthopaedic Implant Infections. *Adv. Drug Del. Rev.* **2012**, *64*, 1165–1176. [CrossRef]
10. Inzana, J.A.; Schwarz, E.M.; Kates, S.L.; Awad, H.A. Biomaterials Approaches to Treating Implant-Associated Osteomyelitis. *Biomaterials* **2016**, *81*, 58–71. [CrossRef]
11. Cong, Y.; Quan, C.; Liu, M.; Liu, J.; Huang, G.; Tong, G.; Yin, Y.; Zhang, C.; Jiang, Q. Alendronate-Decorated Biodegradable Polymeric Micelles for Potential Bone-Targeted Delivery of Vancomycin. *J. Biomater. Sci. Polym. Ed.* **2015**, *26*, 629–643. [CrossRef]
12. Wang, Y.; Li, P.; Xiang, P.; Lu, J.; Yuan, J.; Shen, J. Electrospun Polyurethane/Keratin/Agnp Biocomposite Mats for Biocompatible and Antibacterial Wound Dressings. *J. Mater. Chem. B* **2016**, *4*, 635–648. [CrossRef]
13. Fang, B.; Jiang, Y.; Nüsslein, K.; Rotello, V.M.; Santore, M.M. Antimicrobial Surfaces Containing Cationic Nanoparticles: How Immobilized, Clustered, and Protruding Cationic Charge Presentation Affects Killing Activity and Kinetics. *Colloids Surf. B. Biointerfaces* **2015**, *125*, 255–263. [CrossRef]
14. Fuchs, A.D.; Tiller, J.C. Contact-Active Antimicrobial Coatings Derived from Aqueous Suspensions. *Angew. Chem. Int. Ed.* **2006**, *45*, 6759–6762. [CrossRef]
15. Milović, N.M.; Wang, J.; Lewis, K.; Klibanov, A.M. Immobilized N-Alkylated Polyethylenimine Avidly Kills Bacteria by Rupturing Cell Membranes with No Resistance Developed. *Biotechnol. Bioeng.* **2005**, *90*, 715–722. [CrossRef]

16. Förch, R.; Chifen, A.N.; Bousquet, A.; Khor, H.L.; Jungblut, M.; Chu, L.Q.; Zhang, Z.; Osey-Mensah, I.; Sinner, E.K.; Knoll, W. Recent and Expected Roles of Plasma-Polymerized Films for Biomedical Applications. *Chem. Vap. Deposition* **2007**, *13*, 280–294. [CrossRef]
17. Franklin, A.D. Nanomaterials in Transistors: From High-Performance to Thin-Film Applications. *Science* **2015**, *349*, aab2750. [CrossRef]
18. Montemor, M.d.F. Functional and Smart Coatings for Corrosion Protection: A Review of Recent Advances. *Surf. Coat. Technol.* **2014**, *258*, 17–37. [CrossRef]
19. Zare, M.; Zare, M.; Butler, J.A.; Ramakrishna, S. Nanoscience-Led Antimicrobial Surface Engineering to Prevent Infections. *ACS Appl. Nano Mater.* **2021**, *4*, 4269–4283. [CrossRef]
20. Lishchynskyi, O.; Shymborska, Y.; Stetsyshyn, Y.; Raczkowska, J.; Skirtach, A.G.; Peretiatko, T.; Budkowski, A. Passive Antifouling and Active Self-Disinfecting Antiviral Surfaces. *Chem. Eng. J.* **2022**, *446*, 137048. [CrossRef]
21. Schröder, K.; Finke, B.; Ohl, A.; Lüthen, F.; Bergemann, C.; Nebe, B.; Rychly, J.; Walschus, U.; Schlosser, M.; Liefeith, K. Capability of Differently Charged Plasma Polymer Coatings for Control of Tissue Interactions with Titanium Surfaces. *J. Adhes. Sci. Technol.* **2010**, *24*, 1191–1205. [CrossRef]
22. Cao, N.; Miao, Y.; Zhang, D.; Boukherroub, R.; Lin, X.; Ju, H.; Li, H. Preparation of Mussel-Inspired Perfluorinated Polydopamine Film on Brass Substrates: Superhydrophobic and Anti-Corrosion Application. *Prog. Org. Coat.* **2018**, *125*, 109–118. [CrossRef]
23. Zhou, C.; Shi, Y.; Sun, C.; Yu, S.; Liu, M.; Gao, C. Thin-Film Composite Membranes Formed by Interfacial Polymerization with Natural Material Sericin and Trimesoyl Chloride for Nanofiltration. *J. Membr. Sci.* **2014**, *471*, 381–391. [CrossRef]
24. Gupta, P.; Nayak, K.K. Characteristics of Protein-Based Biopolymer and Its Application. *Polym. Eng. Sci.* **2015**, *55*, 485–498. [CrossRef]
25. Woloszyn, J.D.; Hesse, P.; Hungenberg, K.D.; McAuley, K.B. Parameter Selection and Estimation Techniques in a Styrene Polymerization Model. *Macromol. React. Eng.* **2013**, *7*, 293–310. [CrossRef]
26. Yasuda, H.; Bumgarner, M.; Marsh, H.; Morosoff, N. Plasma Polymerization of Some Organic Compounds and Properties of the Polymers. *J. Polym. Sci., Polym. Chem. Ed.* **1976**, *14*, 195–224. [CrossRef]
27. Kumar, A.; Al-Jumaili, A.; Prasad, K.; Bazaka, K.; Mulvey, P.; Warner, J.; Jacob, M.V. Pulse Plasma Deposition of Terpinen-4-Ol: An Insight into Polymerization Mechanism and Enhanced Antibacterial Response of Developed Thin Films. *Plasma Chem. Plasma Process.* **2020**, *40*, 339–355. [CrossRef]
28. Al-Jumaili, A.; Bazaka, K.; Jacob, M.V. Retention of Antibacterial Activity in Geranium Plasma Polymer Thin Films. *Nanomaterials* **2017**, *7*, 270. [CrossRef]
29. Chan, Y.W.; Siow, K.S.; Ng, P.Y.; Gires, U.; Majlis, B.Y. Plasma Polymerized Carvone as an Antibacterial and Biocompatible Coating. *Mater. Sci. Eng. C* **2016**, *68*, 861–871. [CrossRef]
30. Sharifahmadian, O.; Zhai, C.; Hung, J.; Shineh, G.; Stewart, C.A.C.; Fadzil, A.A.; Ionescu, M.; Gan, Y.; Wise, S.G.; Akhavan, B. Mechanically Robust Nitrogen-Rich Plasma Polymers: Biofunctional Interfaces for Surface Engineering of Biomedical Implants. *Mater. Today Adv.* **2021**, *12*, 100188. [CrossRef]
31. Bhatt, S.; Pulpytel, J.; Arefi-Khonsari, F. Low and Atmospheric Plasma Polymerisation of Nanocoatings for Bio-Applications. *Surf. Innov.* **2015**, *3*, 63–83. [CrossRef]
32. Deng, X.; Leys, C.; Vujosevic, D.; Vuksanovic, V.; Cvelbar, U.; De Geyter, N.; Morent, R.; Nikiforov, A. Engineering of Composite Organosilicon Thin Films with Embedded Silver Nanoparticles Via Atmospheric Pressure Plasma Process for Antibacterial Activity. *Plasma Process. Polym.* **2014**, *11*, 921–930. [CrossRef]
33. Khan, M.; Rehman, N.; Khan, S.; Ullah, N.; Masood, A.; Ullah, A. Spectroscopic Study of Co2 and Co2–N2 Mixture Plasma Using Dielectric Barrier Discharge. *AIP Adv.* **2019**, *9*, 085015. [CrossRef]
34. Lou, B.-S.; Lai, C.-H.; Chu, T.-P.; Hsieh, J.-H.; Chen, C.-M.; Su, Y.-M.; Hou, C.-W.; Chou, P.-Y.; Lee, J.-W. Parameters Affecting the Antimicrobial Properties of Cold Atmospheric Plasma Jet. *J. Clin. Med. Res.* **2019**, *8*, 1930. [CrossRef]
35. Ma, C.; Nikiforov, A.; De Geyter, N.; Morent, R.; Ostrikov, K.K. Plasma for Biomedical Decontamination: From Plasma-Engineered to Plasma-Active Antimicrobial Surfaces. *Curr. Opin. Chem. Eng.* **2022**, *36*, 100764. [CrossRef]
36. Izadjoo, M.; Zack, S.; Kim, H.; Skiba, J. Medical Applications of Cold Atmospheric Plasma: State of the Science. *J. Wound Care* **2018**, *27*, S4–S10. [CrossRef]
37. Jungbauer, G.; Moser, D.; Müller, S.; Pfister, W.; Sculean, A.; Eick, S. The Antimicrobial Effect of Cold Atmospheric Plasma against Dental Pathogens—A Systematic Review of in-Vitro Studies. *Antibiotics* **2021**, *10*, 211. [CrossRef]
38. Gerchman, D.; Bones, B.; Pereira, M.; Takimi, A. Thin Film Deposition by Plasma Polymerization Using D-Limonene as a Renewable Precursor. *Prog. Org. Coat.* **2019**, *129*, 133–139. [CrossRef]
39. Siow, K.S.; Britcher, L.; Kumar, S.; Griesser, H. Qcm-D and Xps Study of Protein Adsorption on Plasma Polymers with Sulfonate and Phosphonate Surface Groups. *Colloids Surf. B. Biointerfaces* **2019**, *173*, 447–453. [CrossRef]
40. Ahmed, N.; Masood, A.; Siow, K.S.; Wee, M.; Haron, F.F.; Patra, A.; Nayan, N.; Soon, C.F.J.P.C.; Processing, P. Effects of Oxygen (O2) Plasma Treatment in Promoting the Germination and Growth of Chili. *Plasma Chem. Plasma Process.* **2022**, *42*, 91–108. [CrossRef]
41. Madkour, A.E.; Tew, G.N. Towards Self-Sterilizing Medical Devices: Controlling Infection. *Polym. Int.* **2008**, *57*, 6–10. [CrossRef]
42. Moyes, R.B.; Reynolds, J.; Breakwell, D.P. Differential Staining of Bacteria: Gram Stain. *Curr. Protoc. Microbiol.* **2009**, *15*, A.3C.1–A.3C.8. [CrossRef] [PubMed]

43. Wickramasinghe, S.; Ju, M.; Milbrandt, N.B.; Tsai, Y.H.; Navarreto-Lugo, M.; Visperas, A.; Klika, A.; Barsoum, W.; Higuera-Rueda, C.A.; Samia, A.C.S. Photoactivated Gold Nanorod Hydrogel Composite Containing D-Amino Acids for the Complete Eradication of Bacterial Biofilms on Metal Alloy Implant Materials. *ACS Appl. Nano Mater.* **2020**, *3*, 5862–5873. [CrossRef]
44. Srinivasan, S.; McKinley, G.H.; Cohen, R.E. Assessing the Accuracy of Contact Angle Measurements for Sessile Drops on Liquid-Repellent Surfaces. *Langmuir* **2011**, *27*, 13582–13589. [CrossRef] [PubMed]
45. Wang, Y.; Sang, D.K.; Du, Z.; Zhang, C.; Tian, M.; Mi, J. Interfacial Structures, Surface Tensions, and Contact Angles of Diiodomethane on Fluorinated Polymers. *J. Phys. Chem. C* **2014**, *118*, 10143–10152. [CrossRef]
46. Belibel, R.; Avramoglou, T.; Garcia, A.; Barbaud, C.; Mora, L. Effect of Chemical Heterogeneity of Biodegradable Polymers on Surface Energy: A Static Contact Angle Analysis of Polyester Model Films. *Mater. Sci. Eng. C* **2016**, *59*, 998–1006. [CrossRef]
47. Fahmy, A.; Mix, R.; Schönhals, A.; Friedrich, J. Surface and Bulk Structure of Thin Spin Coated and Plasma-Polymerized Polystyrene Films. *Plasma Chem. Plasma Process.* **2012**, *32*, 767–780. [CrossRef]
48. Jacob, M.V.; Easton, C.D.; Anderson, L.J.; Bazaka, K. Rf Plasma Polymerised Thin Films from Natural Resources. *Int. J. Mod. Phys. Conf. Ser.* **2014**, *32*, 1460319. [CrossRef]
49. Zhao, C.; Li, L.; Wang, Q.; Yu, Q.; Zheng, J. Effect of Film Thickness on the Antifouling Performance of Poly (Hydroxy-Functional Methacrylates) Grafted Surfaces. *Langmuir* **2011**, *27*, 4906–4913. [CrossRef]
50. Yuan, Y.; Hays, M.P.; Hardwidge, P.R.; Kim, J. Surface Characteristics Influencing Bacterial Adhesion to Polymeric Substrates. *RSC Adv.* **2017**, *7*, 14254–14261. [CrossRef]
51. Derdar, H.; Belbachir, M.; Harrane, A. A Green Synthesis of Polylimonene Using Maghnite-H+, an Exchanged Montmorillonite Clay, as Eco-Catalyst. *Bull. Chem. React. Eng. Catal.* **2019**, *14*, 69–78. [CrossRef]
52. Siow, K.S.; Britcher, L.; Kumar, S.; Griesser, H.J. Plasma Polymers Containing Sulfur and Their Co-Polymers with 1, 7-Octadiene: Chemical and Structural Analysis. *Plasma Process. Polym.* **2017**, *14*, 1600044. [CrossRef]
53. Palaniappan, S.; Narayana, B. Temperature Effect on Conducting Polyaniline Salts: Thermal and Spectral Studies. *J. Polym. Sci. Part A Polym. Chem.* **1994**, *32*, 2431–2436. [CrossRef]
54. Burkey, D.D.; Gleason, K.K. Structure and Thermal Properties of Thin Film Poly (A-Methylstyrene) Deposited Via Plasma-Enhanced Chemical Vapor Deposition. *Chem. Vap. Depos.* **2003**, *9*, 65–71. [CrossRef]
55. Easton, C.; Jacob, M. Ageing and Thermal Degradation of Plasma Polymerised Thin Films Derived from Lavandula Angustifolia Essential Oil. *Polym. Degrad. Stab.* **2009**, *94*, 597–603. [CrossRef]
56. Clouet, F.; Shi, M. Interactions of Polymer Model Surfaces with Cold Plasmas: Hexatriacontane as a Model Molecule of High-Density Polyethylene and Octadecyl Octadecanoate as a Model of Polyester. I. Degradation Rate Versus Time and Power. *J. Appl. Polym. Sci.* **1992**, *46*, 1955–1966. [CrossRef]
57. Bazaka, K.; Jacob, M.V. Post-Deposition Ageing Reactions of Plasma Derived Polyterpenol Thin Films. *Polym. Degrad. Stab.* **2010**, *95*, 1123–1128. [CrossRef]
58. Alancherry, S.; Bazaka, K.; Jacob, M.V. Rf Plasma Polymerization of Orange Oil and Characterization of the Polymer Thin Films. *J. Polym. Environ.* **2018**, *26*, 2925–2933. [CrossRef]
59. Bazaka, K.; Jacob, M.V.; Shanks, R.A. Fabrication and Characterization of Rf Plasma Polymerized Thin Films from 3, 7-Dimethyl-1, 6-Octadien-3-Ol for Electronic and Biomaterial Applications. *Adv. Mater. Res.* **2010**, *123*, 323–326. [CrossRef]
60. Friedrich, J. Mechanisms of Plasma Polymerization–Reviewed from a Chemical Point of View. *Plasma Process. Polym.* **2011**, *8*, 783–802. [CrossRef]
61. Drábik, M.; Polonskyi, O.; Kylián, O.; Čechvala, J.; Artemenko, A.; Gordeev, I.; Choukourov, A.; Slavínská, D.; Matolínová, I.; Biederman, H. Super-Hydrophobic Coatings Prepared by Rf Magnetron Sputtering of Ptfe. *Plasma Process. Polym.* **2010**, *7*, 544–551. [CrossRef]
62. Ahmad, J.; Bazaka, K.; Whittle, J.D.; Michelmore, A.; Jacob, M.V. Structural Characterization of Γ-Terpinene Thin Films Using Mass Spectroscopy and X-Ray Photoelectron Spectroscopy. *Plasma Process. Polym.* **2015**, *12*, 1085–1094. [CrossRef]
63. Park, C.-S.; Jung, E.Y.; Kim, D.H.; Kim, D.Y.; Lee, H.-K.; Shin, B.J.; Lee, D.H.; Tae, H.-S. Atmospheric Pressure Plasma Polymerization Synthesis and Characterization of Polyaniline Films Doped with and without Iodine. *Materials* **2017**, *10*, 1272. [CrossRef]
64. Espina, L.; Gelaw, T.K.; de Lamo-Castellvi, S.; Pagán, R.; Garcia-Gonzalo, D. Mechanism of Bacterial Inactivation by (+)-Limonene and Its Potential Use in Food Preservation Combined Processes. *PLoS ONE* **2013**, *8*, e56769. [CrossRef]
65. Han, Y.; Chen, W.; Sun, Z. Antimicrobial Activity and Mechanism of Limonene against Staphylococcus Aureus. *J. Food Saf.* **2021**, *41*, e12918. [CrossRef]
66. Nikaido, H. Multidrug Efflux Pumps of Gram-Negative Bacteria. *J. Bacteriol.* **1996**, *178*, 5853–5859. [CrossRef]
67. Diu, T.; Faruqui, N.; Sjöström, T.; Lamarre, B.; Jenkinson, H.F.; Su, B.; Ryadnov, M.G. Cicada-Inspired Cell-Instructive Nanopatterned Arrays. *Sci. Rep.* **2014**, *4*, 7122. [CrossRef]
68. Schiffman, J.D.; Elimelech, M. Antibacterial Activity of Electrospun Polymer Mats with Incorporated Narrow Diameter Single-Walled Carbon Nanotubes. *ACS Appl. Mater. Interfaces.* **2011**, *3*, 462–468. [CrossRef]

69. Lee, S.B.; Koepsel, R.R.; Morley, S.W.; Matyjaszewski, K.; Sun, Y.; Russell, A.J. Permanent, Nonleaching Antibacterial Surfaces. 1. Synthesis by Atom Transfer Radical Polymerization. *Biomacromolecules* **2004**, *5*, 877–882. [CrossRef]
70. Kumar, V.; Pulpytel, J.; Giudetti, G.; Rauscher, H.; Rossi, F.; Arefi-Khonsari, F. Amphiphilic Copolymer Coatings Via Plasma Polymerisation Process: Switching and Anti-Biofouling Characteristics. *Plasma Process. Polym.* **2011**, *8*, 373–385. [CrossRef]

polymers MDPI

Article

Nanostructured Polyaniline Films Functionalized through Auxiliary Nitrogen Addition in Atmospheric Pressure Plasma Polymerization

Jae Young Kim [1,†], Hyojun Jang [1,†], Ye Rin Lee [1], Kangmin Kim [2], Habeeb Olaitan Suleiman [1], Choon-Sang Park [3], Bhum Jae Shin [4], Eun Young Jung [1,5,*] and Heung-Sik Tae [1,*]

1 School of Electronic and Electrical Engineering, College of IT Engineering, Kyungpook National University, Daegu 41566, Republic of Korea
2 School of Electronics Engineering, College of IT Engineering, Kyungpook National University, Daegu 41566, Republic of Korea
3 Department of Electrical Engineering, Milligan University, Johnson City, TN 37682, USA
4 Department of Electronics Engineering, Sejong University, Seoul 05006, Republic of Korea
5 The Institute of Electronic Technology, College of IT Engineering, Kyungpook National University, Daegu 41566, Republic of Korea
* Correspondence: eyjung@knu.ac.kr (E.Y.J.); hstae@ee.knu.ac.kr (H.-S.T.); Tel.: +82-940-8863 (E.Y.J.); +82-53-950-6563 (H.-S.T.)
† These authors contributed equally to this work.

Abstract: Polyaniline (PANI) was synthesized from liquid aniline, a nitrogen-containing aromatic compound, through the atmospheric pressure (AP) plasma process using a newly designed plasma jet array with wide spacing between plasma jets. To expand the area of the polymerized film, the newly proposed plasma jet array comprises three AP plasma jet devices spaced 7 mm apart in a triangular configuration and an electrodeless quartz tube capable of applying auxiliary gas in the center of the triangular plasma jets. The vaporized aniline monomer was synthesized into a PANI film using the proposed plasma array device. The effects of nitrogen gas addition on the morphological, chemical, and electrical properties of PANI films in AP argon plasma polymerization were examined. The iodine-doped PANI film was isolated from the atmosphere through encapsulation. The constant electrical resistance of the PANI film indicates that the conductive PANI film can achieve the desired resistance by controlling the atmospheric exposure time through encapsulation.

Keywords: atmospheric pressure plasma; nitrogen addition; plasma polymerization; polyaniline nanostructures

Citation: Kim, J.Y.; Jang, H.; Lee, Y.R.; Kim, K.; Suleiman, H.O.; Park, C.-S.; Shin, B.J.; Jung, E.Y.; Tae, H.-S. Nanostructured Polyaniline Films Functionalized through Auxiliary Nitrogen Addition in Atmospheric Pressure Plasma Polymerization. *Polymers* **2023**, *15*, 1626. https://doi.org/10.3390/polym15071626

Academic Editor: Anton M. Manakhov

Received: 17 February 2023
Revised: 22 March 2023
Accepted: 22 March 2023
Published: 24 March 2023

1. Introduction

Polyaniline (PANI), a representative conductive polymer, is a nitrogen-containing aromatic compound [1,2]. It exhibits excellent conductivity, easy manufacturability, low cost, good biocompatibility, and environmental stability and is used in various applications, such as organic electronics, anticorrosion materials, electrorheological materials, and nanobiomedicines [2–6]. Conventional PANI synthesis mainly involves three processes: chain growth through aniline oxidation, dimerization, and oligomerization/polymerization [6]. It requires an acidic environment and oxidants, and the resulting π-conjugated polymer has low solubility despite the wet process [6–8]. Although monomers with various substituent groups can be employed to improve solubility, they reduce conductivity [9].

Plasma polymerization is a synthetic method that can overcome the limitations of the existing polymerization processes. The plasma used for polymerization has a low ion temperature and a high electron temperature; this type of plasma is called non-thermodynamic-equilibrium plasma or nonthermal plasma [10,11]. Given that its nonthermal property induces high chemical activity without thermal damage, plasma polymerization is particularly advantageous for synthesizing temperature-sensitive organic polymers [11,12].

Recently, many studies have attempted to use atmosphere-generated nonthermal plasma that does not require vacuum equipment for material processing, including polymerization [13–15]. Atmospheric pressure (AP) plasma polymerization is attracting considerable attention as an extremely simple and eco-friendly method because AP plasma can replace chemicals or extreme synthetic environments, such as high temperatures, for the oxidation/reduction of monomers [16,17]. In particular, recent studies have demonstrated that AP plasma polymerization can produce polymer films using only electricity and monomers without the application of additional chemicals and heat [13–16]. In AP plasma polymerization, vaporized monomers are transferred into the plasma region, activated in the plasma medium, and crosslinked while leaving the plasma region [17,18].

Many studies have reported on the effect of applied power on multiple materials synthesized via AP plasma polymerization [19–22]. An important consideration for polymerization using AP plasma is the activation of monomers without breaking the chemical structure during the interaction between monomer molecules and plasma. When high applied power is used to generate a sufficient discharge in AP, the chemical structure of the monomer is weakened and an incomplete structure is synthesized because of excessive plasma energy [21–23]. The resulting degradation of the electrical and optical properties of the conjugated polymer eventually becomes a limitation for the functionality of the conjugated polymer as an electrode, light-emitting, and gas-sensing material in electronic applications [24,25]. AP plasma polymerization allows the precise control of the process and the resulting polymer by changing experimental parameters such as the working gas composition, plasma reactor configuration, discharge initiation/maintenance, and power source type [26–29]. In particular, the nonthermal plasma state during polymerization can be stabilized by controlling the amount of vaporized monomer molecules, the type and flow rate of additional neutral gas, and applied power. By applying additional gas to the AP plasma reactor, the plasma intensity and reactivity can be adjusted, and certain radicals, which can be expected to improve polymer properties, are also supplied.

In this study, PANI was synthesized from liquid aniline, a nitrogen-containing aromatic compound, through the AP plasma process using a newly designed plasma jet array with wide spacing between plasma jets. During PANI synthesis, an auxiliary neutral gas, such as Ar, N_2, or O_2, was introduced in the plasma region to control plasma intensity and generate additional radicals. The effect of the addition of N_2, which improves the quality of the nanostructured PANI film, was examined in detail. The influences of N_2 gas addition on the synthesis process were investigated, and the morphological, chemical, and electrical properties of the synthesized PANI were then examined.

2. Materials and Methods

2.1. AP Plasma Polymerization System

Our previous study reported on aerosol-through-plasma systems that used vaporized liquid monomers, related to AP plasma polymerization techniques, in detail [15]. Figure 1 shows the plasma polymerization equipment. The details of the experimental system, which included a monomer bubbling unit, gas feedline, driving system, and electrical monitoring instrument, are described in a previous work [15], except for the AP plasma generator. In the primary gas supply for polymerization, AP plasma was generated using Ar gas with a flow rate of 2200 sccm, and liquid aniline (MW = 93 $g \cdot mol^{-1}$, Sigma-Aldrich Co., St. Louis, MO, USA) was vaporized using a bubbler and Ar gas with a flow rate of 500 sccm. The additional gas supply was split into three to test Ar, O_2, and N_2 gases. A sinusoidal voltage with a peak value of 10 kV and frequency of 28 kHz was then applied to the AP plasma jet (APPJ) array using an inverter-type power supply.

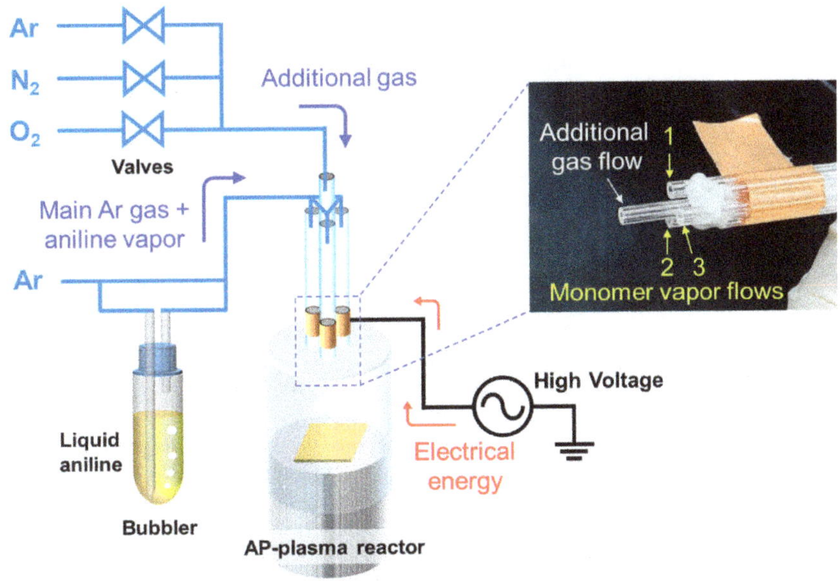

Figure 1. Experimental setup for the AP plasma polymerization system.

2.2. *AP Plasma Reactor with Added Auxiliary Gas*

The proposed AP plasma reactor comprised an APPJ array, guide tube, and substrate stand (Figure 1). The APPJ array was fabricated using four identical quartz tubes with outer and inner diameters of 3 and 1.5 mm, respectively. The APPJ array contained one tube in the center of the structure. Three other tubes were arranged in a triangle around the center tube. In the three quartz tubes arranged in a triangle, the center-to-center distance between the tubes was 7 mm. Cu tape, which was used as a powered electrode, was wrapped around three outer tubes 15 mm apart from the end of the tubes. These three tubes were then combined with each other using copper tape, thereby generating plasma in each quartz tube in the reactor. An electrodeless quartz tube was located in the center of the triangular tubes for auxiliary gas addition, and its end protruded 10 mm from the end of the outer tube to facilitate auxiliary gas supply into the polymerization area. Using the prepared plasma reactor, nanostructured PANI thin films were deposited with a 100 sccm gas flow of Ar, O_2, and N_2 gases for AP Ar-based plasma synthesis.

2.3. *Electrical and Optical Characterization of Generated Plasma*

A voltage probe (P6015A, Tektronix Inc., Beaverton, OR, USA) and a current probe (4100, Pearson Electronics Inc., Palo Alto, CA, USA) were used to examine APPJs during plasma polymerization. The wavelength-unresolved light emission from a photosensor amplifier (C6386-01, Hamamatsu Corp., Hamamatsu, Japan) detecting the region from the visible to the near-infrared bands were displayed on an oscilloscope (TDS3014B, Tektronix Inc., Beaverton, OR, USA). The optical emission spectra (OES) of plasma emission were acquired using a spectrometer with a fiber optic probe (USB-2000+, Ocean Optics Inc., Dunedin, FL, USA).

2.4. *Analysis and Characterization of Nanostructured PANI Films*

The shape and structure of the nanostructured polymer films were observed with a field-emission scanning electron microscope (FE-SEM; SU8220, Hitachi Korea Co., Ltd., Seoul, Korea). Based on the FE-SEM image, the pore area distribution of the film was measured through image analysis using the IMT i-Solution software (IMT i-solution Inc., Burnaby, BC, Canada).

The functional groups of PANI films synthesized by the APPJs were confirmed via attenuated total reflection Fourier transform infrared spectroscopy (ATR-FTIR; Vertex 70, Bruker, Ettlingen, Germany) at the Korea Basic Science Institute (Daegu, Korea). During the ATR-FTIR measurements, spectra ranging from 650 to 4000 cm^{-1} were obtained by averaging 128 scans.

For X-ray photoelectron spectroscopy (XPS; ESCALAB 250XI, Thermo Fisher Scientific, Waltham, MA, USA), a monochromatic Al Kα X-ray source (hυ = 1486.71 eV) operating at 15 kV and 20 mA was used to confirm surface chemical characteristics. XPS measurement was performed over an area of 500 μm × 500 μm. Wide-scan spectra were plotted with data collected every 1 eV. High-resolution measurements were recorded every 0.03 eV. Peak fitting was performed using Gaussian–Lorentzian peak shapes.

2.5. Iodine Doping of PANI Films for the Electrical Conductivity Test

For the electrical conductivity test, PANI films were prepared on a Si substrate with interdigitated electrodes (IDEs). Each IDE comprised 20 pairs of interdigitated and thin electrodes and had a width of 10.8 μm. The distance between IDEs was 2.54 μm. For doping halogen elements, the prepared PANI film sample and 2 g of I_2 (Sigma-Aldrich Co., St. Louis, MO, USA, 99.99%) pellets were placed together in a Petri dish and vacuum-sealed for 30 min. When I_2 was doped in the PANI film through sublimation, the color of the PANI film changed to dark brown.

3. Results and Discussion

3.1. Changes in Glow Discharge and PANI Films Due to Auxiliary Gas Addition during AP Plasma Polymerization

Figure 2 shows the optical images of the glow discharge with the addition of auxiliary gas through the central tube of the APPJ and the resulting PANI films deposited on Si substrates. For AP plasma polymerization, Ar gas containing aniline vapor was added via the three triangular outer tubes to generate the discharge, and a nominal amount of gas was then applied to the central tube. Because each plasma jet in the triangular plasma jet array was separated from the others by a distance of 7 mm, the PANI film could be uniformly deposited on a Si substrate with dimensions of 20 mm × 20 mm without technical difficulties, as shown in Figure 2a. In this study, unlike our previous study [15], the influence of the auxiliary gas during plasma polymerization could be examined without changing the design of the guide tube or substrate stand, because an additional tube can be placed at the center of the triangular plasma jet array. Ar gas containing aniline was introduced into the plasma reactor at a flow rate of 2700 sccm, whereas the added neutral gas had a relatively low flow rate of 100 sccm. The glow emission behavior of the AP plasma and the uniformity of the resulting PANI film were extremely dependent on the type of additive gas.

When a small amount of Ar gas, which was the same as the main gas, was added, the discharge characteristics did not change considerably during plasma polymerization, and a cloudy glow emission attributed to the diffusion of aniline particles with various energy levels was observed (Figure 2a). When this cloudy glow emission was extensive and uniform inside the plasma reactor, a uniform polymer film could be obtained. As shown in Figure 2b,c, the addition of O_2 or N_2 during AP plasma polymerization attenuated the cloudy glow emission. The cloudy glow emissions decreased because the monomers lost energy due to collisions with neutral gases. Such collisions can be attributed to the addition of a small amount of gas different from the discharge gas to the middle of the reactor, wherein many activated monomers were located. In particular, when O_2 was added, the cloudy glow decreased, and only plasma jets were observed. When N_2 was added, the blurred glow emitted a blue color, indicating that some of the neutral nitrogen was transformed into excited species by the plasma energy. The plasma typically generated reactive oxygen species (ROS) and reactive nitrogen species (RNS) when O_2 and N_2 were added during the plasma process, respectively. However, when Ar gas, which was the same as the discharge gas, was used, the reactive byproducts caused morphological and chemical changes in PANI nanostructures.

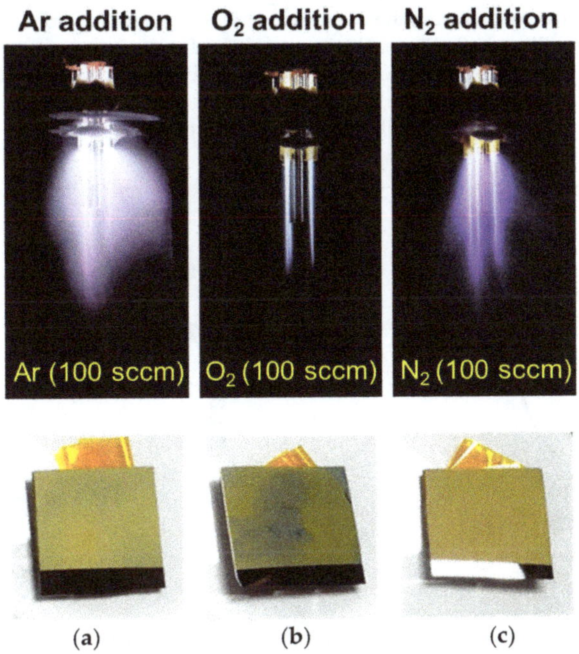

Figure 2. Images of glow plasma during plasma polymerization and PANI films deposited on Si substrates: image of Ar glow discharge during polymerization with the addition of (**a**) Ar, (**b**) O_2, and (**c**) N_2 and images of the resulting PANI films. All additional gases were used at a flow rate of 100 sccm.

The significant suppression of the generated plasma when a small amount of O_2 was applied adversely affected the uniform deposition of the PANI film (Figure 2b). ROS generated by O_2 addition not only hindered the crosslinking of the monomers but also partially etched the resulting PANI film [15]. The PANI film deposited by adding N_2 was uniformly deposited on the Si substrate with dimensions of 20 mm × 20 mm and exhibited a matte beige appearance (Figure 2c). The images of the three PANI films in Figure 2 show that the addition of O_2 considerably degraded the uniformity of the PANI film. Therefore, additional investigations to determine the effect of the additive gas excluded O_2 addition.

3.2. Electrical and Optical Characteristics during AP Plasma Polymerization

The driving voltage, discharge current, and optical intensity were monitored, as shown in Figure 3, to examine the discharge behaviors during plasma polymerization. The driving voltage maintained a constant sinusoidal waveform with a frequency of 28 kHz and was not distorted due to electrical discharge. No change in electrical behaviors due to gas addition could be observed because the addition of a small amount of neutral gas did not fundamentally affect the discharge initiation and maintenance in the plasma reactor. Figure 3a shows the electrical behaviors monitored when Ar was added at a flow rate of 100 sccm. For removing the displacement current caused by the charging and discharging of the capacitive device, the discharge current was obtained by subtracting the current monitored when the operating voltage was applied in the absence of Ar gas from the total current monitored when plasma was normally generated. The discharge current waveform in Figure 3a shows that discharge occurred during the rising and falling periods of the voltage waveform, indicating that the discharges were successful even when the powered electrode acted not only as an anode but also as a cathode. The optical emission of the plasma jets measured near the polymerization area was periodically stable, and the optical intensity during the rising slope of the voltage waveform was higher than that during the falling slope (Figure 3b). The resulting discharge current and optical intensity demonstrated that an intense discharge was produced when the powered electrode served as the anode, demonstrating the typical behavior of dielectric barrier discharge jets generated using a single electrode without a counter electrode [30].

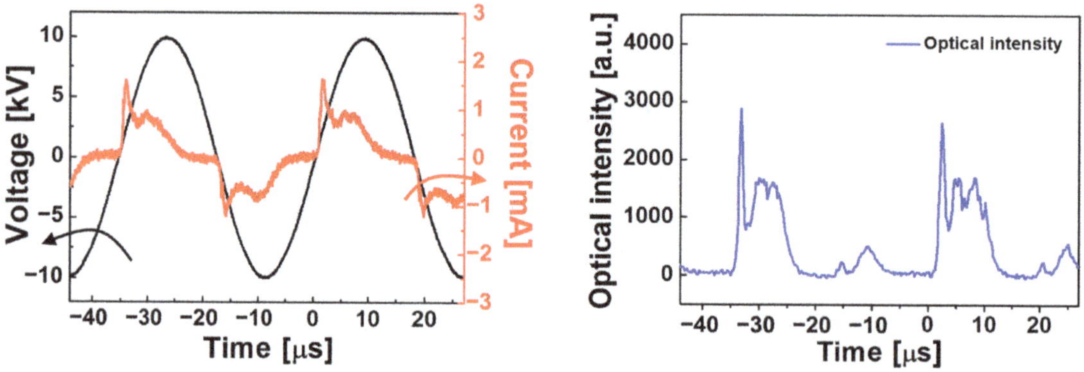

(**a**) (**b**)

Figure 3. Temporal behaviors of (**a**) driving voltage (black line) and discharge current (red line), and (**b**) optical emission from the plasma generated by the proposed APPJ array.

The emission spectra of the plasma jets were obtained during AP plasma polymerization to identify the diverse reactive species created in the plasma medium. OES is a representative diagnostic method for investigating the types and energy levels of excited species in a high-pressure plasma medium and avoids perturbing the plasma medium because it does not involve the use of a diagnostic metal probe [31,32]. Figure 4 shows that emission spectra peaks between 280 and 870 nm were detected in the plasma-generating region during polymerization, demonstrating the presence of excited N_2 as well as Ar species and carbon derivatives in the generated plasma. Many of the emission peaks in the range of 690–860 nm were attributed to the Ar discharge (Figure 4a,b), and those at 300–380 nm were primarily attributed to excited N_2 species. Moreover, multiple carbonaceous peaks (C–N and C–H) were observed (Figure 4c,d). The addition of N_2 gas reduced the intensity of the Ar plasma but substantially increased the N_2 peak, indicating that some of the neutral N_2 species produced RNS.

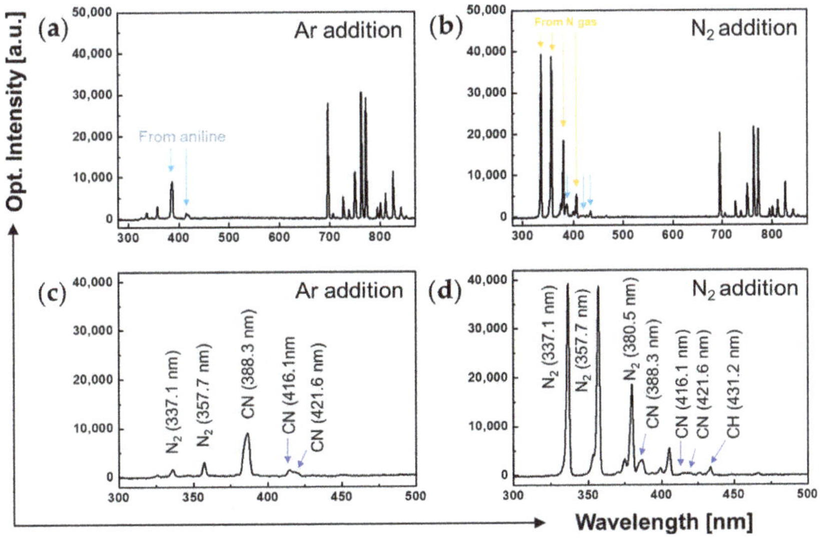

Figure 4. OES of plasma plumes under different gaseous conditions: OES of plasma plumes during AP plasma polymerization with (**a**) Ar addition (100 sccm) and (**b**) N_2 addition (100 sccm). Magnified emission spectra between 300 and 500 nm in the case of (**c**) Ar addition (100 sccm) and (**d**) N_2 addition (100 sccm).

3.3. Changes in the Film Properties of PANI Nanostructures by Nitrogen Addition

PANI films synthesized by adding Ar and N_2 to the central tube of the proposed APPJ during AP polymerization using three triangular plasma jets were observed in detail using the FE-SEM. The PANI film deposited through the proposed AP plasma polymerization process had a porous morphology that comprised crosslinked PANI nanofibers, as shown in Figure 5a,b. Normal PANI has a coarse granular form in which nanoparticles of various sizes are synthesized and aggregated [33,34]. PANI treated with heat or additives possesses a regular and ordered structure in which nanorods and nanofibers are crosslinked [33–35]. During AP plasma polymerization, certain aniline-derived radicals created by plasma energy served as effective additives to promote crosslinking. The crosslinking effects on the nanostructured PANI films appear as an increase in the total pore volume and a decrease in each pore size [33]. Morphological differences were observed between the nanostructured PANI films with the addition of small amounts of Ar and N_2. Figure 5c,d show the pore area distributions obtained by analyzing the high-magnification FE-SEM images in Figure 5a,b. The nanostructured PANI film synthesized with Ar addition had pores on the surface with an area ranging from 200 to 1200 nm^2. However, in the case of N_2 addition, the pore area was distributed between 20 and 300 nm^2, which was smaller and more uniform than the film deposited by Ar gas addition. The top and cross-sectional views of FE-SEM images demonstrated that the PANI film deposited with N_2 addition exhibited higher density and was more uniform than that deposited with Ar addition. Interestingly, the average heights of PANI films with N_2 and Ar addition were 42 and 15 μm, respectively (Figure 6a,b). Furthermore, the PANI films demonstrated better vertical alignment with the addition of N_2 gas. When N_2 was added, aniline derivatives acting as additives were more effectively generated, resulting in improved crosslinking, and the thickness of the PANI film increased because of the high pore volume obtained due to crosslinking [33].

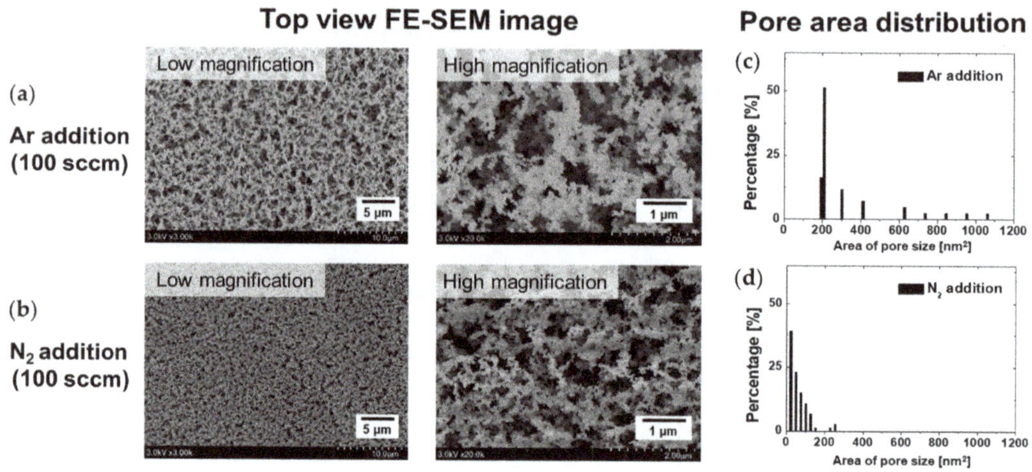

Figure 5. FE-SEM measurements of PANI films synthesized with the addition of (**a**) Ar and (**b**) N_2 (top view). Pore area distribution based on the high-magnification FE-SEM images of PANI films synthesized with the addition of (**c**) Ar and (**d**) N_2.

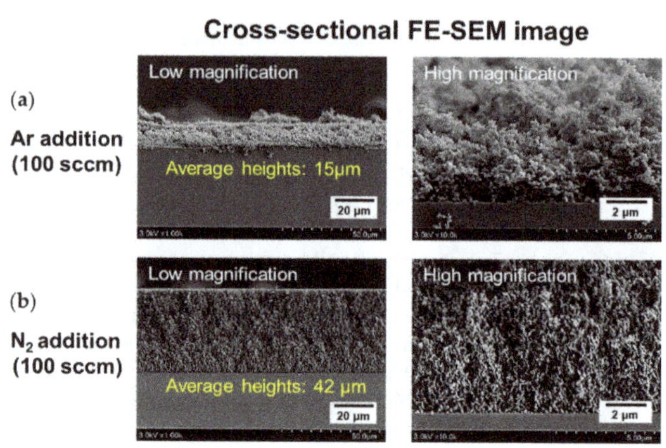

Figure 6. FE-SEM measurements of PANI films synthesized with the addition of (**a**) Ar and (**b**) N_2 (cross-sectional view).

The ATR-FTIR measurements of two PANI films deposited on Si substrates were acquired to examine the chemical characteristics of PANI films in accordance with the use of two different additive gases (Ar and N_2; Figure 7). The ATR-FTIR spectra demonstrated the following distinctive molecular structures of PANI, which demonstrated the successful synthesis of PANI from liquid aniline: C–H deformation from the aromatic ring (763 cm^{-1}), C–N stretching vibrations (1250 and 1313 cm^{-1}), benzenoid stretching vibration (1501 cm^{-1}), quinoid ring stretching vibration (1601 cm^{-1}), C–H asymmetric stretching (2844 and 2959 cm^{-1}), and N–H stretching vibration (3365 cm^{-1}) [36,37]. When 100 sccm N_2 was added during plasma polymerization, the intensity of all major peaks, including nitrogen-related peaks, increased in ATR-FTIR. Hence, the result demonstrated that the strength of the major chemical groups of PANI was enhanced because of the high nanoparticle density and uniform thickness of the nitrogen-added PANI film considering that the FTIR data were obtained with an ATR mode using the total reflection property.

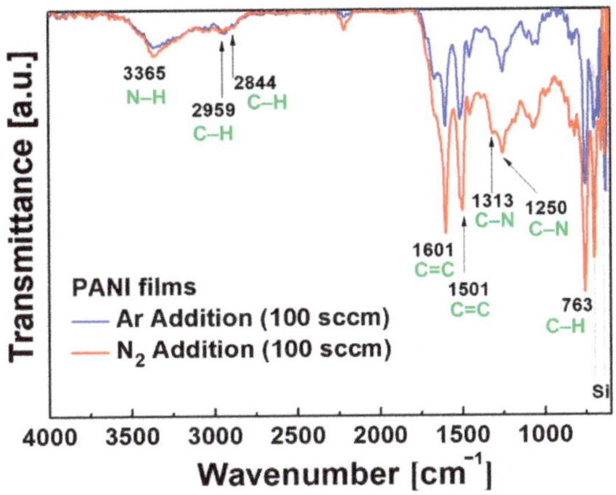

Figure 7. ATR-FTIR spectra of PANI films deposited with Ar and N_2 gases.

The chemical structure of the PANI film surface depending on the introduction of additional gases (Ar and N_2) during polymerization was characterized through XPS measurements (Figure 8). Figure 8a shows the wide-scan XPS spectra of the PANI films synthesized with additional gases (Ar and N_2). These spectra included three primary peaks assigned to oxygen (O 1s), nitrogen (N 1s), and carbon (C 1s). Table 1 summarizes the atomic composition of the PANI surface. Theoretically, a unit of PANI $(C_6H_5N)_n$ had a C/N ratio of 6 and lacked an oxygen component. However, in the case of Ar and N_2 addition, the surface of the synthesized PANI films exhibited similar atomic compositions, such as high amounts of oxygen and C/N ratios above 6. Inevitable exposure to water and oxygen from the air created C–O bonds on the surface of the PANI films [38]. Figure 8b–e show the high-resolution C 1s and N 1s XPS spectra of the synthesized PANI films. The C 1s profiles in Figure 8b,c were fitted to six energetic peaks at 284.6, 285.5, 286.5, 287.2, 288.1, and 289.1 eV, corresponding to the C=C, C–C/C–H, C–N, C–O, C=O, and O–C=O chemical groups, respectively [39]. The N 1s profiles in Figure 8d,e can be decomposed into three peaks centered at 399.0, 400.0, and 401.2 eV, which are assigned to –N=, –NH, and N^+ chemical groups, respectively [38,40]. Detailed information, including chemical states and contributions, is summarized in Table 2.

Table 1. Surface atomic composition ratios of PANI films synthesized with additional gases (Ar and N_2).

Conditions	C 1s (%)	N 1s (%)	O 1s (%)	C/N
Ar addition	76.8	11.3	11.9	6.80
N_2 addition	76.4	11.9	11.7	6.42

Studies related to crosslinked PANI [34,40] have demonstrated that it commonly has chemical properties with fewer quinoid imine groups but more nitrogen-benzenoid groups than normal PANI [34,40,41]. In the present study, these results were observed in the form of reduced –N= groups and increased C–N groups in the XPS results provided in Table 2, indicating that the crosslinking degree of the PANI film synthesized by adding N_2 gas was better than that of the PANI film synthesized by adding Ar. Furthermore, this result demonstrated good agreement with the increase in the number of specific aniline derivatives to promote crosslinking, as shown by FE-SEM measurements. The ratio of quinoid imine and benzenoid amine (–N=)/(–NH–) represents the redox state of PANI [42,43]. A ratio close to 0 indicates that PANI in the reduction state is dominant [40,42]. The conductivity of PANI can be improved through the oxidation of an amine (–NH–) with a dopant, such as I_2 [44,45]. When PANI is clearly in the reduced state or the value of (–N=)/(–NH–) is low, improved electrical properties can be expected after I_2 doping [44,45].

Therefore, as shown in Table 2, the electrical properties of the PANI film synthesized with the addition of N_2 can be expected to improve after doping.

Table 2. Summary of the contribution of each group in Figure 8b–e.

	Group	Binding Energy (eV)	Composition (%)	
			Ar Addition	N_2 Addition
C 1s	C=C	284.6	12.8	21.3
	C–C/C–H	285.5	52.7	43.0
	C–N	286.5	15.9	24.1
	C–O	287.2	11.5	6.4
	C=O	288.1	5.2	3.7
	O–C=O	289.1	1.9	1.5
N 1s	–N=	399.0	17.2	14.4
	–NH–	400.0	60.1	79.7
	N^+	401.2	22.7	5.9
	(–N=)/(–NH–)	-	0.29	0.18

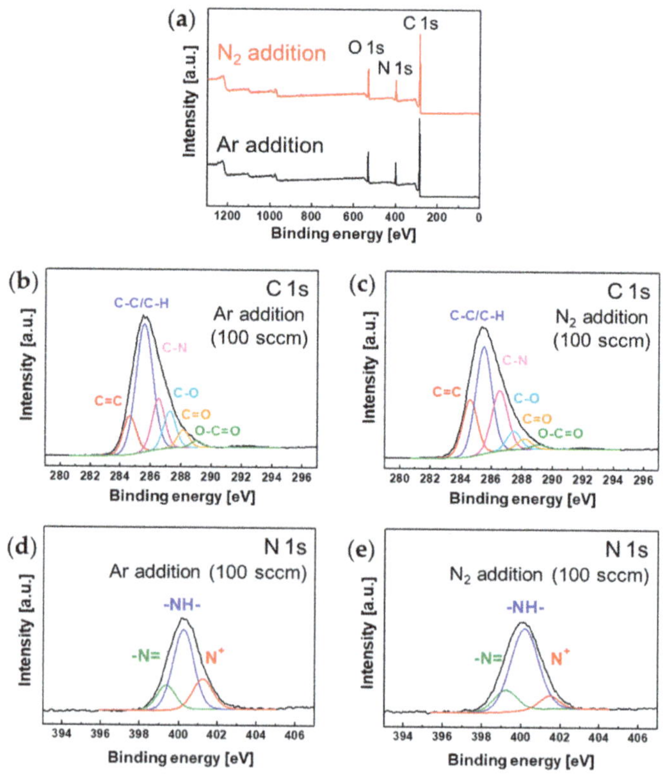

Figure 8. (**a**) Wide-scan XPS spectra of PANI films synthesized with additional gases (Ar and N_2). (**b**,**c**) C 1s core-level spectra of PANI films synthesized with Ar and N_2 gases. (**d**,**e**) N 1s core-level spectra of PANI films synthesized with Ar and N_2 gases.

3.4. Electrical Properties of Conductive PANI Films

Because conjugated nanostructured polymer films are fabricated for use as electronic devices, they must have electrical properties. A facile method for imparting conductivity to conjugated polymers is to dope the polymers with halogen elements [44,46]. Among halogen elements, I_2 is popular for the conductive functionalization of conjugated polymers because it is relatively easy to handle [47,48]. Therefore, I_2 was doped into PANI films fabricated with the addition of different auxiliary gases to examine the electrical properties of the prepared PANI films.

Figure 9 shows the resistance change of I_2-doped PANI films synthesized with the addition of Ar and N_2 in accordance with the exposure time to the ambient air. For a detailed comparison of the effect of gas addition, the resistance change of the I_2-doped PANI film synthesized by adding O_2 was recorded on the graph. The resistance measurement limit was 50 MΩ. If this limit is exceeded, the resistance is considered to be infinite. When an Ar flow of 100 sccm was added to the polymerization process, the resistance of the conductive PANI film continuously increased in the air and reached the measurement limit of 50 MΩ in 3 h. When N_2 was added, the resistance increased more slowly and reached 50 MΩ in 5 h. The I_2-doped PANI film obtained with the addition of N_2 was less affected by atmospheric hydration and had better resistance stability.

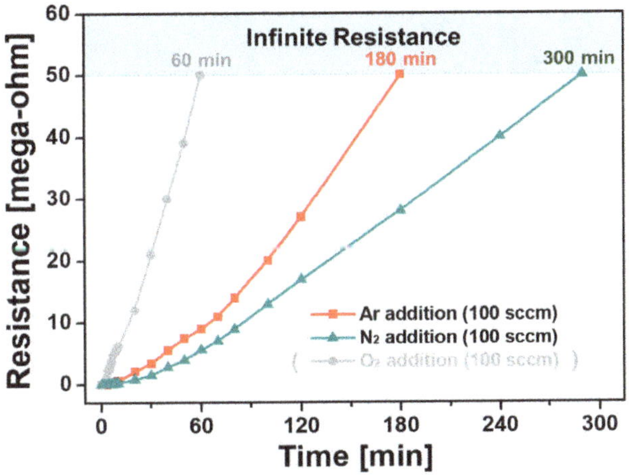

Figure 9. Electrical resistance of I_2-doped PANI films prepared with different additional gases.

This significant difference in resistance stability is closely related to the density and regularity of the PANI nanocomposite, as shown in the FE-SEM images (Figures 5 and 6). The PANI nanostructured film prepared by adding N_2 had excellent vertical orientation such that its nanostructure was extremely dense and uniform and was less affected by hydration in the ambient air, thereby improving its resistance stability. However, the PANI film prepared by adding O_2 was significantly affected by atmospheric hydration because it was irregular even when viewed with the naked eye; moreover, its resistance stability inevitably deteriorated. Furthermore, the resistance measurement confirmed that the conductive PANI film synthesized by adding N_2 exhibited better electrical properties than that synthesized by adding Ar, as predicted based on the XPS results.

I_2-doped PANI films are expected to possess excellent electrical resistance stability when isolated from moisture and oxygen in the external environment by sealing. Thus, the fabricated PANI film was encapsulated with sealing tape and a film to secure stable electrical properties. The encapsulation test was performed using the PANI film polymerized with the addition of 100 sccm N_2. This film had the best conductive performance. The change in resistance was measured over several days (Figure 10) to examine the resistance behavior of the encapsulated I_2-doped PANI film. When the I_2-doped PANI film was left in the atmosphere for 80 min after doping and the electrical resistance reached 10 MΩ, the PANI film on the IDE area was sealed by using polyimide tape (Kapton® tape, DuPont, Wilmington, DE, USA) and an elastic sealing film (PARAFILM® M, Bemis Company, Neenah, WI, USA) as well as isolated from the outside. After encapsulation, the increase in the electrical resistance of the PANI film was greatly attenuated, and the electrical resistance finally reached saturation at 15 MΩ after doping for 10 h (Encapsulation 1; red line in Figure 10). In another encapsulation test, the PANI film was encapsulated when the electrical resistance was 22 MΩ at 150 min. The resistance reached saturation at 30 MΩ 10 h after doping (Encapsulation 2; blue line in Figure 10). Monitoring the different resistances of the two encapsulated PANI films over three days revealed that both films had extremely consistent resistance values without even a change of 1 MΩ. The results indicated that the electrical resistance of the conductive PANI film can be manipulated by actively controlling the atmospheric exposure time of the polymer film via encapsulation. The long-term monitoring of the resistance of the PANI films with encapsulation revealed that three weeks were required to reach the measurement limit of 50 MΩ. The change in the resistance value of the PANI film is expected to be permanently avoided if encapsulation technology that is more advanced than the use of polyimide tape and sealing film can be employed.

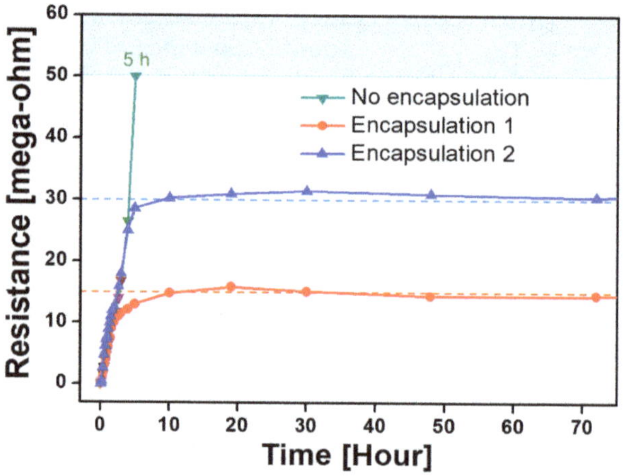

Figure 10. Changes in the electrical resistance of I_2-doped PANI films subjected to different encapsulation processes over three days. Encapsulation process examined with PANI films polymerized with N_2 gas added only.

4. Conclusions

In this study, changes in the morphological, chemical, and electrical properties of nanostructured PANI films with the addition of neutral gas for AP plasma polymerization were investigated. A separate gas tube was added to the center of the APPJ array device, and an auxiliary gas could be injected into the middle of the generated plasma. Adding a small amount of N_2 for AP plasma polymerization not only improved the density and uniformity of the resulting nanostructured PANI film but also increased the growth rate of the PANI film. The latter effect is beneficial for rapid polymerization. Moreover, in contrast to the addition of Ar, the addition of a small amount of N_2 could improve the resistance stability of the conductive PANI film. This property can be exploited to obtain constant electrical conductivity by isolating the conductive polymer film from moisture and oxygen in the ambient air via encapsulation. Adding a small amount of N_2 gas was experimentally demonstrated to be a facile approach for improving the properties of conjugated polymer films without considerably affecting the maintenance of AP plasma undergoing polymerization and deposition simultaneously. Further careful investigations are required to understand exactly how nitrogen molecules/radicals increase the deposition rate of PANI films or the density of PANI nanoparticles in the film. N_2 addition during plasma polymerization can be a practical approach for easily controlling the surface roughness of conductive polymers, thereby improving the capture of gas molecules when conductive polymers are used as the detection layer of gas sensors.

Author Contributions: Conceptualization, J.Y.K., H.J. and H.-S.T.; methodology, H.J., K.K., H.O.S., C.-S.P. and E.Y.J.; validation, J.Y.K., Y.R.L., C.-S.P. and E.Y.J.; formal analysis, J.Y.K., H.J., Y.R.L., H.O.S., B.J.S. and E.Y.J.; investigation, J.Y.K., H.J., K.K., H.O.S., C.-S.P., B.J.S. and H.-S.T.; resources, J.Y.K. and H.-S.T.; data curation, J.Y.K., H.J., K.K., H.O.S., E.Y.J. and H.-S.T.; writing—original draft preparation, H.J., J.Y.K. and H.-S.T.; writing—review and editing, J.Y.K., H.J., E.Y.J. and H.-S.T.; visualization, J.Y.K., H.J. and E.Y.J.; supervision, H.-S.T.; project administration, J.Y.K.; funding acquisition, J.Y.K. and H.-S.T. All authors have read and agreed to the published version of the manuscript.

Funding: This study was supported by two National Research Foundation of Korea (NRF) grants funded by the Korean government (No. 2020R1I1A3071693 and No. 2021R1I1A3049028).

Institutional Review Board Statement: Not applicable.

Data Availability Statement: Not applicable.

Conflicts of Interest: The authors declare no conflict of interest.

References

1. Laha, S.; Luthy, R.G. Oxidation of aniline and other primary aromatic amines by manganese dioxide. *Environ. Sci. Technol.* **1990**, *24*, 363–373. [CrossRef]
2. Zare, E.N.; Makvandi, P.; Ashtari, B.; Rossi, F.; Motahari, A.; Perale, G. Progress in conductive polyaniline-based nanocomposites for biomedical applications: A review. *J. Med. Chem.* **2020**, *63*, 1–22. [CrossRef]
3. Lawal, A.T.; Wallace, G.G. Vapor phase polymerization of conducting and non-conducting polymers: A review. *Talanta* **2014**, *119*, 133–143. [CrossRef] [PubMed]
4. Travaglini, L.; Micolich, A.P.; Cazorla, C.; Zeglio, E.; Lauto, A.; Mawad, D. Single-material OECT-based flexible complementary circuits featuring polyaniline in both conducting channels. *Adv. Funct. Mater.* **2021**, *31*, 2007205. [CrossRef]
5. Gao, F.; Mu, J.; Bi, Z.; Wang, S.; Li, Z. Recent advances of polyaniline composites in anticorrosive coatings: A review. *Prog. Organ. Coat.* **2021**, *151*, 106071. [CrossRef]
6. Ćirić-Marjanović, G. Recent advances in polyaniline research: Polymerization mechanisms, structural aspects, properties and applications. *Synth. Met.* **2013**, *177*, 1–47. [CrossRef]
7. Sapurina, I.; Stejskal, J. The mechanism of the oxidative polymerization of aniline and the formation of supramolecular polyaniline structures. *Polym. Int.* **2008**, *57*, 1295–1325. [CrossRef]
8. Bhadra, S.; Khastgir, D.; Singha, N.K.; Lee, J.H. Progress in preparation, processing and applications of polyaniline. *Prog. Polym. Sci.* **2009**, *34*, 783–810. [CrossRef]
9. Udum, Y.A.; Pekmez, K.; Yıldız, A. Electrochemical preparation of a soluble conducting aniline–thiophene copolymer. *Eur. Polym. J.* **2005**, *41*, 1136–1142. [CrossRef]
10. Scholtz, V.; Pazlarova, J.; Souskova, H.; Khun, J.; Julak, J. Nonthermal plasma—A tool for decontamination and disinfection. *Biotechnol. Adv.* **2015**, *33*, 1108–1119. [CrossRef]
11. Khelifa, F.; Ershov, S.; Habibi, Y.; Snyders, R.; Dubois, P. Free-radical-induced grafting from plasma polymer surfaces. *Chem. Rev.* **2016**, *116*, 3975–4005. [CrossRef] [PubMed]
12. Bárdos, L.; Baránková, H. Cold atmospheric plasma: Sources, processes, and applications. *Thin Solid Film* **2010**, *518*, 6705–6713. [CrossRef]
13. Pandiyaraj, K.N.; Ramkumar, M.C.; Kumar, A.A.; Vasu, D.; Padmanabhan, P.V.A.; Tabaei, P.S.E.; Cools, P.; Geyter, N.D.; Morent, R.; Jaganathan, S.K. Development of phosphor containing functional coatings via cold atmospheric pressure plasma jet—Study of various operating parameters. *Appl. Surf. Sci.* **2019**, *488*, 343–350. [CrossRef]
14. Yang, J.; Pu, Y.; Miao, D.; Ning, X. Fabrication of durably superhydrophobic cotton fabrics by atmospheric pressure plasma treatment with a siloxane precursor. *Polymers* **2018**, *10*, 460. [CrossRef] [PubMed]
15. Kim, J.-Y.; Jang, H.-J.; Jung, E.; Bae, G.; Lee, S.; Park, C.-S.; Shin, B.; Tae, H.-S. Improvement of the uniformity and electrical properties of polyaniline nanocomposite film by addition of auxiliary gases during atmospheric pressure plasma polymerization. *Nanomaterials* **2021**, *11*, 2315. [CrossRef] [PubMed]
16. Vazirinasab, E.; Jafari, R.; Momen, G. Evaluation of atmospheric-pressure plasma parameters to achieve superhydrophobic and self-cleaning HTV silicone rubber surfaces via a single-step, eco-friendly approach. *Surf. Coat. Technol.* **2019**, *375*, 100–111. [CrossRef]

17. Peran, J.; Ražić, S.E. Application of atmospheric pressure plasma technology for textile surface modification. *Text. Res. J.* **2020**, *90*, 1174–1197. [CrossRef]
18. Hegemann, D.; Hossain, M.M.; Körner, E.; Balazs, D.J. Macroscopic description of plasma polymerization. *Plasma Process. Polym.* **2007**, *4*, 229–238. [CrossRef]
19. Deynse, A.V.; Cools, P.; Leys, C.; Geyter, N.D.; Morent, R. Surface activation of polyethylene with an argon atmospheric pressure plasma jet: Influence of applied power and flow rate. *Appl. Surf. Sci.* **2015**, *328*, 269–278. [CrossRef]
20. Morent, R.; Geyter, N.D.; Jacobs, T.; Vlierberghe, S.V.; Dubruel, P.; Leys, C.; Schacht, E. Plasma-polymerization of HMDSO using an atmospheric pressure dielectric barrier discharge. *Plasma Process. Polym.* **2009**, *6*, S537–S542. [CrossRef]
21. Bashir, M.; Rees, J.M.; Zimmerman, W.B. Plasma polymerization in a microcapillary using an atmospheric pressure dielectric barrier discharge. *Surf. Coat. Technol.* **2013**, *234*, 82–91. [CrossRef]
22. Yang, P.; Zhang, J.; Guo, Y. Synthesis of intrinsic fluorescent polypyrrole nanoparticles by atmospheric pressure plasma polymerization. *Appl. Surf. Sci.* **2009**, *255*, 6927–6929. [CrossRef]
23. Kim, S.Y.; Kim, S.-Y.; Choi, J.; Lee, S.; Jo, S.M.; Joo, J.; Lee, H.-S. Two step microwave plasma carbonization including low plasma power pre-carbonization for polyacrylonitrile based carbon fiber. *Polymer* **2015**, *69*, 123–128. [CrossRef]
24. McQuade, D.T.; Pullen, A.E.; Swager, T.M. Conjugated polymer-based chemical sensors. *Chem. Rev.* **2000**, *100*, 2537–2574. [CrossRef]
25. Smela, E. Conjugated polymer actuators for biomedical applications. *Adv. Mater.* **2003**, *15*, 481–494. [CrossRef]
26. Choukourov, A.; Pleskunov, P.; Nikitin, D.; Titov, V.; Shelemin, A.; Vaidulych, M.; Kuzminova, A.; Solař, P.; Hanuš, J.; Kousal, J.; et al. Advances and challenges in the field of plasma polymer nanoparticles. *Beilstein J. Nanotechnol.* **2017**, *8*, 2002–2014. [CrossRef]
27. Hegemann, D.; Nisol, B.; Watson, S.; Wertheimer, M.R. Energy conversion efficiency in plasma polymerization—A comparison of low- and atmospheric-pressure processes. *Plasma Process. Polym.* **2016**, *13*, 834–842. [CrossRef]
28. Malinowski, S.; Herbert, P.A.F.; Rogalski, J.; Jaroszyńska-Wolińska, J. Laccase enzyme polymerization by soft plasma jet for durable bioactive coatings. *Polymers* **2018**, *10*, 532. [CrossRef]
29. Jalaber, V.; Del Frari, D.; De Winter, J.; Mehennaoui, K.; Planchon, S.; Choquet, P.; Detrembleur, C.; Moreno-Couranjou, M. Atmospheric aerosol assisted pulsed plasma polymerization: An environmentally friendly technique for tunable catechol-bearing thin films. *Front. Chem.* **2019**, *7*, 183. [CrossRef]
30. Park, C.-S.; Kim, D.H.; Shin, B.J.; Tae, H.-S. Synthesis and characterization of nanofibrous polyaniline thin film prepared by novel atmospheric pressure plasma polymerization technique. *Materials* **2016**, *9*, 39. [CrossRef]
31. Zaplotnik, R.; Primc, G.; Vesel, A. Optical emission spectroscopy as a diagnostic tool for characterization of atmospheric plasma jets. *Appl. Sci.* **2021**, *11*, 2275. [CrossRef]
32. Naz, M.Y.; Shukrullah, S.; Rehman, S.U.; Khan, Y.; Al-Arainy, A.A.; Meer, R. Optical characterization of non-thermal plasma jet energy carriers for effective catalytic processing of industrial wastewaters. *Sci. Rep.* **2021**, *11*, 2896. [CrossRef] [PubMed]
33. Ayad, M.; Zaghlol, S. Nanostructured crosslinked polyaniline with high surface area: Synthesis, characterization and adsorption for organic dye. *Chem. Eng. J.* **2012**, *204–206*, 79–86. [CrossRef]
34. Wang, X.; Deng, J.; Duan, X.; Liu, D.; Guo, J.; Liu, P. Crosslinked polyaniline nanorods with improved electrochemical performance as electrode material for supercapacitors. *J. Mater. Chem. A* **2014**, *2*, 12323. [CrossRef]
35. Lv, X.; Wang, Y.; Wang, Y.-A.; Lin, X.; Ni, Y. Crosslinked polyaniline nanorods coupled with molybdenum disulfide on functionalized carbon cloth for excellent electrochemical performance. *J. Solid State Electrochem.* **2021**, *25*, 1871–1880. [CrossRef]
36. Sharma, A.K.; Bhardwaj, P.; Dhawan, S.K.; Sharma, Y. Oxidative synthesis and electrochemical studies of poly(aniline-co-pyrrole)-hybrid carbon nanostructured composite electrode materials for supercapacitor. *Adv. Mater. Lett.* **2015**, *6*, 414–420. [CrossRef]
37. Srinivasan, P.; Gottam, R. Infrared Spectra: Useful technique to identify the conductivity level of emeraldine form of polyaniline and indication of conductivity measurement either two or four probe technique. *Mat. Sci. Res. India* **2018**, *15*, 209–217. [CrossRef]
38. Golczak, S.; Kanciurzewska, A.; Fahlman, M.; Langer, K.; Langer, J.J. Comparative XPS surface study of polyaniline thin films. *Solid State Ion.* **2008**, *179*, 2234–2239. [CrossRef]
39. Elmas, S.; Beelders, W.; Nash, J.; Macdonald, T.J.; Jasieniak, M.; Griessera, H.J.; Nann, T. Photo-doping of plasma-deposited polyaniline (PAni). *RSC Adv.* **2016**, *6*, 70691. [CrossRef]
40. Cruz-Silva, R.; Romero-García, J.; Angulo-Sáncheza, J.L.; Flores-Loyola, E.; Farías, M.H.; Castillón, F.F.; Díaz, J.A. Comparative study of polyaniline cast films prepared from enzymatically and chemically synthesized polyaniline. *Polymer* **2004**, *45*, 4711–4717. [CrossRef]
41. Lee, Y.M.; Kim, J.H.; Kang, J.S.; Ha, S.Y. Annealing effects of dilute polyaniline/NMP solution. *Macromolecules* **2000**, *33*, 7431–7439. [CrossRef]
42. Bhadra, S.; Singha, N.K.; Khastgir, D. Polyaniline by new miniemulsion polymerization and the effect of reducing agent on conductivity. *Synth. Met.* **2006**, *156*, 1148–1154. [CrossRef]
43. Li, Z.F.; Kang, E.T.; Neoh, K.G.; Tan, K.L. Effect of thermal processing conditions on the intrinsic oxidation states and mechanical properties of polyaniline films. *Synth. Met.* **1997**, *87*, 45–52. [CrossRef]
44. Pron, A.; Rannou, P. Processible conjugated polymers: From organic semiconductors to organic metals and superconductors. *Prog. Polym. Sci.* **2002**, *27*, 135–190. [CrossRef]

45. Lee, Y.W.; Do, K.; Lee, T.H.; Jeon, S.S.; Yoon, W.J.; Kim, C.; Ko, J.; Im, S.S. Iodine vapor doped polyaniline nanoparticles counter electrodes for dye-sensitized solar cells. *Synth. Met.* **2013**, *174*, 6–13. [CrossRef]
46. Fan, L.; Xu, X. A simple strategy to enhance electrical conductivity of nanotube-conjugate polymer composites via iodine-doping. *RSC Adv.* **2015**, *5*, 78104–78108. [CrossRef]
47. Wang, J.; Neoh, K.G.; Kang, E.T. Comparative study of chemically synthesized and plasma polymerized pyrrole and thiophene thin films. *Thin Solid Films* **2004**, *446*, 205–217. [CrossRef]
48. Silverstein, M.S.; Visoly-Fisher, I. Plasma polymerized thiophene: Molecular structure and electrical properties. *Polymer* **2002**, *43*, 11–20. [CrossRef]

polymers

Article

Development of an Atmospheric Pressure Plasma Jet Device Using Four-Bore Tubing and Its Applications of In-Liquid Material Decomposition and Solution Plasma Polymerization

Gyu Tae Bae [1,†], Hyo Jun Jang [1,†], Eun Young Jung [1,2], Ye Rin Lee [1], Choon-Sang Park [3], Jae Young Kim [1,*] and Heung-Sik Tae [1,4,*]

1 School of Electronic and Electrical Engineering, College of IT Engineering, Kyungpook National University, Daegu 41566, Republic of Korea
2 The Institute of Electronic Technology, College of IT Engineering, Kyungpook National University, Daegu 41566, Republic of Korea
3 Department of Electrical Engineering, Milligan University, Johnson City, TN 37682, USA
4 School of Electronics Engineering, College of IT Engineering, Kyungpook National University, Daegu 41566, Republic of Korea
* Correspondence: jyk@knu.ac.kr (J.Y.K.); hstae@ee.knu.ac.kr (H.-S.T.); Tel.: +82-53-950-6563 (H.-S.T.)
† These authors contributed equally to this work.

Abstract: In this study, we describe an atmospheric pressure plasma jet (APPJ) device made of four-bore tubing operable in inhospitable humid environments and introduce two potential applications of liquid material processing: decomposition of aqueous phosphorus compounds and solution-plasma polymerization. A four-bore tube was used as the plasma transfer conduit and two diagonal bores contained metal wires. In the proposed APPJ device, the metal wires serving as electrodes are completely enclosed inside the holes of the multi-bore glass tube. This feature allows the APPJ device to operate both safely and reliably in humid environments or even underwater. Thus, we demonstrate that the proposed electrode-embedded APPJ device can effectively decompose aqueous phosphorus compounds into their phosphate form by directly processing the solution sample. As another application of the proposed APPJ device, we also present the successful synthesis of polypyrrole nanoparticles by solution plasma polymerization in liquid pyrrole.

Keywords: atmospheric pressure plasma; multi-bore tube; phosphorus compound decomposition; plasma polymerization; plasma processing; plasma treatment

Citation: Bae, G.T.; Jang, H.J.; Jung, E.Y.; Lee, Y.R.; Park, C.-S.; Kim, J.Y.; Tae, H.-S. Development of an Atmospheric Pressure Plasma Jet Device Using Four-Bore Tubing and Its Applications of In-Liquid Material Decomposition and Solution Plasma Polymerization. *Polymers* **2022**, *14*, 4917. https://doi.org/10.3390/polym14224917

Academic Editor: Petr Smolka

Received: 3 November 2022
Accepted: 9 November 2022
Published: 14 November 2022

1. Introduction

Nonthermal atmospheric pressure (AP) plasma is a weakly ionized gas medium with ionized charged particles, exciting species with varying energy levels, highly reactive but short-lived radicals, and free electrons [1–5]. Because these byproducts of nonthermal AP plasma are effective agents for various materials, plasma technology has been effectively applied in material processing over the past few decades, despite the difficulty in diagnosing whether AP plasma is in contact with materials [6,7]. The non-equilibrium discharge behavior owing to partial ionization allows to attain high electron energies in the plasma medium while retaining ions and neutral species at room temperature [8]. The presence of various radicals and highly energetic electrons at low gas temperatures indicates that nonthermal AP plasma is particularly advantageous when processing heat-sensitive materials, such as organic materials, polymers, and volatiles [9–13]. In addition, because there are no bulky chambers or complicated vacuum components in the plasma generation system, and no chemical waste is generated after the process, material processing with AP plasma is generally considered an eco-friendly process with the advantages of having a simple overall experimental setup, fast processing, and easy maintenance. Regarding using nonthermal AP plasma, material decomposition and synthesis, two representative material processing methods, are gradually increasing in popularity, and research on various AP plasma device designs is attracting attention as well.

The AP plasma jet (APPJ) device, which can create a nonthermal AP plasma having the simple configuration of an electrode and a conduit, ignites the discharge gas flowing through the conduit by a powered electrode and deliver the plasma plume to the outside of the conduit [14–17]. APPJ has the advantage of being able to easily generate plasma at AP using a discharge gas, such as argon (Ar) or helium that ignites the electrical discharge better than air and delivers it close to the target material. To effectively utilize APPJ for material processing, various efforts have been made to transfer the plasma and its byproducts as close to the target as possible. There are several reports on the fabrication of various types of APPJ devices using flexible plastic tubes, as well as glass, quartz, or ceramic conduits [18–21]. Nevertheless, the difficulties related to the effectiveness of APPJ in the synthesis, decomposition, or functionalization of organic materials include the precise delivery of the reactive radicals in the plasma medium to the target and the risk of unwanted electric shock around the powered electrode owing to utilizing high voltages. In particular, when processing liquid materials using this method, it is necessary to pay attention to the malfunction of APPJ due to the undesirable electrical breakdown occurring at the electrode exposed to high-humidity environments. If APPJs can be guaranteed to operate stably and safely in inhospitable humidity conditions, applications, where APPJs can be used for liquid-material processing, will be greatly expanded. Our research group has attempted to ensure that APPJs function well in humid environments without discharge failure [22,23]. In particular, the fabrication of an APPJ device using multi-bore tubing is a promising approach to exhibiting plasma generation even in an inhospitable environment; however, there was no follow-up study except for a photo image of plasma generation [23].

Therefore, in this study, we demonstrate an APPJ device with built-in electrodes inside a plasma transfer conduit to facilitate the plasma treatment of organic materials in the liquid state. To avoid operational failure in humid environments and improve the targeted delivery of plasma, a multi-bore glass tube was employed as the plasma delivery tube. The multi-bore tube has four identical holes. Two of them are for openings in the nonthermal plasma jets and the other two are for the wire electrodes. Our research group demonstrated that the proposed APPJ can safely and stably generate plasma plumes not only in an atmospheric environment but also in water, underlining its suitability for the decomposition of aqueous phosphorus compounds related to wastewater treatment and monitoring. Furthermore, this APPJ device can be applied to plasma polymerization by generating in-liquid plasma in the monomer solution.

2. Materials and Methods

2.1. Atmospheric Pressure Plasma Jet Device and Plasma Operation System

The APPJ device with a four-bore glass tube and plasma generation system employed in this study was described in our previous report [23] and is shown in Figure 1. The four-bore borosilicate glass tube has four equal holes with a diameter of 1.5 mm and a total outer diameter of 6.35 mm. Two ends of the four holes were sealed with epoxy resin. Stainless steel wires with a 1.2 mm diameter were inserted into the sealed holes. The overall length of the APPJ device was 25 cm and the width was 6.35 mm, which spatially separated the resulting plasma columns and the electrical feeder by a distance of 25 cm.

A schematic of the experimental setup employed in this study is shown in Figure 1. High-purity (HP) grade Ar gas with 99.999% purity was used as the carrier gas, and a sinusoidal voltage waveform was used to power the APPJ device. A voltage probe (P6015A, Tektronix Inc., Beaverton, OR, USA) and current probe (4100, Pearson Electronics Inc., Palo Alto, CA, USA) were used along with a fiber-optic spectrometer (USB-2000+, Ocean Optics Inc., Dunedin, FL, USA) to measure the electrical and optical characteristics of the plasma plumes. The instantaneous waveforms of voltage, current and optical emission were displayed in real time on an oscilloscope (TDS3014B, Tektronix Inc., Beaverton, OR, USA).

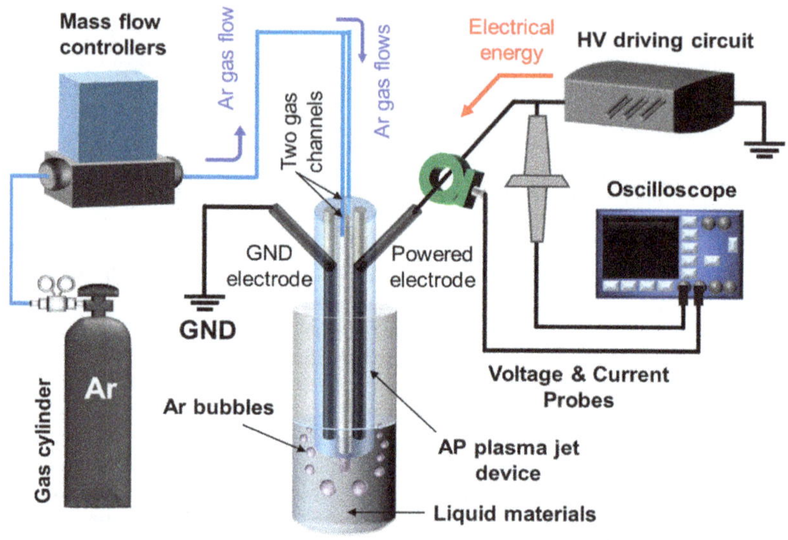

Figure 1. Schematic of experimental setup including an atmospheric pressure plasma jet (APPJ) device with two embedded electrodes.

2.2. Preparing a Phosphorus Compound Solution for Assessing Plasma Decomposition

A mass of 1.98 mg of β-glycerol phosphate disodium salt pentahydrate (BGP; $C_3H_7Na_2O_6P{\cdot}5H_2O$, Sigma-Aldrich Inc., St. Louis, MO, USA) was dissolved in 200 mL of deionized (DI) water to prepare a diluted BGP solution having a phosphorus concentration of 1.0 mg/L. The BGP solution (10 mL) was placed in a glass vial and irradiated with nonthermal plasma for up to 30 min using the proposed APPJ device. A sinusoidal voltage with a peak voltage of 6.5 kV and a frequency of 25 kHz was applied, and an Ar flow rate of 2 standard liters per minute (slm) was used to generate a nonthermal plasma to be used in the examination. The orthophosphate concentrations of the BGP samples with and without plasma treatment were measured using ion chromatography system (ICS3000, Dionex Corp., Sunnyvale, CA, USA) at the Korea Basic Science Institute (KBSI, Busan, Korea). The measurements were taken at an injection loop volume of 20 μL and column temperature of 30 °C.

2.3. Ascorbic Acid Reduction Method Using Phosphate Standard Solutions

1.1 mg of potassium phosphate monobasic (KH_2PO_4, Sigma-Aldrich Inc., St. Louis, MO, USA) was dissolved in 250 mL of DI water to prepare a phosphate standard solution having a phosphorus concentration of 1.0 mg/L. Then, DI water was further added to the standard solution to prepare phosphate standard samples containing 0.2, 0.4, 0.6, and 0.8 mg/L of phosphorus, respectively. 0.18 g of a colorimetric reagent for the ascorbic acid method (HI736, Hanna Instruments Inc., Woonsocket, RI, USA) was added to 10 mL of all standard solutions and reacted for 10 min. The absorption spectra in the visible-near infrared region of the prepared phosphate standard solutions were recorded using a spectrometer (USB-4000 UV-vis, Ocean Optics Inc., Dunedin, FL, USA). Calibration curves for the standard solutions were plotted based on the absorbance characteristics at 710 nm [24,25]. The decomposition of the BGP solution into orthophosphate (PO_4^{3-}) by plasma treatment was investigated comparatively based on these calibration curves.

2.4. Solution Plasma Polymerization for Examination of Plasma Synthesis

Polypyrrole (PPy) nanomaterials were synthesized from a pyrrole monomer solution using the proposed electrode-embedded APPJ device. The amount of liquid pyrrole per treatment was 25 mL, and an Ar flow rate of 500 standard cubic centimeters per minute was used. A bipolar pulse with an amplitude of 7 kV and frequency of 5 kHz was applied to the APPJ device using a high-voltage power amplifier (20/20C-HS, Trek, Inc., Lockport, NY, USA) and function generator (AFG-3102, Tektronix Inc., Beaverton, OR, USA). The pulse duration of positive and negative polarity in bipolar pulse was equal to 100 μs and the polymerization duration time was up to 6 h.

2.5. Preparation of Polypyrrole Nanoparticles

To separate the PPy nanoparticles synthesized from the pyrrole monomer, the processed pyrrole solution was mixed with ethanol and precipitated using a centrifuge at 13,500 rpm for 20 min. Then, the upper part of the purified solution, except for settled PPy nanoparticles, was removed using a micropipette. The settled PPy nanoparticles were added to distilled water and rinsed twice with a centrifuge under the same conditions. Finally, solid PPy nanoparticle powders were obtained after drying in an oven at 60 °C for 12 h, and then collected and fixed on copper tape.

2.6. Analysis and Characterization of Polypyrrole Nanoparticles

The ultraviolet-visible (UV-vis) absorption spectra of the pyrrole solution processed using the proposed APPJ were recorded over a spectral range of 250–500 nm using a UV-vis spectrophotometer (LAMBDA 950, Perkin Elmer, Inc., Waltham, MA, USA) at KBSI (Daegu, Korea).

Field-emission scanning electron microscopy (FE-SEM; SU8220, Hitachi Korea Co. Ltd., Seoul, Korea) was used to observe the shape and size of the PPy nanoparticles.

The functional groups of the nanoparticles synthesized by the plasma process were identified using Fourier-transform infrared spectroscopy (FT-IR; Vertex 70, Bruker, Ettlingen, Germany) at KBSI (Daegu, Korea). The attenuated total reflection (ATR) FT-IR spectra were measured by the average of 128 scans in the range 650–4000 cm^{-1} at a resolution of 0.6 cm^{-1}.

2.7. Statistical Analysis

All quantitative data related to phosphorus compound decomposition are presented as mean ± standard deviation (SD). The mean and SD values were obtained from triplicate measurements of each experiment (n = 3).

3. Results and Discussion

3.1. Electrode-Embedded Atmospheric Pressure Plasma Jet Device

As shown in Figure 2a, the two hollow ends containing wire electrodes were sealed to ensure that the device functioned properly with respect to both safety and stability. Figure 2b shows a photograph of the end of the 4-bore tube with the wire electrodes inserted. Because the diameter of this metal wire was 1.2 mm, there was only a spatial margin of 300 μm between the inner diameter of the hole and the outer diameter of the metal wire. Because of this tight spatial margin, the proposed APPJ device has the advantage that the gap between the two metal wires, the most important experimental parameter for initiating and maintaining the plasma, hardly changes regardless of the length of the plasma device. Consequently, the proposed AP plasma device can be manufactured in various lengths for processing purposes by determining the overall length of the 4-bore tube. In addition, the two-gas nozzle configuration using a 4-bore tube has the great advantage of doubling the spatial expansion of the plasma plume without any difficulty. The spatial expansion of the plasma plume can increase the plasma treatment area, resulting in faster material decomposition and polymerization processes. In the proposed APPJ device, two hollows with wire electrodes and two hollows from which plasma plumes are emitted are positioned symmetrically. Thus, creating two plasma plumes does not require twice as much power as generating one, because it still uses the same two wire electrodes. The use of two gas nozzles is an important design of the proposed APPJ device and can provide a key clue for the large-area treatment of AP plasma.

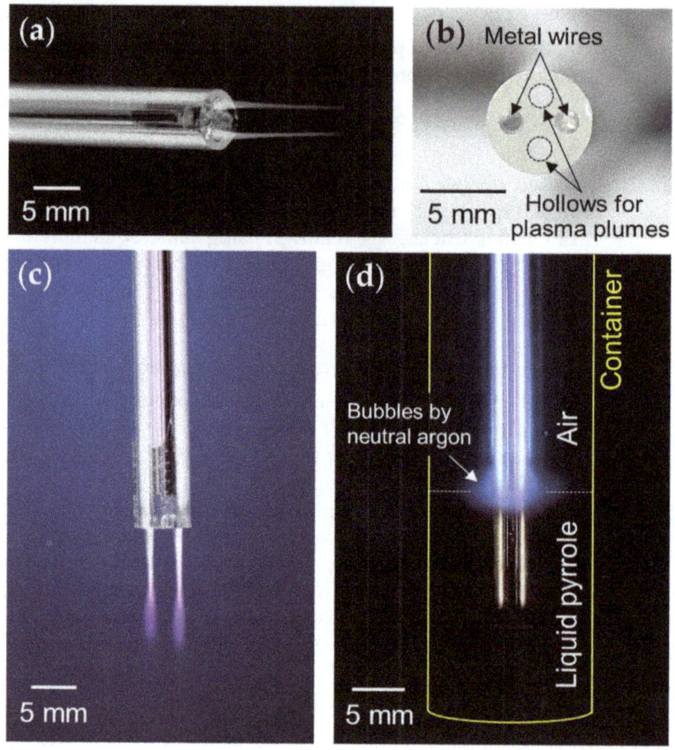

Figure 2. Details of the proposed APPJ device: (**a**) An APPJ device with two wire electrodes embedded in the four-bore tube. (**b**) Photograph of the four-bore glass tube with wire electrodes inserted. (**c**) Plasma plumes generated by the proposed plasma device in the air, and (**d**) plasma processing of liquid pyrrole.

To demonstrate plasma generation in air shown in Figure 2c, an Ar gas flow rate of 1.5 slm was applied to the electrodeless hollows; subsequently, the APPJ device produced two plasma plumes of equal length of 2 cm. In this case, the operating voltage was a sinusoidal waveform with an amplitude of 5 kV and a frequency of 25 kHz. The discharge was initiated inside the 4-bore tube, through which the Ar gas flowed, and the plasma afterglow was transferred to the solution according to the gas flow. Even though the plasma plumes were observed to shorten to less than 1 mm in liquid pyrrole, the plasma device could still operate in a stable state (Figure 2d). The shorter plasma jet length in liquid than in air is due to the different densities of the media [23]. In this AP plasma device, Ar plasma is ignited inside the glass tube via two built-in electrodes, and the afterglow of the Ar plasma is emitted to the outside through two nozzles as neutral argon gas flows. When the external medium is fairly dense, such as water or liquid monomers, it is difficult for the plasma to flow out of the plasma device, resulting in very short plasma jet lengths. Therefore, using higher Ar flow rates to increase the length of the Ar plasma plume works to some extent in air, but does not work well in dense media such as liquids. However, even though the plasma jet length was reduced to less than 1 mm, numerous radicals were still present in the plasma medium and came out through the device along with the Ar bubbles.

3.2. Optical and Electrical Characteristics of Atmospheric Pressure Plasma Jet

Compared with conventional APPJs, which only work well under normal air conditions, the proposed APPJ ensures consistent and stable operation not only under normal air conditions, but also in humid or liquid environments. The electrical characteristics measured in the proposed APPJ device operating in liquid pyrrole are depicted in Figure 3. These electrical characteristics are important because they show the general electrical properties of dielectric barrier discharges, even under liquid environmental conditions where normal APPJs cannot sustain stable driving. All measured data were averaged over 16 periods of waveforms for data reliability. Figure 3a,b presents the applied voltage and measured total current flow when alternating-current (AC) plasma is sustained. The current observed during the plasma-on state, shown in Figure 3b, consists of the discharge and displacement currents. The discharge current was acquired by subtracting the current obtained when the operating voltage was applied without discharge gas from the total current displayed on the oscilloscope when the discharge occurred. From Figure 3c, the discharge current ($I_{Plasma\ ON}–I_{Plasma\ OFF}$) appears to be periodic, according to that of the driving voltage. This periodicity indicated that the device operated in a stable state. It is also observed that the discharge current is evenly distributed in the positive and negative periods of the voltage waveforms because the two identical metal wires inserted into the 4-bore tube alternatively assume the roles of the anode and cathode during the AC discharges.

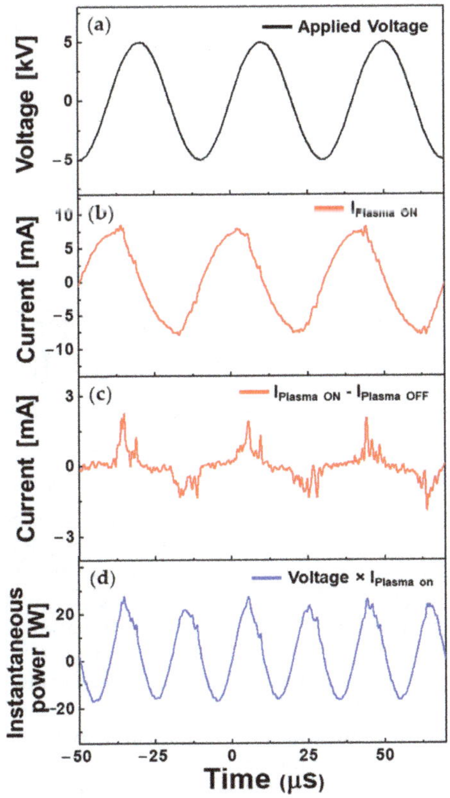

Figure 3. Temporal electrical behavior of plasma jet device operating in liquid pyrrole. Profiles of (**a**) driving voltage, (**b**) total current, (**c**) discharge current, and (**d**) instantaneous power as functions of time.

The instantaneous power consumption is shown in Figure 3d, and the average power was calculated according to the following formula:

$$P = \frac{1}{T}\int_0^T U(t) \times I(t)dt$$

where T is the period of the applied voltage, $U(t)$ is the voltage signal, $I(t)$ is the acquired current, and t is the time. The integrated value of the waveform of the instantaneous power during one period (0–40.0 μs) is 1.041×10^{-4} J. Therefore, the average power during one period is approximately 2.602 W, which is low energy consumption.

The optical emission spectra of the proposed APPJ were measured using a miniature spectrometer to identify the various reactive species produced by the plasma plume in the air. In this optical measurement, the temperature of the plasma plumes emitted into the air was measured to be 41.2 °C. Figure 4 shows that although Ar gas played an integral part in plasma jet production, many gaseous species, such as excited OH, N_2, and O present in ambient air, also served as catalysts in the plasma medium. In particular, the presence of reactive oxygen and nitrogen species (ROS and RNS) is beneficial because they have been associated with the compound decomposition response.

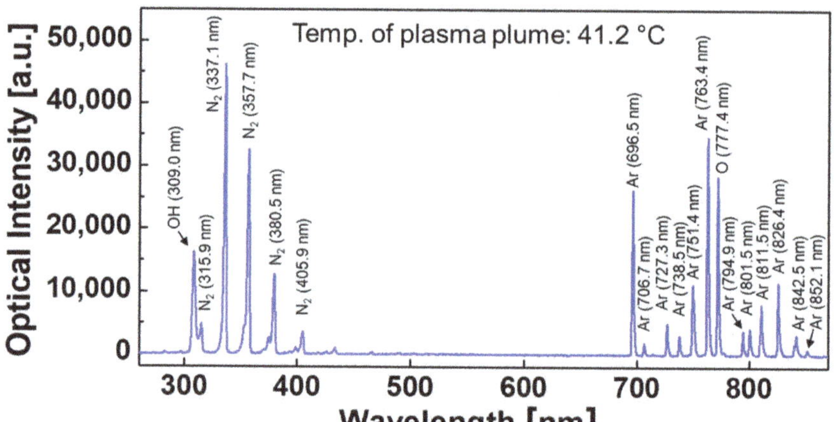

Figure 4. Optical emission spectra of the generated plasma plume, which was monitored using a fiber-optic spectrometer.

3.3. Decomposition of Aqueous Phosphorus Compounds by Atmospheric Pressure Plasma Jet

The proposed APPJ device can function safely and reliably even in harsh environments such as extremely humid environments because the ends of the two holes containing the wire electrodes are completely sealed. As shown in Figure 1, thin wire electrodes inserted into the two hollows of the four-bore tubes are completely isolated from the outside. Because the electrode has no contact with the external environment at the end of the APPJ device, two plasma jets can be maintained not only in ambient air but also in water, which greatly expands the application field of plasma processing.

Recently, several studies on the decomposition of organic/inorganic compounds using charged particles and reactive species generated by plasma have been reported [26–28]. Among these topics, we are interested in the decomposition of aqueous phosphorus compounds related to water quality monitoring. Because phosphorus in aquatic ecosystems exists as a variety of phosphorus compounds, the total phosphorus amount in water can be determined only when phosphorus compounds are decomposed into orthophosphate, PO_4^{3-}. The ascorbic acid method, which is widely known as the standard method for monitoring water quality, is also based on determining the amount of phosphate decomposed from diverse aqueous phosphorus compounds [29,30].

We tested the measurement of phosphate decomposed by in-liquid plasma in a phosphorus compound solution as a potential application of the proposed APPJ device. Because the proposed APPJ device can generate nonthermal plasma not only in the air but also in water, the surface treatment of a phosphorus compound solution using plasma plumes generated in air and treatment by immersing the APPJ device in the phosphorus compound solution were investigated and compared (Figure 5a). A BGP solution containing 1.0 mg/L phosphorus was selected as an example phosphorus compound and treated with nonthermal plasma for up to 30 min using the proposed 4-bore tube-based APPJ device. Figure 5b plots the temperature changes in the BGP solution by Ar plasma jet in air and liquid Ar plasma treatments. Because the heat generated by the APPJ device is directly transferred to the solution during the in-liquid plasma treatment, the saturation temperature of the BGP solution was observed to be approximately 6 °C higher for the in-liquid plasma treatment than that for the plasma treatment on the solution surface. However, even when the BGP solution was treated with the two types of plasma for 30 min, the solution temperature was saturated to just below 40 °C, indicating that the heat generated by plasma processing did not affect the decomposition of the phosphorus compound in this experiment.

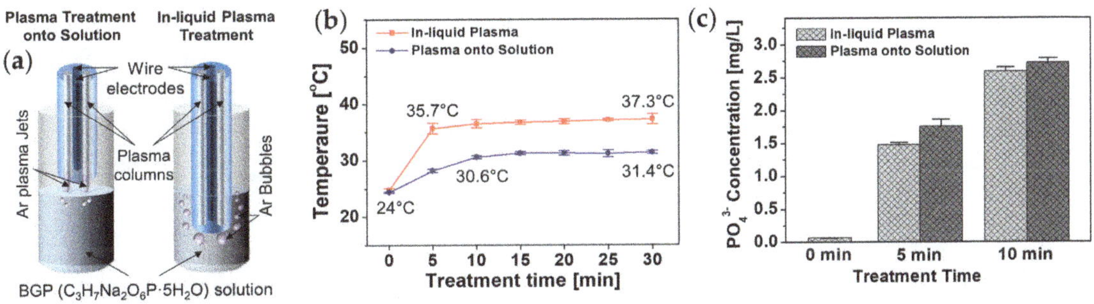

Figure 5. Plasma decomposition of phosphorus compounds: (**a**) Schematics of two different plasma treatments for a BGP solution by the proposed APPJ device. (**b**) Temperature change in BGP solution treated by plasma treatments. (**c**) Changes in orthophosphate concentration before and after two plasma treatments of BGP solution via ion chromatography. Data are presented as the mean ± SD of three repeated experiments.

Ion chromatography measurements demonstrated that the orthophosphate concentration in the BGP solution considerably increased after 5 and 10 min of plasma exposure (Figure 5c). This indicated that the nonthermal plasma generated by the proposed AP plasma device was effective in decomposing BGP in the form of $C_3H_7Na_2O_6P \cdot 5H_2O$ into orthophosphate in the form of PO_4^{3-}. In particular, when the Ar plasma jet generated in the air was irradiated on the surface of the BGP solution, the amount of orthophosphate in the BGP solution was higher than that obtained when the Ar plasma jet was generated in water. This is because the nitrogen and oxygen present in the air participate in the discharge process and produce more ROS and RNS. When discharge occurs in water, ROS can be generated by water molecules, but RNS cannot be generated owing to the lack of nitrogen. Accordingly, we demonstrated that the RNS generated affected the decomposition of the BGP solution into orthophosphate and plasma the jet generated in the air was more effective in decomposing the BGP solution into orthophosphate.

However, under the condition of plasma treatment on the solution surface, it was also noticed that the plasma device should be very close to the solution target, less than 1 cm, for reliable treatment. Plasma operation using conventional APPJ generators at such high environmental humidity can create unwanted discharge at the exposed powered electrode and sometimes cause discharge failure; however, the proposed APPJ device can completely avoid this experimental risk. Moreover, in the case of the plasma treatment of the solution, because the distance between the device and the solution surface may be changed due to sample evaporation, the plasma treatment effect may change with an increase in the plasma processing time. Therefore, if a precise and reliable plasma process that does not change the plasma influence as a function of time is required, in-liquid plasma treatment is preferable.

Figure 6a depicts the absorption spectra in the visible-near infrared region of BGP solutions treated with in-liquid plasma as a function of plasma treatment time. Based on the ascorbic acid method [31], the amount of phosphate can be quantified by comparing the absorption of the prepared phosphate standard solution at 710 nm with that of the plasma-treated samples [24,32]. Figure 6b shows the results of the decomposition efficiency of the BGP samples into orthophosphate by in-liquid plasma according to treatment time, which indicates that most of the phosphorus compounds were degraded into phosphate after in-liquid plasma treatment using the APPJ device for 30 min. Similarly, the absorption spectra and decomposition efficiency of the BGP sample with plasma treatment on its surface are shown in Figure 6c,d. In this case, observe that the processing time for most of the phosphorus compounds to be decomposed into phosphates in the plasma-treated BGP sample is shorter than 10 min. This means that the Ar plasma jet generated in the air is more effective in decomposing the BGP solution into orthophosphates than the in-liquid Ar plasma, but it is also noted that both plasma treatment approaches are sufficiently effective for decomposing BGP solutions. This result agrees well with the ion chromatography results shown in Figure 5c. The experimental results demonstrated that the plasma decomposition method available in a humid environment can be effectively combined with the ascorbic acid reduction to measure the total aqueous phosphorus concentration.

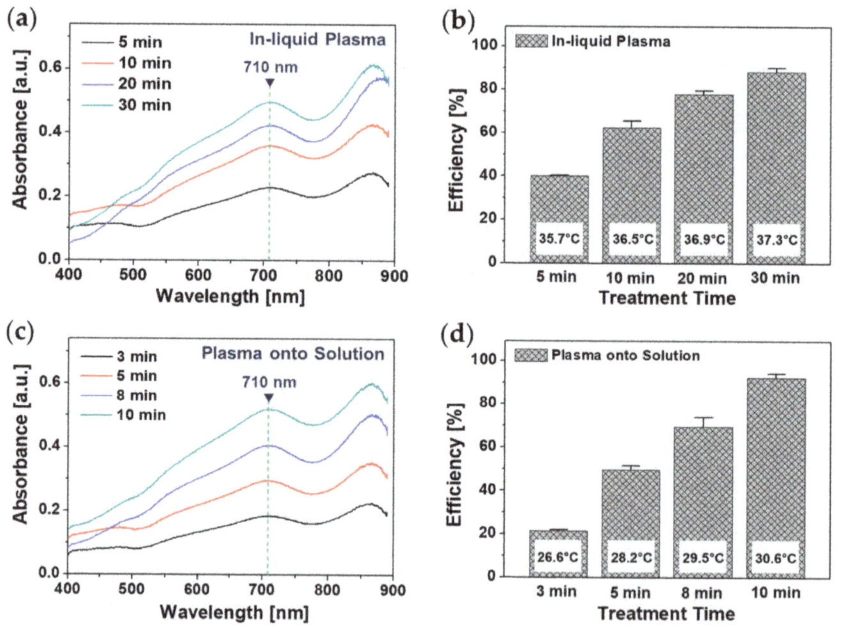

Figure 6. Results of ascorbic acid method: (**a**) Visible-NIR absorption spectrum curves of BGP solution treated by in-liquid plasma for 5, 10, 20, and 30 min and (**b**) efficiency of phosphorus decomposition to phosphate form via in-liquid plasma. (**c**) Visible-NIR absorption spectrum curves of Ar plasma treatment onto BGP solution for 3, 5, 8, and 10 min and (**d**) efficiency of phosphorus decomposition into phosphate form via Ar plasma treatment onto BGP solution. Phosphorus decomposition efficiencies are presented as the mean ± SD of three repeated experiments.

3.4. Plasma Polymerization in Liquid Monomer Using Electrode-Embedded Atmospheric Pressure Plasma Device

Another benefit of the proposed APPJ device is that the solution plasma process (SPP) can be applied without any operational difficulty. An SPP mainly uses a spark discharge generated in a liquid by applying an electrical impulse or high voltage between two pin-shaped electrodes facing each other with a small gap [33–35]. The liquid near the metal electrode is locally vaporized by Joule heating to generate a strong spark discharge, and radicals generated from liquid molecules react with each other to form nanoparticles and cool rapidly in the liquid. Because of this nanoparticle formation mechanism by solution plasma, it is difficult to synthesize organic nanoparticles that require low-temperature plasma. Our research group introduced a gas channel and dielectric barrier to synthesize polymeric nanomaterials using SPP in a previous study [36]. The use of an Ar bubble channel can significantly reduce the operating voltage while stably generating a streamer discharge, and the dielectric barrier effectively controls the discharge current.

The proposed electrode-embedded APPJ device using Ar gas channels can easily generate low-temperature discharge, even in a solution. Accordingly, the proposed APPJ device was applied to PPy synthesis through SPP. Figure 7 shows the solution color change in liquid pyrrole during SPP with various processing times. The color of the liquid pyrrole gradually changed from yellowish brown to dark brown, representing that particles were formed from the liquid pyrrole. When liquid pyrrole with the initial temperature of 24 °C was treated with SPP using the proposed APPJ, the temperature of the liquid pyrrole was saturated within 30 min and maintained at approximately 38 °C until 6 h of SPP. This low saturation temperature implies that the heating of the liquid phase by SPP does not contribute to nanoparticle synthesis. Because the Ar plasma jet was generated during immersion in liquid pyrrole, evaporation and liquid disturbance by Ar bubbling inevitably occurred, which could affect the solution plasma process. Liquid pyrrole (25 mL of liquid pyrrole was evaporated by Ar bubbling, leaving only 15 mL after 6 h of SPP, and the liquid pyrrole sample was well stirred by Ar bubbling during the SPP. The pH and conductivity of the original pyrrole solution were 8.25 and 0.331 μS/cm, respectively. After 6 h of SPP, the pH was changed to 7.08 and conductivity to 0.882 μS/cm.

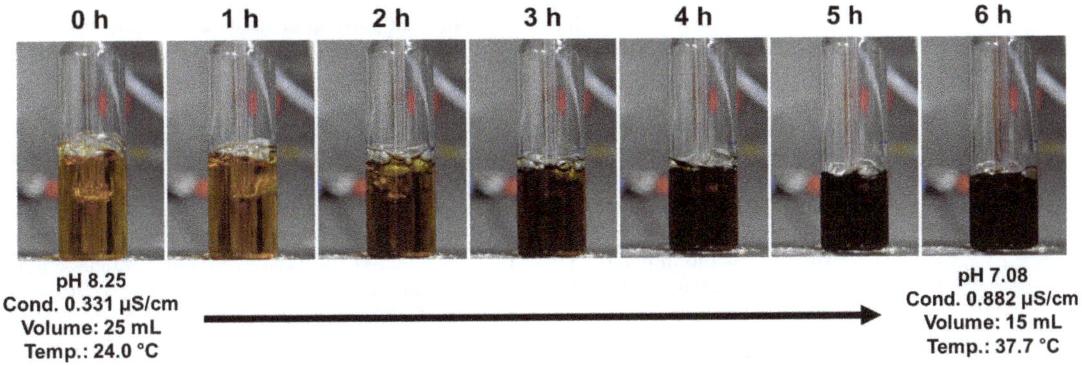

Figure 7. Color change in liquid pyrrole with various process times during solution plasma process (SPP) using the proposed APPJ device.

Oxidation of pyrrole monomers is generally known to produce chemically active pyrrole cations [37,38]. Pyrrole cations created this way polymerize with pyrrole monomers or other cations; consequently, H^+ ions are released from the pyrrole monomer [37,38]. SPP using the proposed APPJ induced the oxidation of pyrrole, leading to the formation of pyrrole cations. The generated pyrrole cations chemically reacted with other neutral pyrrole molecules or pyrrole cations during SPP to first form pyrrole oligomers and finally form PPy particles in liquid pyrrole. Based on the conductivity and pH data of the SP-treated liquid pyrrole (Figure 7), it can be inferred that pyrrole ions, particularly cations, were formed during the SPP.

Figure 8 shows UV-vis spectra of pure liquid pyrrole and pyrrole solution treated with the proposed APPJ for 6 h. The pyrrole samples with and without plasma treatment exhibit significant differences in absorbance over the 275–400 nm region, owing to the presence of pyrrole oligomers in the plasma-treated pyrrole sample. In particular, absorptions near 275 and 320 nm indicate the presence of bipyrrole and terpyrrole, respectively [39]. Thus, the UV-vis results demonstrate that pyrrole monomers were synthesized in larger units by SPP using the proposed APPJ. In addition, it is generally known that the pH of the solution decreases during PPy synthesis owing to liberated H^+ ions [40]. Therefore, the experimental results depicted in Figures 7 and 8 indicate that PPy is synthesized in the liquid pyrrole by SPP using the proposed APPJ device.

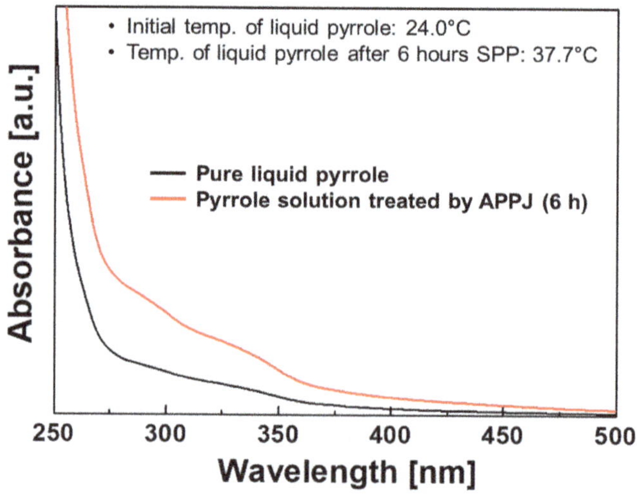

Figure 8. Ultraviolet-visible (UV-vis) spectra of the pure liquid pyrrole and pyrrole solution treated by the proposed APPJ device for 6 h.

Figure 9a shows a sample of PPy nanoparticles synthesized for 6 h and then collected on copper tape for morphological analysis using FE-SEM images. As depicted in the FE-SEM image in Figure 9b, the synthesized nanoparticles exhibited a small spherical-like structure with a size distribution of approximately tens of nanometers to 250 nm. Figure 10 plots the FT-IR spectra of the pyrrole monomer (upper graph) and PPy nanoparticles (lower graph) synthesized by SPP using the proposed APPJ device. In the FT−IR spectrum of the pyrrole monomer, the peaks in the 3500–3300 and 1150−1000 cm^{-1} ranges originated from the stretching of the N−H bond and in-plane bending of the C-H bond, respectively. The aromatic ring stretching in the C=C/C−C and C−N bonds were observed as peaks in the 1750−1500 and 1350−1200 cm^{-1} ranges, respectively [41]. In the FT−IR spectrum of PPy nanoparticles, the band around 3300 cm^{-1} is attributed to the stretching vibration of the N−H bonds. The peaks at 2889 and 1045 cm^{-1} are attributed to the stretching and in-plane bending of the C−H bonds, respectively. The peak at 1681 cm^{-1} can be attributed to C=C/C=O stretching. C−C/C=C and C−N stretching in the pyrrole ring have been observed as the peaks at 1556 and 1211 cm^{-1}, respectively. From the FE-SEM and FT-IR results, it can be concluded that PPy nanoparticles were successfully synthesized under low-temperature plasma.

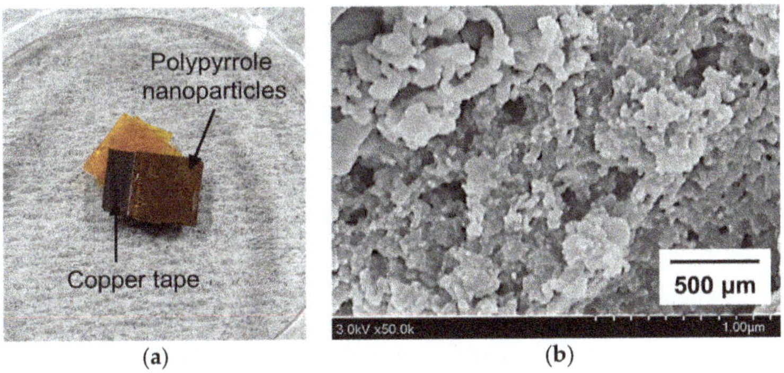

Figure 9. Sample of polypyrrole (PPy) nanoparticles: (**a**) Photograph of PPy nanoparticles collected on copper tape for FE-SEM measurements and (**b**) FE-SEM image of PPy nanoparticles synthesized by the proposed APPJ device.

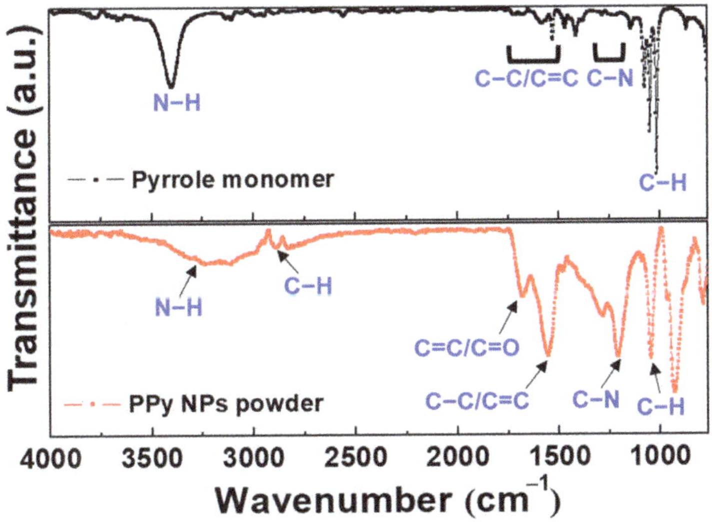

Figure 10. Fourier transformation infrared spectroscopy (FT-IR) spectra of pure pyrrole monomer (upper graph) and PPy nanoparticles synthesized by SPP using the proposed APPJ device (lower graph).

4. Conclusions

In summary, in this study, we proposed an electrode-embedded APPJ device made of 4-bore glass tubing for use in humid environments. The device configuration, namely, having electrodes isolated from the environment and spatial separation of the plasma jet and electrical input, allows stable and safe operation in a humid environment or even in water. The proposed APPJ device has Ar gas bubble channels and a dielectric barrier inside the plasma delivery conduit; therefore, it can stably generate low-temperature plasma regardless of the external environment. We demonstrated that aqueous phosphorus compounds can be effectively decomposed into phosphate forms by plasma treatment using the proposed APPJ device. Preliminary studies on the decomposition of phosphorus compounds reported that the decomposition process using nonthermal plasma could be used as a new alternative to the pretreatment of the ascorbic acid method to verify the total phosphorus content in freshwater. As another potential application of the proposed APPJ device, PPy nanoparticles were synthesized in liquid pyrrole via solution plasma polymerization. By applying the proposed APPJ device to synthesize PPy nanoparticles using SPP, thermal damage such as pyrrole carbonization and electrode erosion can be avoided. SPP using the electrode-embedded APPJ device is expected to improve the quality, purity, and uniformity of the resulting organic nanomaterials. In addition, the proposed APPJ has a low energy consumption during operation, which is reasonable enough to transfer this technology from the laboratory to the industry. An excellent approach to effectively increase the plasma processing/processing capacity for transferring this technology to the industry is to consider using multiple plasma devices as an array.

Author Contributions: Conceptualization, J.Y.K. and H.-S.T.; methodology, G.T.B., H.J.J., E.Y.J. and C.-S.P.; validation, E.Y.J., Y.R.L. and J.Y.K.; formal analysis, G.T.B., H.J.J., E.Y.J., Y.R.L. and J.Y.K.; investigation, G.T.B., H.J.J., C.-S.P., J.Y.K. and H.-S.T.; resources, J.Y.K. and H.-S.T.; data curation, G.T.B., H.J.J., E.Y.J., C.-S.P., J.Y.K. and H.-S.T.; writing—original draft preparation, G.T.B., H.J.J., J.Y.K. and H.-S.T.; writing—review and editing, G.T.B., H.J.J., J.Y.K. and H.-S.T.; visualization, G.T.B., H.J.J., E.Y.J. and J.Y.K.; supervision, H.-S.T.; project administration, J.Y.K.; funding acquisition, J.Y.K. All authors have read and agreed to the published version of the manuscript.

Funding: This study was supported by two National Research Foundation of Korea (NRF) grants funded by the Korean government (No. 2020R1I1A3071693 and No. 2021R1I1A3049028).

Institutional Review Board Statement: Not applicable.

Data Availability Statement: Not applicable.

Conflicts of Interest: The authors declare no conflict of interest.

References

1. Becker, K.H.; Schoenbach, K.H.; Eden, J.G. Microplasmas and applications. *J. Phys. D Appl. Phys.* **2006**, *39*, R55–R70. [CrossRef]
2. Fridman, A.; Chirokov, A.; Gutsol, A. Non-thermal atmospheric pressure discharges. *J. Phys. D Appl. Phys.* **2005**, *38*, R1–R24. [CrossRef]
3. Schutze, A.; Jeong, J.Y.; Babayan, S.E.; Park, J.; Selwyn, G.S.; Hicks, R.F. The atmospheric-pressure plasma jet: A review and comparison to other plasma sources. *IEEE Trans. Plasma Sci.* **1998**, *26*, 1685–1694. [CrossRef]
4. Laroussi, M.; Akan, T. Arc-free atmospheric pressure cold plasma jets: A review. *Plasma Process. Polym.* **2007**, *4*, 777–788. [CrossRef]
5. Tendero, C.; Tixier, C.; Tristant, P.; Desmaison, J.; Leprince, P. Atmospheric pressure plasmas: A review. *Spectrochim. Acta B* **2006**, *61*, 2–30. [CrossRef]
6. Bhatt, S.; Pulpytel, J.; Aref-Khonsari, F. Low and atmospheric plasma polymerisation of nanocoatings for bio-applications. *Surf. Innov.* **2015**, *3*, 63–83. [CrossRef]
7. Penkov, O.V.; Khadem, M.; Lim, W.-S.; Kim, D.-E. A review of recent applications of atmospheric pressure plasma jets for materials processing. *J. Coat. Technol. Res.* **2015**, *12*, 225–235. [CrossRef]
8. Jang, H.J.; Jung, E.Y.; Parsons, T.; Tae, H.-S.; Park, C.-S. A review of plasma synthesis methods for polymer films and nanoparticles under atmospheric pressure conditions. *Polymers* **2021**, *13*, 2267. [CrossRef]
9. Graves, D.B. The emerging role of reactive oxygen and nitrogen species in redox biology and some implications for plasma applications to medicine and biology. *J. Phys. D Appl. Phys.* **2012**, *45*, 263001. [CrossRef]
10. Jalaber, V.; Del Frari, D.; De Winter, J.; Mehennaoui, K.; Planchon, S.; Choquet, P.; Detrembleur, C.; Moreno-Couranjou, M. Atmospheric aerosol assisted pulsed plasma polymerization: An environmentally friendly technique for tunable catechol-bearing thin films. *Front. Chem.* **2019**, *7*, 183. [CrossRef] [PubMed]
11. Lisco, F.; Shaw, A.; Wright, A.; Walls, J.M.; Iza, F. Atmospheric-pressure plasma surface activation for solution processed photovoltaic devices. *Sol. Energy* **2017**, *146*, 287–297. [CrossRef]
12. Galmiz, O.; Tucekova, Z.K.; Kelar, J.; Zemanek, M.; Stupavska, M.; Kovacik, D.; Cernak, M. Effect of atmospheric pressure plasma on surface modification of paper. *AIP Adv.* **2019**, *9*, 105013. [CrossRef]
13. Jiang, C.; Mohamed, A.-A.H.; Stark, R.H.; Yuan, J.H.; Schoenbach, K.H. Removal of volatile organic compounds in atmospheric pressure air by means of direct current glow discharges. *IEEE Trans. Plasma Sci.* **2005**, *33*, 1416–1425. [CrossRef]
14. Walsh, J.L.; Kong, M.G. Room-temperature atmospheric argon plasma jet sustained with submicrosecond high-voltage pulses. *Appl. Phys. Lett.* **2007**, *91*, 221502. [CrossRef]
15. Xiong, Q.; Lu, X.P.; Jiang, Z.H.; Tang, Z.Y.; Hu, J.; Xiong, Z.L.; Pan, Y. An atmospheric pressure nonequilibrium plasma jet device. *IEEE Trans. Plasma Sci.* **2008**, *36*, 986–987. [CrossRef]
16. Park, H.S.; Kim, S.J.; Joh, H.M.; Chung, T.H.; Bae, S.H.; Leem, S.H. Optical and electrical characterization of an atmospheric pressure microplasma jet with a capillary electrode. *Phys. Plasmas* **2010**, *17*, 033502. [CrossRef]
17. Barman, K.; Mudgal, M.; Rane, R.; Bhattacharjee, S. Effect of magnetic field on optical emission from cold atmospheric pressure micro-plasma jet. *Phys. Plasmas* **2021**, *28*, 123503. [CrossRef]

18. Kim, J.Y.; Kim, S.-O.; Wei, Y.; Li, J. A flexible cold microplasma jet using biocompatible dielectric tubes for cancer therapy. *Appl. Phys. Lett.* **2010**, *96*, 203701. [CrossRef]
19. Sato, T.; Furuya, O.; Ikeda, K.; Nakatani, T. Generation and transportation mechanisms of chemically active species by dielectric barrier discharge in a tube for catheter sterilization. *Plasma Process. Polym.* **2008**, *5*, 606–614. [CrossRef]
20. Schweigert, I.; Zakrevsky, D.; Gugin, P.; Yelak, E.; Golubitskaya, E.; Troitskaya, O.; Koval, O. Interaction of cold atmospheric argon and helium plasma jets with bio-target with grounded substrate beneath. *Appl. Sci.* **2019**, *9*, 4528. [CrossRef]
21. Kostov, K.G.; Nishime, T.M.C.; Machida, M.; Borges, A.C.; Prysiazhnyi, V.; Koga-Ito, C.Y. Study of cold atmospheric plasma jet at the end of flexible plastic tube for microbial decontamination. *Plasma Process. Polym.* **2015**, *12*, 1383–1391. [CrossRef]
22. Kim, J.Y.; Ballato, J.; Foy, P.; Hawkins, T.; Wei, Y.; Li, J.; Kim, S.-O. Apoptosis of cultured tumor cells treated with 200 μm-sized flexible microplasma jet. *IEEE Trans. Plasma Sci.* **2011**, *39*, 2974–2975. [CrossRef]
23. Kim, J.Y.; Kim, J.H.; Tae, H.-S.; Moon, D.W. Electrode-embedded atmospheric pressure plasma jet device for humid environment. *IEEE Trans. Plasma Sci.* **2014**, *42*, 2476–2477. [CrossRef]
24. Bae, G.T.; Kim, J.Y.; Kim, D.Y.; Jung, E.Y.; Jang, H.J.; Park, C.-S.; Jang, H.; Lee, D.H.; Lee, H.-K.; Tae, H.-S. Potential application of pin-to-liquid dielectric barrier discharge structure in decomposing aqueous phosphorus compounds for monitoring water quality. *Materials* **2021**, *14*, 7559. [CrossRef] [PubMed]
25. Zhang, J.-Z.; Fischer, C.J.; Ortner, P.B. Continuous flow analysis of phosphate in natural waters using hydrazine as a reductant. *Intern. J. Environ. Chem.* **2001**, *80*, 61–73. [CrossRef]
26. Takeuchi, N.; Yasuoka, K. Review of plasma-based water treatment technologies for the decomposition of persistent organic compounds. *Jpn. J. Appl. Phys.* **2021**, *60*, SA0801. [CrossRef]
27. Sano, N.; Kawashima, T.; Fujikawa, J.; Fujimoto, T.; Kitai, T.; Kanki, T.; Toyoda, A. Decomposition of organic compounds in water by direct contact of gas corona discharge: influence of discharge conditions. *Ind. Eng. Chem. Res.* **2002**, *41*, 5906–5911. [CrossRef]
28. Barjasteh, A.; Dehghani, Z.; Lamichhane, P.; Kaushik, N.; Choi, E.H.; Kaushik, N.K. Recent progress in applications of non-thermal plasma for water purification, bio-sterilization, and decontamination. *Appl. Sci.* **2021**, *11*, 3372. [CrossRef]
29. Zhou, M.; Struve, D.M. The effects of post-persulfate-digestion procedures on total phosphorus analysis in water. *Water Res.* **2004**, *38*, 3893–3898. [CrossRef] [PubMed]
30. Gross, A.; Boyd, C.E. A digestion procedure for the simultaneous determination of total nitrogen and total phosphorus in pond water. *J. World Aquac. Soc.* **1998**, *29*, 300–303. [CrossRef]
31. To, Y.S.; Randall, C.W. Evaluation of ascorbic acid method for determination of orthophosphates. *J. Water Pollut. Control Fed.* **1977**, *49*, 689–692.
32. Ma, J.; Yuan, Y.; Zhou, T.; Yuan, D. Determination of total phosphorus in natural waters with a simple neutral digestion method using sodium persulfate. *Limnol. Oceanogr. Methods* **2017**, *15*, 372–380. [CrossRef]
33. Morishita, T.; Ueno, T.; Panomsuwan, G.; Hieda, J.; Yoshida, A.; Bratescu, M.A.; Saito, N. Fast formation routes of nanocarbons in solution plasma processes. *Sci. Rep.* **2016**, *6*, 36880. [CrossRef] [PubMed]
34. Saito, N.; Bratescu, M.A.; Hashimi, K. Solution plasma: A new reaction field for nanomaterials synthesis. *Jpn. J. Appl. Phys.* **2018**, *57*, 0102A4. [CrossRef]
35. Kim, H.-J.; Shin, J.-G.; Park, C.-S.; Kum, D.S.; Shin, B.J.; Kim, J.Y.; Park, H.-D.; Choi, M.; Tae, H.-S. In-liquid plasma process for size- and shape-controlled synthesis of silver nanoparticles by controlling gas bubbles in water. *Materials* **2018**, *11*, 891. [CrossRef] [PubMed]
36. Shin, J.-G.; Shin, B.J.; Jung, E.Y.; Park, C.-S.; Kim, J.Y.; Tae, H.-S. Effects of a dielectric barrier discharge (DBD) on characteristics of polyaniline nanoparticles synthesized by a solution plasma process with an Ar gas bubble channel. *Polymers* **2020**, *12*, 1939. [CrossRef] [PubMed]
37. Qian, R.; Pei, Q.; Huang, Z. The role of H^+ ions in the electrochemical polymerization of pyrrole. *Makromol. Chem.* **1991**, *192*, 1263–1273. [CrossRef]
38. Katritzky, A.R.; Pozharskii, A.F. Reactivity of heterocycles. In *Handbook of Heterocyclic Chemistry*, 2nd ed.; Academic Press: Cambridge, MA, USA, 2000; pp. 377–379.
39. Zotti, G.; Martina, S.; Wegner, G.; Schlüter, A.-D. Well-Defined Pyrrole Oligomers: Electrochemical and UV/vis Studies. *Adv. Mater.* **1992**, *4*, 798–801. [CrossRef]
40. Vernitskaya, T.V.; Efimov, O.N. Polypyrrole: A conducting polymer; its synthesis, properties and applications. *Russ. Chem. Rev.* **1997**, *66*, 443–457. [CrossRef]
41. Coates, J. Interpretation of infrared spectra, a practical approach. *Encycl. Anal. Chem.* **2006**, *20*, 10815–10837.

polymers

MDPI

Article

Study of a Conical Plasma Jet with a Cloth-Covered Nozzle for Polymer Treatment

Felipe Vicente de Paula Kodaira [1,*], Ana Carla de Paula Leite Almeida [1], Thayna Fernandes Tavares [1], Antje Quade [2], Luis Rogério de Oliveira Hein [1] and Konstantin Georgiev Kostov [1]

1 School of Engineering, São Paulo State University (UNESP), Guaratinguetá 12516-410, SP, Brazil
2 Leibniz Institute for Plasma Science and Technology—INP, 17489 Greifswald, Germany; quade@inp-greifswald.de
* Correspondence: kodaira.felipe@gmail.com

Abstract: Although atmospheric pressure plasma jets (APPJs) have been widely employed for materials modification, they have some drawbacks, such as the small treatment area (couple of cm^2). To overcome this limitation, a funnel-like APPJ with a wide exit has been proposed. In this work, a gas-permeable cotton cloth covered the nozzle of the device to improve the gas flow dynamics and increase its range of operation. The funnel jet was flushed with Ar, and the plasma was ignited in a wide range of gas flow rates and the gap distances between the exit nozzle and the sample holder. The device characterization included electric measurements and optical emission spectroscopy (OES). To evaluate the size of the treatment and the degree of surface modification, large samples of high-density polyethylene (PE) were exposed to plasma for 5 min. Afterward, the samples were analyzed via water contact angle WCA measurements, scanning electron microscopy (SEM), and X-ray photoelectron spectroscopy (XPS). It was found that surface modification occurs simultaneously on the top and bottom faces of the samples. However, the treatment incorporated different functional groups on each side.

Keywords: APPJ; polymer treatment; conical APPJ; surface modification; atmospheric plasma

Citation: Kodaira, F.V.d.P.; Almeida, A.C.d.P.L.; Tavares, T.F.; Quade, A.; Hein, L.R.d.O.; Kostov, K.G. Study of a Conical Plasma Jet with a Cloth-Covered Nozzle for Polymer Treatment. *Polymers* **2023**, *15*, 3344. https://doi.org/10.3390/polym15163344

Academic Editors: Francesco Paolo La Mantia and Choon-Sang Park

Received: 27 June 2023
Revised: 21 July 2023
Accepted: 7 August 2023
Published: 9 August 2023

1. Introduction

Atmospheric pressure plasma jets (APPJs) emerged by the end of the 1990s as an alternative to the conventional dielectric barrier discharge (DBD) systems that are prone to processing thin films and flat substrates, but are inadequate for treating 3D structures. APPJs are low-cost devices where electric discharge is ignited in a noble gas that flows through a dielectric capillary and the resulting plasma is ejected into the surrounding environment (usually ambient air) [1]. There, the energetic plasma species (electrons, photons, and metastables) interact with air molecules, creating reactive oxygen and nitrogen species (RONS) that are driven by the gas flow to a target where surface modification/decontamination can take place [2]. Normally, APPJs generate several cm long plasma plumes that can be easily adapted to treat irregular 3D objects and internal surfaces of narrow tubes or cavities [3,4]. In particular, the RONS generated by atmospheric plasma jets have shown to be very effective in improving the wettability of polymers [5,6]. Plasma jets are very versatile devices that can be driven by pulsed DC, AC, RF, and microwave sources, with diverse designs and operation parameters [1]. Moreover, the gas temperature of the plasma plume can be kept quite low (under 40 °C in some cases), which makes the plasma jets suitable for medical applications and the treatment of thermo-sensitive materials [4]. In summary, the simple construction, flexible geometry, and low cost make APPJ very attractive for use in diverse fields, ranging from material synthesis, surface modification, gas conversion, to treatment of liquids through biomedical and agricultural applications [1,2]. However, the plasma plume diameter is determined by the inner diameter of the capillary [7], (normally in the order of a few mm), which combined with the limited lifetime of reactive species results in a small area (typically several cm^2) that can be covered by RONS [8], and eventually modified. Moreover, within the plasma-treated area (usually with a circular shape), the degree of material surface modification is not uniform, being more intense at the center (where the plasma jet hits the target) and less pronounced at the edges [9]. These issues may not be a problem in the case of a localized treatment, like plasma medicine, but they are serious drawbacks in the field of material processing. To remedy these disadvantages, different approaches, like target/jet manipulation [10], employing arrays of many plasma jets [11], different electrode configurations [12–15], high discharge frequency [16,17], shielding gases [9,18], the cooperative merging of several small plasma jets [19–21], and the use of a larger dielectric tubes [22] with different geometries [23,24] have been proposed. These methods help extend the plasma modification effect over a larger area (up to tens of cm^2), but all of them have drawbacks. For instance, one common disadvantage of the large tube diameter [22], the cooperative jets merging [19,20], and the plasma jet arrays [11] is that those devices usually rely upon excessively high flow rates (~10 L/min) of helium gas. Moreover, due to the complex gas flow dynamics in irregular tubes [24] and the complex interaction between individual jets in a plasma array [25], ensuring uniform surface modification is very hard. In some excitation schemes like in [5,12,15,16], a long (up to 10 cm), brush-type plasma jet was generated, which, combined with a one-dimensional target movement, can provide uniform treatment over a large rectangular area [16]. Nevertheless, as shown in [12,16], the discharge is prone to instabilities that perturb plasma plume distribution along the gas duct, thus compromising the treatment uniformity. In [10], the authors report on an especially designed mechanical system for simultaneous jet rotation and tilting, which can provide uniform decontamination over a circular area (~20 cm^2). Overall, the jet/target manipulation systems help achieve a larger treatment area, but they also add more process parameters and increment the device cost. To increase the size of plasma-modified surfaces, some authors proposed different shielding gases [9] and lateral He gas streams [18]. Once again, these new components increase the process cost and system complexity. In previous work, we proposed a simple plasma jet system terminating with a conical horn exit, which can perform uniform surface modification over the entire area covered by the horn [26], without any target/jet manipulation. The device also has low power and moderate gas consumption (several L/min) and provides surface modification on both sides of planar polymer samples [27]. Moreover, 3D objects (like seeds or small samples) can be placed inside the funnel and treated on all sides simultaneously [26,27]. The beneficial properties of this jet device can be explained by the effective trapping of the reactive species inside the conical horn that can be achieved at very short (few mm) sample-to-horn exit distances and moderate gas flow rates [28].

In this work, we report on the investigation of an atmospheric plasma jet terminating with a 70 mm diameter conical nozzle that is covered by a woven fabric. The latter improves the device's operation so that it can operate at a wide range of Ar flow rates and gap distances, and produces a stable plasma under the entire area of the funnel exit. Therefore, the proposed plasma jet device is suitable for applications that require uniform surface modification over a relatively large area.

The polymer chosen in this work was polyethylene (PE), a versatile polymer, that finds numerous applications across various industries due to its desirable properties such as high chemical resistance, low electrical conductivity, lightweight nature, and cost-effectiveness [29]. As such a popular polymer, many studies are conducted to develop or improve its processing, including plasma treatments, both at low pressure and atmospheric pressure, using various types of devices such as, for example, DBD discharges and APPJs [29–34].

2. Materials and Methods

A schematic drawing of the plasma jet configuration employed in this work is shown in Figure 1. It consists of a pin electrode (2.4 mm thick tungsten rod) centered inside a

2.5 mm thick quartz conical funnel, which has inner diameters of 4 mm at the straight part and 70 mm and at the funnel exit, respectively. The device was installed vertically, with its wider orifice pointing downward. The bottom of the funnel exit was completely covered by a 0.5 mm thick plain-woven fabric composed of intertwined (90°) cotton fibers, with a density of 150 treads/square inch (from Indústria Têxtil Nossa Senhora do Belém LTDA, Itatiba, Brazil). The cotton fabric allows for gas to pass through it also, and helps improve the gas flow distribution across the funnel exit. Beneath the funnel, a grounded metal electrode (Ø 155 mm) covered by a 3 mm thick glass slab was placed, on top of which the PE samples were placed. The distance from the funnel bottom to the glass slab varied from 0 to 10 mm. Argon gas (99.996% purity from AirLiquide, Cuiabá, Brazil) was admitted into the system through a Teflon holder on the funnel top end. The gas flow rate was adjusted ranging from 0 to 4 SLM by a Horiba model STEC N100 (Horiba, Ltd., Kyoto, Japan) mass-flow controller. Plasma was excited by a commercial AC power supply (Minipuls 6, GBS Elektronik, Dresden, Germany), which was operated in burst mode. The generated voltage waveform was an amplitude-modulated sinewave, i.e., a sequence of N high-voltage oscillations is followed by a voltage off-period. The number of high-voltage cycles (N = 12), the signal frequency (25.0 kHz), and the burst repetition period (Tr = 2.0 ms) were kept constant, while the voltage magnitude and discharge current changed depending on the discharge parameters.

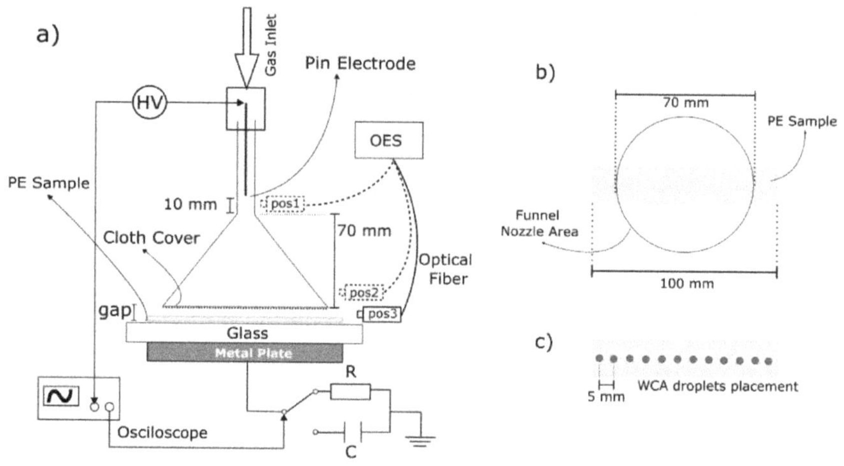

Figure 1. (**a**) Schematic drawing of the plasma jet configuration; (**b**) Top view scheme of the PE sample position during treatment; (**c**) Scheme of droplet distribution for WCA measurements.

The charge transferred to the target and the discharge current were obtained by measuring the voltage drop across a serial capacitor of 10 nF or a serial resistor of 100 Ω, respectively. The high-voltage signal applied to the pin electrode was measured by a P6015A Tektronix voltage divider (1×1000). The signals were monitored on a digital oscilloscope (TDS3032C model from Tektronix, Beaverton, OR, USA). To determine the plasma jet mean power, we used the Q-V Lissajous method, which was adapted for amplitude-modulated voltage signals, as reported in previous work [26].

For all operation conditions, plasma was generated in the form of randomly distributed filaments that commenced from the pin electrode, propagated down the quartz funnel, and after passing through the cloth, reached the glass plate over the grounded electrode. Although permeable, the cloth keeps an Ar-enriched atmosphere inside the conical horn, which facilitates the discharge ignition. Therefore, the cloth cover allows for device operation with bigger gaps between the exit nozzle and the target. For instance, in previous works [26,27], where a similar device was operated without the cloth cover, plasma could not be excited at gaps bigger than 3 mm.

Optical emission spectroscopy was performed in three regions: on the edge of the pin electrode, above the cloth inside the quartz funnel, and right below the cloth (as shown in Figure 1, pos1, pos2, and pos3, respectively), to investigate how the cover interferes with the propagation of the species generated by the plasma. For these measurements, an Avantes (Apeldoorn, Netherlands) spectrometer model AvaSpec-ULS-RS-TEC was used.

It is known from the literature that the treatment of polymers by atmospheric plasmas results in enhanced wettability and adhesion, properties that are very much desirable for coating, printing, painting, and dying [35,36]. All these applications require plasma sources that can generate uniform plasma over a large area. For initial tests, we employed polyethylene (PE) as a common engineering polymer. The plasma treatments were performed on 0.960 g/cm^3, 1.0 mm thick commercial PE (from LAMIEX INDUSTRIA DE PLASTICOS LTDA., Pinhais, Brazil). The samples were cut in rectangular shapes with a 100 × 20 mm size. The sample was placed below the device's nozzle, as shown in Figure 1b. It is important to note that the length of the samples is greater than the diameter of the funnel so that measurements could be taken along the sample to check the homogeneity of the treatment and verify whether any changes occur beyond the funnel exit geometric area. To remove organic contaminants, all samples were ultrasonically cleaned first in distilled water for 15 min, and after that in isopropyl alcohol for 10 min and finally left to dry at room temperature overnight.

To assess the surface modification of polymer samples induced by the conical plasma jet, we employed water contact angle (WCA) measurements. They were performed using the sessile drop method on a Rame-Hart goniometer (300 F1) with an automated drop dispenser. Deionized water was used as a test liquid and the volume of each drop was set to 1.0 μL. For each sample, a sequence of equidistant droplets was deposited on the sample surface to determine the treatment area and the radial distribution of the WCA, as shown in Figure 1c. The WCA measurements were performed within an interval of 10 min after the plasma treatment, and both faces of the samples, the top (facing the funnel nozzle) and bottom (facing the glass dielectric), were analyzed; the measurement along each surface took about 3 min to be performed.

To evaluate the functionalization of the sample surface, XPS analysis was performed with Kratos AXIS Supra (Kratos Analytical, Manchester, UK). For these measurements, it was assumed that the treatment is symmetrical, so after being treated, the samples were cut in half (50 × 20 mm) and one half was analyzed at the top face and the other half at the bottom. Longitudinal scans with a 2 mm sampling interval of these samples gave us the values of their elemental composition from the edge to the center.

To evaluate the possible modifications of surface morphology, SEM measurements were performed by a Carl Zeiss (Oberkochen, Baden-Württemberg, Germany) EVO LS 15 at a low pressure (10^{-3} Pa), and electron high tension (EHT) of 5.0 kV. Before the measurements, the samples were covered by a 6 nm thick gold layer deposited via sputtering. Topography images were taken from both the top and bottom faces of the samples, at the center and the border of the treated samples.

3. Results

The porous fabric at the funnel exit greatly improves the gas flow dynamics as well as discharge homogeneity, thus allowing for device operation over a wide range of gas flow rates (1–4 SLM) and distances (0–10 mm). Only in extreme cases, like the combinations of the largest gaps with very low Ar flows, the discharge is not stable. The device operating with 4.0 SLM of argon with two different gaps of 3 mm and 8 mm is shown in Figure 2a,b, respectively. It can be observed that by increasing the gap, the discharge filaments tend to stop running along the walls of the funnel and go straight to the glass plate that covers the second electrode.

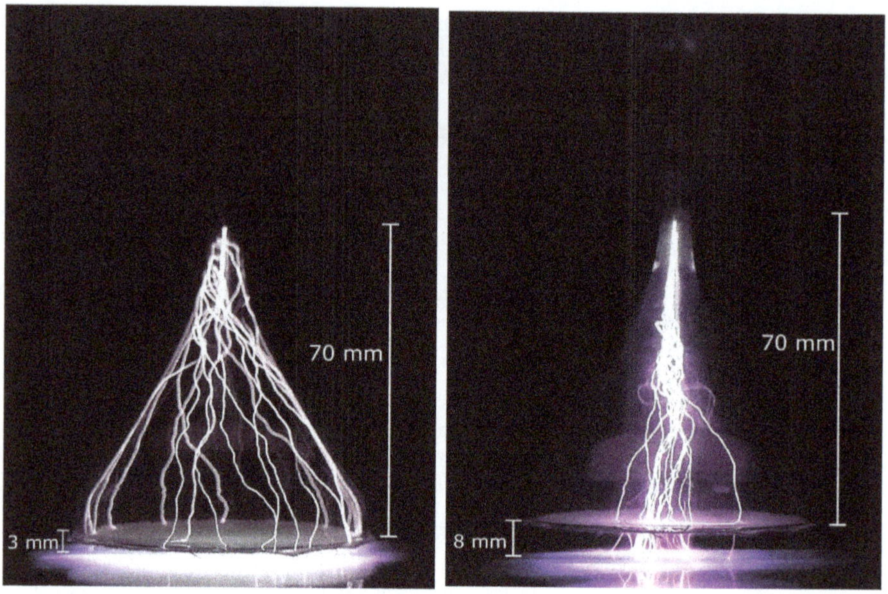

(a) (b)

Figure 2. Photos of the funnel device operating with an argon flow of 4.0 SLM and gap distance of (**a**) 3.0 mm, and (**b**) 8.0 mm.

The distance to the target, as well as the gas flow rate can affect the plasma jet operation, and more specifically, its current waveform and mean power. The typical current and voltage waveforms at different gas flow rates are shown in Figure 3a–c for a gap distance of 5.0 mm, and in Figure 4a,b for an 8 mm gap. In the latter case, it was not possible to ignite the discharge at a gas flow of 2.0 SLM. It can be seen from them that higher gap values require a higher argon flow to obtain stable voltage/current waveforms, which is evidenced by little variations in the signal amplitude within the burst. Moreover, as can be seen in Figures 3 and 4, the current amplitude increases with the argon flow, while conversely, the voltage magnitude first decreases rapidly and after that, it tends to saturate at higher flows.

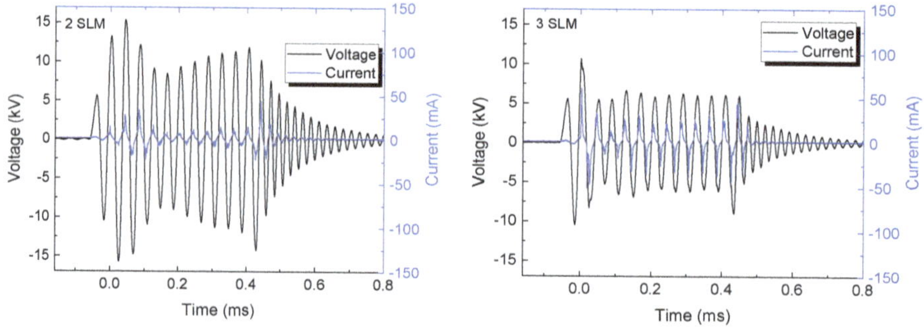

(a) (b)

Figure 3. *Cont.*

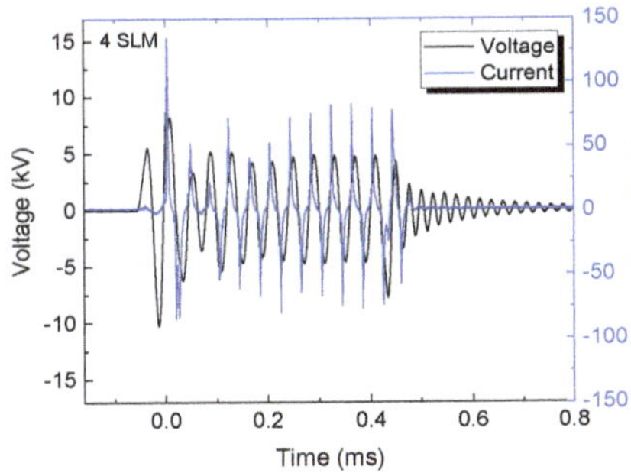

(c)

Figure 3. Typical current/voltage waveforms for discharge with a 5.0 mm gap and argon flow of (**a**) 2.0 SLM, (**b**) 3.0 SLM, and (**c**) 4.0 SLM.

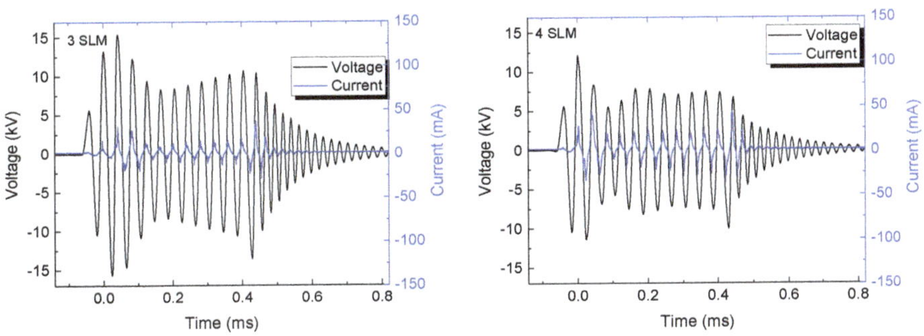

(a) (b)

Figure 4. Typical current/voltage waveforms for discharge with an 8.0 mm gap and argon flow of (**a**) 3.0 SLM and (**b**) 4.0 SLM.

Figure 5 depicts the influence of gap distance for different Ar flow rates on (a) the RMS discharge current, and (b) the discharge power. It is possible to observe that for each gas flow rate, there is an optimum gap distance where the power reaches its maximum. Also, for the gas flow rates in Figure 5a, the RMS current is relatively stable up to a certain gap value, beyond which it decreases sharply, and the power maximum occurs right after this fall. This shift in the maximum of the RMS current and voltage signals is probably due to the nonmonotonic behavior of the applied voltage signal at different gaps.

The results of optical emission spectroscopy for an Ar flow of 4.0 SLM and gap distance of 4 mm are presented in Figure 6. the optical fiber position for each measurement is depicted in Figure 1 indicated as pos1, pos2, and pos3, corresponding, respectively, to Figure 6a–c. To evaluate the effect of permeable cotton fabric on the composition of the generated species, the spectra were acquired in two places inside the funnel, in the funnel narrow part close to the Ar admission (Figure 6a), and right above the cloth in the funnel large part (Figure 6b). Figure 6c shows the OES spectrum collected right below the cloth. As can be seen, the spectra inside the funnel are dominated by the intense argon spectral line, which means that the cloth can hold an argon-enriched atmosphere inside the device. This is an important role that facilitated the discharge ignition and allowed for the device operation with bigger gaps. Alongside the Ar lines, the spectrum in Figure 6a presents a few molecular lines, mostly OH emissions coming from moisture inside the gas line. However, the spectrum taken close to the permeable cloth (Figure 6b) shows the characteristic emission bands of excited N_2 molecules, which indicate Ar mixing with air. On the other hand, the spectrum under the cloth shown in Figure 6c exhibits some weaker atomic lines coming from Ar expelled through the cloth, and is dominated by the typical molecular emissions coming from excited N_2, OH, and NO molecules that are commonly generated in atmospheric pressure plasma jets [37].

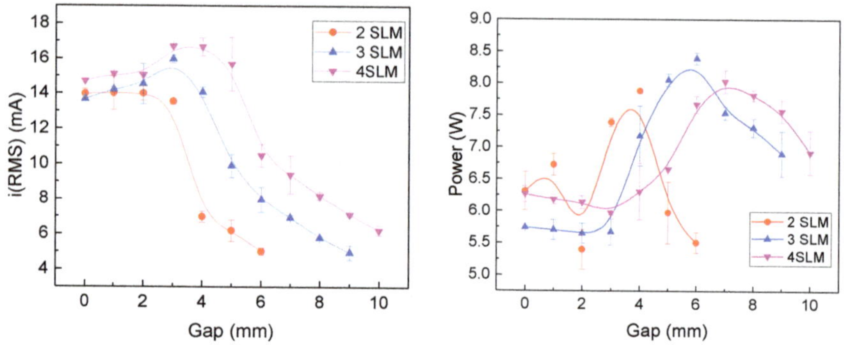

(a) (b)

Figure 5. Influence of the gap distance for different Ar flow rates on (**a**) rms currents, and (**b**) discharge power.

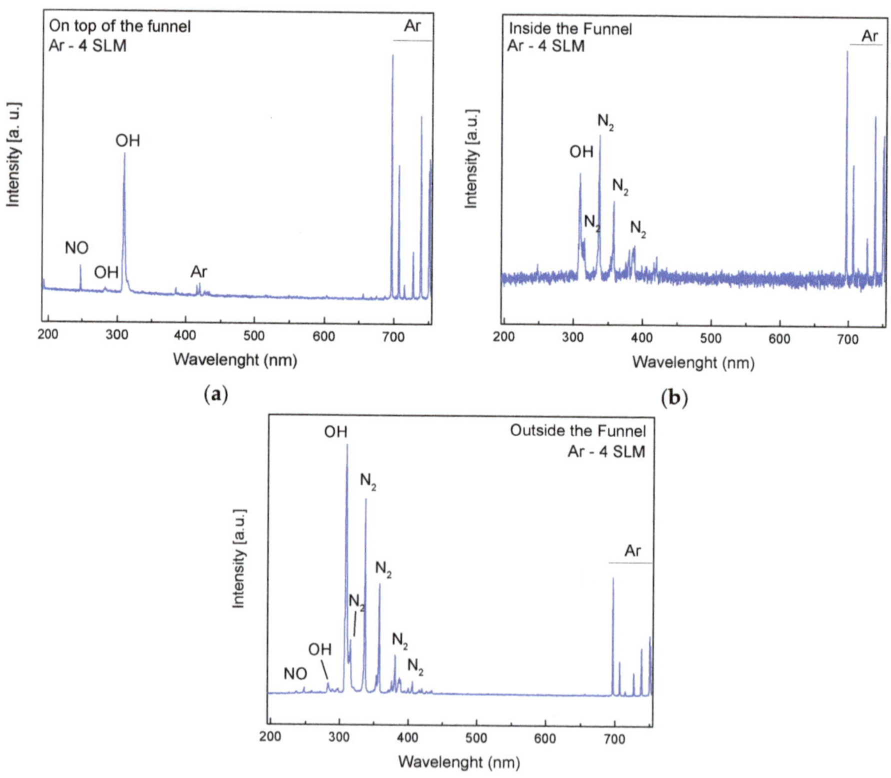

(a) (b) (c)

Figure 6. OES for discharges with a 4.0 mm gap (**a**) inside, on top of the funnel with 4.0 SLM of argon flow, (**b**) inside the funnel (right above the cloth) with 4.0 SLM of argon flow, and (**c**) outside the funnel with 4.0 SLM of argon flow.

The PE samples were exposed to plasma for 5.0 min and after that subjected to wettability analysis. A series of WCA measurements were performed with a 5 mm spacing between the consecutive drops to obtain the treatment profile along the target's long side, as shown in Figure 1c. The top and bottom faces of each sample were measured. As noted in previous work [27], when applying a plasma jet to a thin dielectric sample, a secondary DBD-type discharge is generated between the sample bottom and the sample holder. Figure 7 depicts the water contact angle distributions for different gap values, while keeping an argon flow of 4.0 SLM. The samples were larger than the funnel diameter and stretched beyond the funnel edge. As can be seen in Figure 7, the plasma modification effect happened within the entire funnel geometrical area, as well as in its close vicinity. Also, as in [27], both sides of the samples were treated. Except for the big gap of 8 mm, which is discussed separately, within the funnel region, the bottom and top sides of the samples exhibit similarly wide plateau-like WCA distributions. As the sample's top face is in direct contact with the Ar plasma jet, the degree of surface modification is higher at a shorter gap distance (d = 3 mm). On the other hand, the bottom side of the sample has approximately the same WCA for gap distances of 3.0 and 5.0 mm. As shown in Figure 2b, for bigger gaps, the discharge filaments tend to concentrate in the middle of the funnel, thus causing more intense treatment in that region. Therefore, the contact angle distribution obtained for the 8.0 mm gap (blue triangles in Figure 7a,b) is not uniform along the sample, exhibiting a triangular shape with a minimum roughly at the funnel center. Another observation is that on the top side of the sample, the wettability remains relatively unchanged for about another 10 mm beyond the funnel geometrical region, while for the bottom side, the WCA distribution starts decreasing more abruptly. This finding can be explained by the fact that reactive species formed on the top side of the sample are carried by the Ar flow outside the funnel, where they can still alter the polymer surface characteristics. On the other hand, the reactive species produced under the sample by the secondary DBD discharge can expand only due to diffusion, thus covering a smaller area.

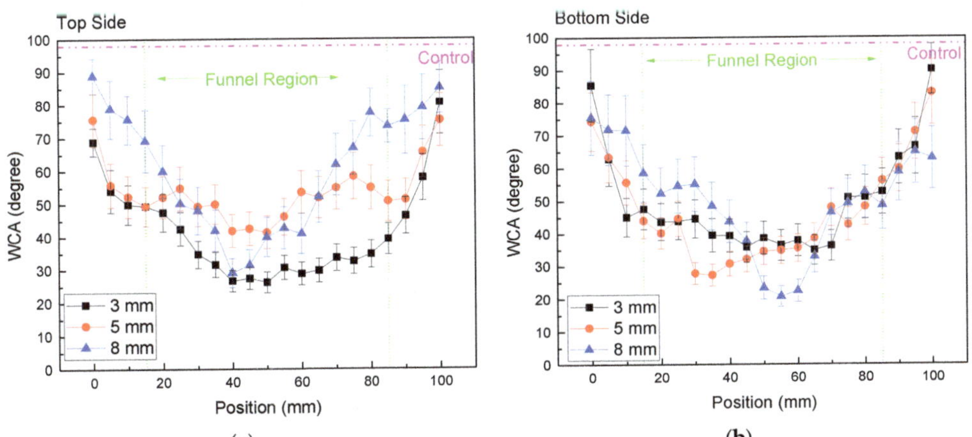

Figure 7. WCA distribution of PE samples treated with an argon flow of 4.0 SLM and different gap values for (**a**) the top side of the sample, and (**b**) the bottom side of the sample.

The results from XPS analysis of PE samples treated for 5 min using a 5.0 mm gap and argon flow of 4.0 SLM are presented in Figure 8. The treated samples were cut in half and the left-hand part of the sample was analyzed on its top face, while the right-hand part was measured on its bottom face. The spatially resolved XPS measurements were taken along the sample's central line with a 2.0 mm separation, starting from the edge of the sample to its center, which coincides with the center of the funnel. The elemental composition of the untreated PE sample is 96 at% C and 4 at% O due to surface oxidation.

In Figure 9, an example of an XPS C 1s spectrum is given. This measurement was taken from position 19 mm on the top side, and the C 1s peaks were fitted using four components: C–C/C–H @ BE 285.0 eV (calibr.), C–OH/R @ BE 286.5 ± 0.2 eV, C=O @ BE 287.8 eV, and COOH/R @ BE 289.2 ± 0.2 eV [38–40]. Figure 8a,b shows the distribution of the elemental composition of both faces of the treated PE sample. It is possible to see a reduction in the carbon content and the appearance of oxygen on both sides of the treated sample. Traces of Si (due to sample contamination) and a small amount of N (<3 at%) were also detected. However, a significant percentage of nitrogen atoms (up to 8 at%) was found only on the bottom side of the sample (Figure 8a). This is probably due to the nature of the secondary air DBD discharge formed between the sample and the sample holder. As shown in the literature [32], plasma treatments of polymers in the air usually result in the incorporation of oxygen and nitrogen atoms into the material surface [41].

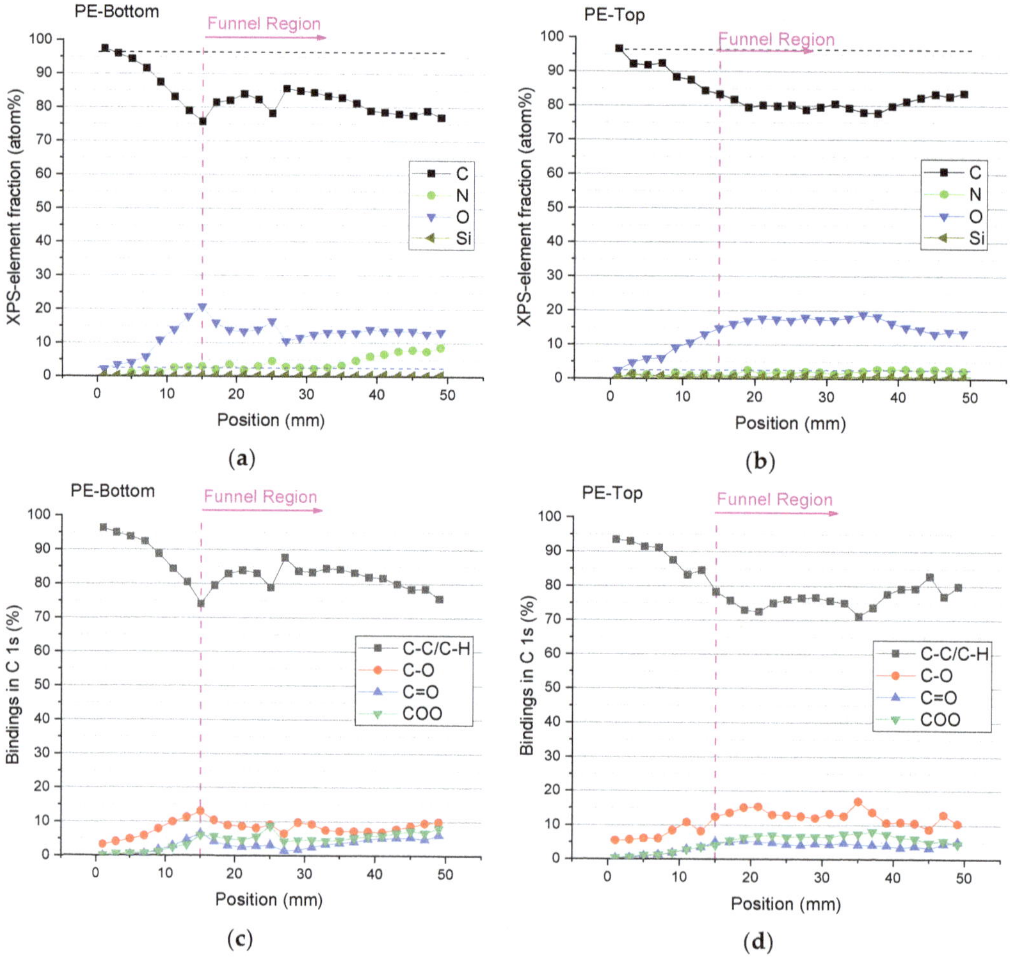

Figure 8. Spatially resolved XPS results of PE samples treated for 5 min with a gap of 5 mm and argon flow of 4.0 SLM. Element fractions on the (**a**) bottom side and (**b**) top side of the sample. Carbon bindings on (**c**) the bottom side and (**d**) top side of the sample.

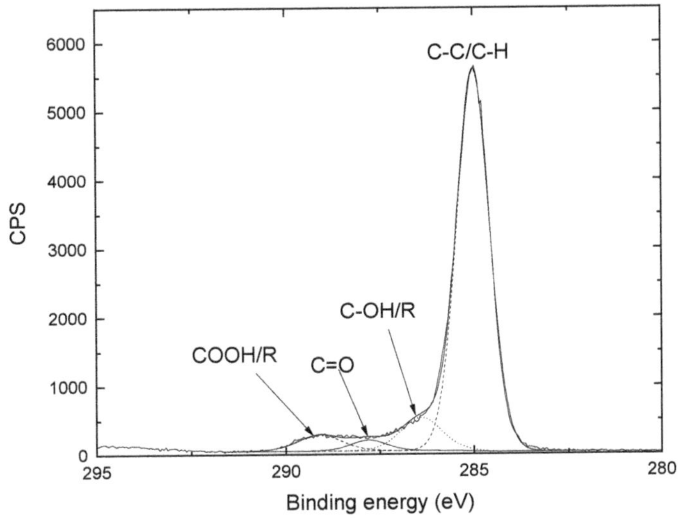

Figure 9. Example of an XPS spectra deconvolution, taken in the position of 19 mm on the top side. C 1s peaks were fitted using four components: C–C/C–H @ BE 285.0 eV (calibr.), C–OH/R @ BE 286.5 ± 0.2 eV, C=O @ BE 287.8 eV, and COOH/R @ BE 289.2 ± 0.2 eV.

In Figure 8c,d, the binding in the C 1s peaks on both sample faces is represented. It is possible to observe that as one approaches the funnel region, there is a gradual reduction in the C–C and C–H bonds and the emergence of oxygenated groups such as C–O, C=O, and COO on the surface. When entering into the funnel region, the amount of these functional groups tends to remain practically unchanged throughout the sample. It is worth mentioning that the reduction in functional groups, as well as the O/C atom contents outside the funnel region is more abrupt on the bottom side of the sample than the ones on the top samples side. This finding also corroborates with the WCA distributions.

The plasma treatment can modify the chemistry of polymeric surfaces by adding functional groups [42]. This functionalization is observed mainly with the addition of oxygen and nitrogen groups. The extent of PE surface functionalization was determined using XPS, and it is depicted in Figure 10, where the degree of functionalization is shown. The degree of functionalization provides information about the ratio of the sum of functional groups (-OH/R, =O, -COOH/R) to C–C/C–H [38–40]. Again, it is possible to see in Figure 10a that at the end of the funnel region, there is a sharp decay in the degree of functionalization for the bottom of the sample. On the upper face of the sample, the corresponding functionalization decay is smoother (Figure 10b). Moreover, within the funnel area, the top PE side exhibits higher functionalization levels of approximately 35%, whereas the corresponding values for the bottom sample face are around 25%. This is in good agreement with the WCA measurements that exhibited similar trends. It should be noted, however, that the functionalization of the upper sample side is predominantly attributed to oxygen groups, whereas in addition to oxygen groups, the lower side also exhibits the presence of nitrogen functional groups.

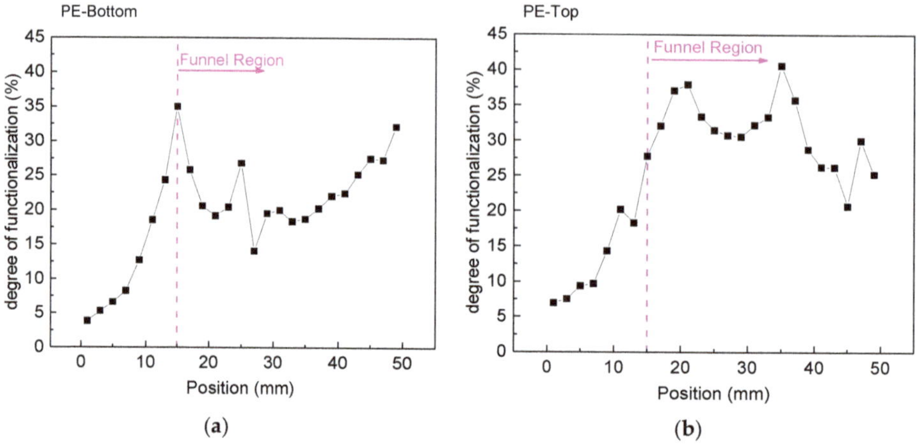

Figure 10. Degree of surface functionalization of the PE samples obtained via XPS (**a**) on the bottom side of the sample and (**b**) the top side. The samples were treated for 5 min with a gap of 5.0 mm and an argon flow of 4.0 SLM.

The results from the SEM analyses are presented in Figure 11. It shows typical micrographs of the bottom and top side of the PE sample taken in the central region (Figure 10a,b), respectively, and the edge region (Figure 11c,d) after a 5 min plasma treatment with an argon flow of 5.0 SLM and an 8.0 mm gap. For comparison, Figure 11e depicts the surface of the untreated PE. First, in the edge region on both sides of the sample (Figure 11c,d), no changes in the surface morphology can be observed when compared to the untreated sample (Figure 11e). This observation agrees with the finding that at large gaps, the discharge is mostly concentrated at the funnel central part, and consequently, less intense surface modification happens at the funnel edge. On the other hand, in the central region of the treated sample (Figure 11a,b), some significant changes can be seen. On the top side of the treated sample (Figure 11b), only mild morphological alterations, probably due to the etching and removal of surface contaminants, can be seen. However, Figure 11a clearly shows the formation of small round structures in the central part on the bottom face of the PE sample. These structures consist of loosely bonded short polymer fragments that are known as low molecular-weight oxidized materials (LMWOMs). Their formation is typically observed on the surface of polymers treated in air plasma [41,43,44], thus corroborating the existence of the secondary DBD discharge formed under the PE sample.

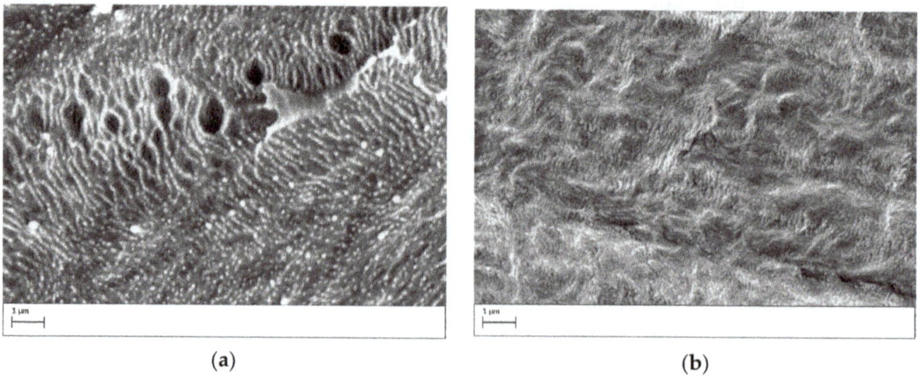

Figure 11. *Cont.*

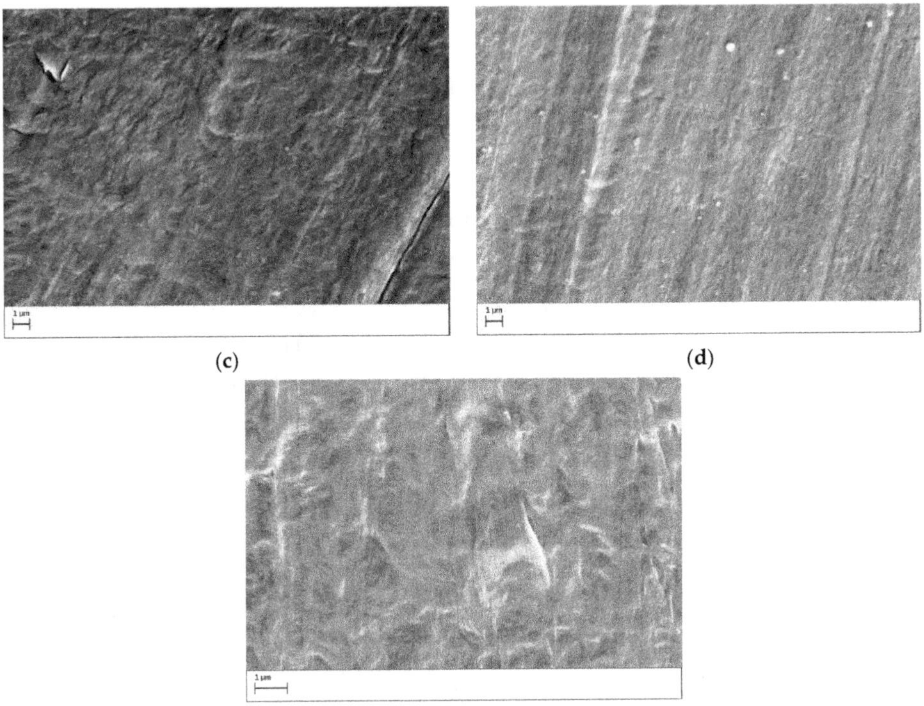

(c) (d) (e)

Figure 11. SEM images of the PE samples treated for 5 min with a gap of 5.0 mm and an argon flow of 4.0 SLM. The analysis was performed at (**a**) the central region on the bottom side of the sample, (**b**) the central region on the top side of the sample, (**c**) the border on the bottom side of the sample, (**d**) the border on the top side of the sample, and (**e**) the center of the untreated sample.

4. Conclusions

The funnel-shaped plasma jet has demonstrated potential for uniform treatment over large areas when compared to conventional plasma jets, without the requirement of high gas flow rates. By employing a cloth cover on the exit nozzle, the device has been able to operate at increased distances from the sample holder, thereby expanding its potential applications in material treatments. The device was characterized by electrical and optical measurements, showing that device power consumption and proportions between the excited reactive species can be optimized through the careful choosing of the operation parameters, such as gas flow rate and gap distance. It has been observed that the device operation is influenced by variations in the flow rate and gap, with greater distances leading to a reduction in the uniform part of the treated area.

Simultaneous treatment of both sides of the sample has been successfully achieved, which offers a time-saving advantage in situations where the treatment of both surfaces is necessary. Specifically, the upper face is subjected to direct treatment from the discharge emanating from the funnel, while the lower face is treated by a secondary DBD discharge. However, despite the simultaneous treatment, differences exist between the surface modification effect on the two sides of the PE sample. The bottom side presented the incorporation of nitrogen groups and the formation of LMWOM, while the upper sample face exhibited a more intense incorporation of O-containing groups. Furthermore, it is worth noting that the treatment of the upper side of the samples occurs over a broader area than that covered by the funnel, while the treatment of the lower side is confined to a few millimeters beyond this region. These observations contribute to a deeper understanding of the operational characteristics and limitations of the funnel-shaped plasma jet device, facilitating its implementation in various material treatment scenarios.

Author Contributions: Conceptualization, F.V.d.P.K. and K.G.K.; methodology, F.V.d.P.K., A.C.d.P.L.A., T.F.T., A.Q. and L.R.d.O.H.; formal analysis, F.V.d.P.K., A.C.d.P.L.A., T.F.T., A.Q. and L.R.d.O.H.; investigation, F.V.d.P.K., A.C.d.P.L.A. and T.F.T.; resources, K.G.K.; data curation, F.V.d.P.K. and T.F.T.; writing—original draft preparation, F.V.d.P.K. and K.G.K.; writing—review and editing, F.V.d.P.K. and K.G.K.; supervision, F.V.d.P.K. and K.G.K.; project administration, K.G.K.; funding acquisition, K.G.K. All authors have read and agreed to the published version of the manuscript.

Funding: The authors acknowledge a postdoctoral fellowship from the UNESP under project 4235, as well as financial support from the CAPES-PrInt and the São Paulo Research Foundation FAPESP under grant 2019/05856-7.

Institutional Review Board Statement: Not applicable.

Conflicts of Interest: The authors declare no conflict of interest.

References

1. Winter, J.; Brandenburg, R.; Weltmann, K.D. Atmospheric pressure plasma jets: An overview of devices and new directions. *Plasma Sources Sci. Technol.* **2015**, *24*, 064001. [CrossRef]
2. Viegas, P.; Slikboer, E.; Bonaventura, Z.; Guaitella, O.; Sobota, A.; Bourdon, A. Physics of plasma jets and interaction with surfaces: Review on modelling and experiments. *Plasma Sources Sci. Technol.* **2022**, *31*, 053001. [CrossRef]
3. Geng, J.; Yin, S.; Huang, S.; Tang, Q.; Luo, H.; Chen, F. Flexible cold plasma jet with controllable length and temperature for hydrophilic modification. *Phys. Plasmas* **2018**, *25*, 083508. [CrossRef]
4. Corbella, C.; Portal, S.; Keidar, M. Flexible Cold Atmospheric Plasma Jet Sources. *Plasma* **2023**, *6*, 72–88. [CrossRef]
5. Shaw, D.; West, A.; Bredin, J.; Wagenaars, E. Mechanisms behind surface modification of polypropylene film using an atmospheric-pressure plasma jet. *Plasma Sources Sci. Technol.* **2016**, *25*, 065018. [CrossRef]
6. Ma, C.; Nikiforov, A.; Hegemann, D.; De Geyter, N.; Morent, R.; Ostrikov, K. Plasma-controlled surface wettability: Recent advances and future applications. *Int. Mater. Rev.* **2023**, *68*, 82–119. [CrossRef]
7. Birer, Ö. Reactivity zones around an atmospheric pressure plasma jet. *Appl. Surf. Sci.* **2015**, *354*, 420–428. [CrossRef]
8. Cho, G.; Kim, Y.; Uhm, H.S. The jet-stream channels of gas and plasma in atmospheric-pressure plasma jets. *J. Korean Phys. Soc.* **2016**, *69*, 525–535. [CrossRef]
9. Narimisa, M.; Onyshchenko, Y.; Morent, R.; De Geyter, N. Improvement of PET surface modification using an atmospheric pressure plasma jet with different shielding gases. *Polymer* **2021**, *215*, 123421. [CrossRef]
10. Li, H.; Li, M.; Zhu, H.; Zhang, Y.; Du, X.; Chen, Z.; Xiao, W.; Liu, K. Realizing high efficiency and large-area sterilization by a rotating plasma jet device. *Plasma Sci. Technol.* **2022**, *24*, 045501. [CrossRef]
11. Zhou, R.; Zhang, B.; Zhou, R.; Liu, F.; Fang, Z.; Ostrikov, K. Linear-field plasma jet arrays excited by high-voltage alternating current and nanosecond pulses. *J. Appl. Phys.* **2018**, *124*, 033301. [CrossRef]
12. Wu, K.; Liu, J.; Wu, J.; Chen, M.; Ran, J.; Pang, X.; Jia, P.; Li, X.; Ren, C. A double-mode planar argon plume produced by varying the distance from an atmospheric pressure plasma jet. *High Volt.* **2023**. [CrossRef]
13. Joh, H.M.; Kang, H.R.; Chung, T.H.; Kim, S.J. Electrical and Optical Characterization of Atmospheric-Pressure Helium Plasma Jets Generated with a Pin Electrode: Effects of the Electrode Material, Ground Ring Electrode, and Nozzle Shape. *IEEE Trans. Plasma Sci.* **2014**, *42*, 3656–3667. [CrossRef]
14. Wang, S.; Zhang, J.; Li, G.; Wang, D. Cold large-diameter plasma jet near atmospheric pressure produced via a triple electrode configuration. *Vacuum* **2014**, *101*, 317–320. [CrossRef]
15. Lv, X.; Ren, C.; Ma, T.; Feng, Y.; Wang, D. An Atmospheric Large-Scale Cold Plasma Jet. *Plasma Sci. Technol.* **2012**, *14*, 799–801. [CrossRef]
16. Polášková, K.; Nečas, D.; Dostál, L.; Klíma, M.; Fiala, P.; Zajíčková, L. Self-organization phenomena in cold atmospheric pressure plasma slit jet. *Plasma Sources Sci. Technol.* **2022**, *31*, 125014. [CrossRef]
17. Omran, A.V.; Busco, G.; Ridou, L.; Dozias, S.; Grillon, C.; Pouvesle, J.-M.; Robert, E. Cold atmospheric single plasma jet for RONS delivery on large biological surfaces. *Plasma Sources Sci. Technol.* **2020**, *29*, 105002. [CrossRef]
18. Johnson, M.J.; Boris, D.R.; Petrova, T.B.; Walton, S.G. Extending the volume of atmospheric pressure plasma jets through the use of additional helium gas streams. *Plasma Sources Sci. Technol.* **2020**, *29*, 015006. [CrossRef]
19. Li, H.-P.; Nie, Q.-Y.; Yang, A.; Wang, Z.-B.; Bao, C.-Y. An Atmospheric Cold Plasma Jet with a Good Uniformity, Robust Stability, and High Intensity over a Large Area. *IEEE Trans. Plasma Sci.* **2014**, *42*, 2470–2471. [CrossRef]
20. Nizard, H.; Gaudy, T.; Toutant, A.; Iacono, J.; Descamps, P.; Leempoel, P.; Massines, F. Influence of discharge and jet flow coupling on atmospheric pressure plasma homogeneity. *J. Phys. D Appl. Phys.* **2015**, *48*, 415204. [CrossRef]
21. O'Connor, N.; Humphreys, H.; Daniels, S. Cooperative Merging of Atmospheric Pressure Plasma Jet Arrays. *IEEE Trans. Plasma Sci.* **2014**, *42*, 756–758. [CrossRef]

22. O'Neill, F.T.; Twomey, B.; Law, V.J.; Milosavljevic, V.; Kong, M.G.; Anghel, S.D.; Dowling, D.P. Generation of Active Species in a Large Atmospheric-Pressure Plasma Jet. *IEEE Trans. Plasma Sci.* **2012**, *40*, 2994–3002. [CrossRef]
23. Kang, H.R.; Chung, T.H.; Joh, H.M.; Kim, S.J. Effects of Dielectric Tube Shape and Pin-Electrode Diameter on the Plasma Plume in Atmospheric Pressure Helium Plasma Jets. *IEEE Trans. Plasma Sci.* **2017**, *45*, 691–697. [CrossRef]
24. Yamamoto, A.; Kawano, Y.; Nakai, M.; Nakagawa, T.; Sakugawa, T.; Hosseini, H.; Akiyama, H. Investigation of gas flow dependence of plasma jet produced by pulsed power. *IEEE Trans. Plasma Sci.* **2015**, *43*, 3451–3455. [CrossRef]
25. Ghasemi, M.; Olszewski, P.; Bradley, J.W.; Walsh, J.L. Interaction of multiple plasma plumes in an atmospheric pressure plasma jet array. *J. Phys. D Appl. Phys.* **2013**, *46*, 052001. [CrossRef]
26. Mui, T.S.M.; Mota, R.P.; Quade, A.; Hein, L.R.d.O.; Kostov, K.G. Uniform surface modification of polyethylene terephthalate (PET) by atmospheric pressure plasma jet with a horn-like nozzle. *Surf. Coat. Technol.* **2018**, *352*, 338–347. [CrossRef]
27. Kodaira, F.V.d.P.; Leal, B.H.S.; Tavares, T.F.; Quade, A.; Hein, L.R.d.O.; Chiappim, W.; Kostov, K.G. Simultaneous Treatment of Both Sides of the Polymer with a Conical-Shaped Atmospheric Pressure Plasma Jet. *Polymers* **2023**, *15*, 461. [CrossRef]
28. Punith, N.; Singh, A.K.; Ananthanarasimhan, J.; Boopathy, B.; Chatterjee, R.; Hemanth, M.; Chakravortty, D.; Rao, L. Generation of neutral pH high-strength plasma-activated water from a pin to water discharge and its bactericidal activity on multidrug-resistant pathogens. *Plasma Process. Polym.* **2023**, *20*, 2200133. [CrossRef]
29. Knuuttila, H.; Lehtinen, A.; Nummila-Pakarinen, A. Advanced Polyethylene Technologies—Controlled Material Properties. In *Long Term Properties of Polyolefins*; Advances in Polymer Science; Springer: Berlin/Heidelberg, Germany, 2004; pp. 13–28. [CrossRef]
30. Van Deynse, A.; Cools, P.; Leys, C.; Morent, R.; De Geyter, N. Surface modification of polyethylene in an argon atmospheric pressure plasma jet. *Surf. Coat. Technol.* **2015**, *276*, 384–390. [CrossRef]
31. Organski, L.; Wang, X.; Myers, A.; Chen, Y.-C.; Park, K.; Horava, S.D.; Richard, C.A.; Yeo, Y.; Shashurin, A. Inner surface modification of polyethylene tubing induced by dielectric barrier discharge plasma. *J. Vac. Sci. Technol. A* **2022**, *40*, 063005. [CrossRef]
32. Neto, J.F.d.M.; de Souza, I.A.; Feitor, M.C.; Targino, T.G.; Diniz, G.F.; Libório, M.S.; Sousa, R.R.M.; Costa, T.H.d.C. Study of High-Density Polyethylene (HDPE) Kinetics Modification Treated by Dielectric Barrier Discharge (DBD) Plasma. *Polymers* **2020**, *12*, 2422. [CrossRef]
33. Šourková, H.J.; Weberová, Z.; Antoň, J.; Špatenka, P. Wettability and Adhesion of Polyethylene Powder Treated with Non-Equilibrium Various Gaseous Plasma in Semi-Industrial Equipment. *Materials* **2022**, *15*, 686. [CrossRef]
34. Murthy, V.S.M.D.; Vaidya, U. Improving the adhesion of glass/polypropylene (glass-PP) and high-density polyethylene (HDPE) surfaces by open air plasma treatment. *Int. J. Adhes. Adhes.* **2019**, *95*, 102435. [CrossRef]
35. Penkov, O.V.; Khadem, M.; Lim, W.-S.; Kim, D.-E. A review of recent applications of atmospheric pressure plasma jets for materials processing. *J. Coat. Technol. Res.* **2015**, *12*, 225–235. [CrossRef]
36. Cvelbar, U.; Walsh, J.L.; Černák, M.; de Vries, H.W.; Reuter, S.; Belmonte, T.; Corbella, C.; Miron, C.; Hojnik, N.; Jurov, A.; et al. White paper on the future of plasma science and technology in plastics and textiles. *Plasma Process. Polym.* **2019**, *16*, 1700228. [CrossRef]
37. Luan, P.; Kondeti, V.S.S.K.; Knoll, A.J.; Bruggeman, P.J.; Oehrlein, G.S. Effect of water vapor on plasma processing at atmospheric pressure: Polymer etching and surface modification by an Ar/H_2O plasma jet. *J. Vac. Sci. Technol. A* **2019**, *37*, 031305. [CrossRef]
38. Chemello, G.; Knigge, X.; Ciornii, D.; Reed, B.P.; Pollard, A.J.; Clifford, C.A.; Howe, T.; Vyas, N.; Hodoroaba, V.; Radnik, J. Influence of the Morphology on the Functionalization of Graphene Nanoplatelets Analyzed by Comparative Photoelectron Spectroscopy with Soft and Hard X-rays. *Adv. Mater. Interfaces* **2023**, *10*, 2300116. [CrossRef]
39. Thiha, A.; Ibrahim, F.; Muniandy, S.; Madou, M.J. Microplasma direct writing for site-selective surface functionalization of carbon microelectrodes. *Microsyst. Nanoeng.* **2019**, *5*, 62. [CrossRef]
40. Merenda, A.; Ligneris, E.D.; Sears, K.; Chaffraix, T.; Magniez, K.; Cornu, D.; Schütz, J.A.; Dumée, L.F. Assessing the temporal stability of surface functional groups introduced by plasma treatments on the outer shells of carbon nanotubes. *Sci. Rep.* **2016**, *6*, 31565. [CrossRef]
41. Booth, J.-P.; Mozetic, M.; Nikiforov, A.Y.; Oehr, C. Foundations of plasma surface functionalisation of polymers for industrial and biological applications. *Plasma Sources Sci. Technol.* **2022**, *31*, 103001. [CrossRef]
42. Williamson, J.B.; Lewis, S.E.; Johnson, R.R.; Manning, I.M.; Leibfarth, F.A. C−H Functionalization of Commodity Polymers. *Angew. Chem. Int. Ed.* **2019**, *58*, 8654–8668. [CrossRef] [PubMed]
43. Dell'Orto, E.C.; Vaccaro, A.; Riccardi, C. Morphological and chemical analysis of PP film treated by Dielectric Barrier Discharge. *J. Phys. Conf. Ser.* **2014**, *550*, 012032. [CrossRef]
44. Kehrer, M.; Rottensteiner, A.; Hartl, W.; Duchoslav, J.; Thomas, S.; Stifter, D. Cold atmospheric pressure plasma treatment for adhesion improvement on polypropylene surfaces. *Surf. Coat. Technol.* **2020**, *403*, 126389. [CrossRef]

polymers MDPI

Article

Antibacterial Effects of a Carbon Nitride (CN) Layer Formed on Non-Woven Polypropylene Fabrics Using the Modified DC-Pulsed Sputtering Method

Young-Soo Sohn [1], Sang Kooun Jung [2], Sung-Youp Lee [3] and Hong Tak Kim [3,*]

[1] Department of Biomedical Engineering, Daegu Catholic University, Gyeongsan 38439, Republic of Korea; sohnys@cu.ac.kr
[2] Econet Korea Ltd., Gumi 39373, Republic of Korea; econetkorea@naver.com
[3] Department of Physics, Kyungpook National University, Daegu 41566, Republic of Korea; physylee@knu.ac.kr
* Correspondence: zam89blue@gmail.com

Abstract: In the present study, the surface of non-woven polypropylene (NW-PP) fabric was modified to form CN layers using a modified DC-pulsed (frequency: 60 kHz, pulse shape: square) sputtering with a roll-to-roll system. After plasma modification, structural damage in the NW-PP fabric was not observed, and the C–C/C–H bonds on the surface of the NW-PP fabric converted into C–C/C–H, C–N(CN), and C=O bonds. The CN-formed NW-PP fabrics showed strong hydrophobicity for H_2O (polar liquid) and full-wetting characteristics for CH_2I_2 (non-polar liquid). In addition, the CN-formed NW-PP exhibited an enhanced antibacterial characteristic compared to NW-PP fabric. The reduction rate of the CN-formed NW-PP fabric was 89.0% and 91.6% for *Staphylococcus aureus* (ATCC 6538, Gram-positive) and *Klebsiella pneumoniae* (ATCC4352, Gram-negative), respectively. It was confirmed that the CN layer showed antibacterial characteristics against both Gram-positive and Gram-negative bacteria. The reason for the antibacterial effect of CN-formed NW-PP fabrics can be explained as the strong hydrophobicity due to the CH_3 bond of the fabric, enhanced wetting property due to CN bonds, and antibacterial activity due to C=O bonds. Our study presents a one-step, damage-free, mass-productive, and eco-friendly method that can be applied to most weak substrates, allowing the mass production of antibacterial fabrics.

Keywords: antibacterial effect; carbon nitride; staphylococcus aureus; klebsiella pneumonia; non-woven fabric; sputtering

Citation: Sohn, Y.-S.; Jung, S.K.; Lee, S.-Y.; Kim, H.T. Antibacterial Effects of a Carbon Nitride (CN) Layer Formed on Non-Woven Polypropylene Fabrics Using the Modified DC-Pulsed Sputtering Method. *Polymers* **2023**, *15*, 2641. https://doi.org/10.3390/polym15122641

Academic Editor: Choon-Sang Park

Received: 27 April 2023
Revised: 5 June 2023
Accepted: 8 June 2023
Published: 10 June 2023

1. Introduction

Non-woven (NW) fabrics are fabric-like materials made of short and long polymer fibers, which are held together with chemical, mechanical, and thermal treatments. Polypropylene (PP) is currently the most popular material to manufacture NW fabrics due to its low cost, light weight, good chemical stability, excellent moisture resistance, and insulation properties. In addition, PP is an eco-friendly material that is naturally decomposed outdoors. NW fabrics, including PP, have been widely used in various applications, such as medical supplies, food packing materials, sanitary products, fashion textiles, civil engineering products, and building materials [1–3]. In principle, the materials used for biomedical applications should be biocompatible and have antibacterial properties. Pathogens such as bacteria and viruses cause fatal injuries to the human body, and if they spread widely, they cause great social and economic losses. Until now, removing bacteria and viruses has been considered a problem to be solved in the category of medicine and pharmacy. Recently, it has been reported in various research institutes around the world that virus removal can be performed even by using the antibacterial properties of metals known for a long time. Typically, antibacterial metals, such as silver, platinum, and copper, are commonly applied to various materials in the form of nanomaterials and thin films [4–6].

The basic mechanisms of antibacterial activity for antibacterial metals and metal oxides are the enzyme interference of metal ions, the production of reactive oxygen species (ROS), the destruction of cell membranes, the direct genotoxic activity for some metals, and the prevention of absorption of important microelements by microbes [7]. However, these metals can be easily oxidated, and their antibacterial effect can be rapidly reduced. Metal oxides, including ZnO nanoparticles, can promote the generation of ROS, and this can lead to antibacterial effects [7,8]. However, studies on the harmfulness of nanoparticles to the human body are needed, and the mass production of nanoparticles is also difficult. Compared to these materials, carbon and carbon-related materials are relatively safe for the human body and have the advantage of being inexpensive. Recently, antibacterial effects of carbon-related materials, such as diamond-like carbon, diamond, and graphitic carbon nitride (g-C_3N_4), have been investigated, and several studies confirmed their antibacterial effect [9–13]. For example, activated carbon (AC) has a larger surface area due to its porous structure, and this property can efficiently promote antibacterial effects. In addition, antibacterial effects can also be promoted if we manipulate the surface of these materials. More specifically, super-hydrophilic and -hydrophobic surfaces have been reported to prevent bacterial attachment to surfaces, resulting in low bacterial adhesion [9,14,15]. However, these properties are difficult to achieve on weak materials such as fabrics, NW fabrics, paper, and textiles. Previous studies have shown that wettability can be controlled through carbon surface treatment on paper, which is a soft and porous substrate, and it has been confirmed that there is no damage to the paper [16]. This study investigated the antibacterial effect induced by the formation of a carbon-nitride (CN) layer on the NW-PP fabric. The CN layer was formed on the NW-PP fabrics using a modified DC-pulsed sputtering, which has been shown to minimize substrate damage [16,17]. The main strategy for acquiring the antibacterial effect was to use the porous structure of NW-PP fabrics and modify their surfaces to convert a C–C bond to a C–N bond. It is expected that this design can increase the surface area and promote the reaction with pathogens.

2. Materials and Methods

The NW-PP fabrics were treated by DC-pulsed magnetron sputtering. The sputtering device is equipped with a roll-to-roll device, suitable for the flexible substrate process. The configuration of the magnet was alternately repeated N-polarity lines and S-polarity lines. The magnet arrays of the sputtering gun were configured to acquire a magnetic field parallel to the substrate using magnetic simulation software (FEMM) [18]. The size of the carbon target was 0.416 m × 0.298 m, and the distance between the target and the NW-PP fabrics was 6 cm. The chamber was evacuated up to a base pressure of 6.7×10^{-4} Pa before the modification process of the NW-PP fabrics, and the working pressure was maintained at 0.23 Pa (Ar: 180 sccm, N_2: 20 sccm) during the deposition process. The plasma was generated by a DC-pulsed power (SJ power, Bucheon, Republic of Korea, SPF-2). The applied DC pulse was a negative square shape with a frequency of 60 kHz and a duty ratio of 50%. During the plasma treatment, a power of 450 W (voltage: 450 V, current: 1 A, current density: 124 mA/m^2) was applied to the sputtering source. The CN layer formations were performed at room temperature, and the treatment time was 20 s (roll speed: 100 mm/min). The schematic diagram of the modified DC-pulsed sputtering system is shown in Figure 1. One disadvantage of sputtering deposition is a line-of-sight process, and this means that conformal deposition is difficult on complex-shaped substrates, including porous structural materials. In the modified sputtering, it is possible to widen the incident angle of particles into the substrate because of the modified sputtering source and the roll-to-roll system. Although highly uniform deposition through porous materials is not easy using this system, conformal processing is possible to a certain extent [16].

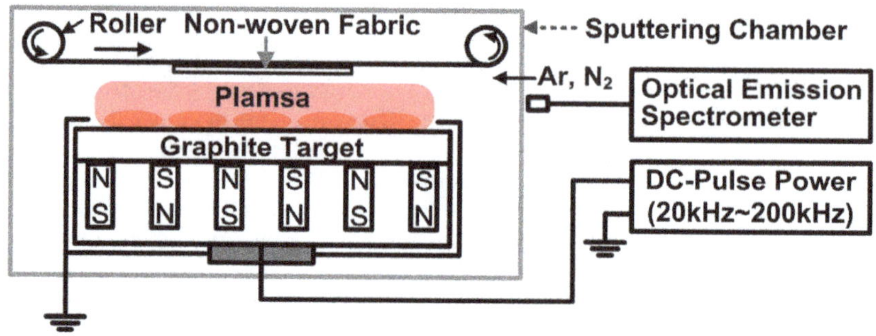

Figure 1. A schematic diagram of a DC-pulsed sputtering system for the formation of a carbon nitride (CN) layer on non-woven PP (NW-PP) fabrics.

The emission spectrum of the Ar/N_2-C plasma was measured by an optical emission spectrometer (Avantes, Gelderland, Netherlands, AvaSpec-256) with a blazed grating (groove density: 600 line/mm, blaze: 250 nm) in the range of 200–900 nm. The surface morphology of the CN-formed NW-PP fabrics was investigated using a scanning electron microscope (SEM, Hitachi, S-4800) operating at the acceleration voltage of 5 kV. The elemental distribution of the fabrics was analyzed at the acceleration voltage of 15 kV using an energy-dispersive X-ray spectroscope (EDX) with an X-ray detector (Horiba, Kyoto, Japan, X-Max^N). The chemical bonds on the surface of the fabrics were analyzed by an X-ray photoelectron spectrometer (XPS, ThermoFisher, Waltham, MA, USA, Nexsa) using Al-K_α radiation (1486.6 eV). The contact angle was measured by a contact angle goniometer with a digital microscope camera (Amscope, Irvine, CA, USA, MU1000,). The test solutions were distilled water (DI, H_2O) and diiodomethane (CH_2I_2, 99%, Sigma-Aldrich, St, Louis, MO, USA), which were representative polar and dispersive solutions, respectively. The contact angle was evaluated using image analysis software (ImageJ) [19], and the calculations of surface energy were performed by the Owen–Wendt method [20]. The antibacterial tests of the CN-formed NW-PP fabrics were performed by appointing an authorized agency (Korea Far Infrared Association Co., Seoul, Republic of Korea). The KS K 0693 test was employed as the antibacterial test method [21]. A brief summary of the KS K 0693 test was as follows: first, reference and fabric samples were sterilized at a high temperature of 120 °C and a high pressure of 103 kPa for 20 min before the test; second, the bacteria culture was injected into the sterilized samples; then, the samples were incubated at a temperature of 37 °C for 18 h. *Staphylococcus aureus* (ATCC 6538, *S. aureus*) and *Klebsiella pneumoniae* (ATCC 4352, *K. pneumoniae*) were used as the representative strains of Gram-positive and Gram-negative bacteria, respectively. The reduction rate (R) was obtained through a comparison of bacterial growth in reference and NW-PP fabrics. The reference sample for the antibacterial test was a cotton fiber, which had a mass per unit area of 115 $\pm$ 5 g/m^2 and a whiteness index of 70 $\pm$ 5 measured by a reflectometer under a CIE standard light source of D_{65} condition (KS K ISO 105-F02:2001).

3. Results and Discussion

Optical emission spectroscopic analysis is a useful technique for analyzing active species in processing plasma, and Figure 2 demonstrates the optical emission spectrum of Ar/N_2–C plasma in the DC-pulsed sputtering system. For Ar emissions, we observed neutral argon lines (Ar I) in the range of 580–850 nm and single ionized argon lines (Ar II) in the range of 400–520 nm [16,22,23]. Some Ar I lines overlapped with Ar II lines at a wavelength of 415.9, 420.1, and 427.2 nm. For N_2 emissions, we noted that all the emission lines of N_2 were attributed to molecular transitions and overlapped with the Ar emission lines mostly in the 400–790 nm range. The first positive lines of N_2, between 590 and 790 nm, were attributed to the transition of $B^3\Pi_g - A^3\Sigma_u$. The second positive lines of N_2 in the range of 340 and 400 nm were ascribed as the transition of $C^3\Pi_u - B^3\Pi_g$, and the first negative line of N_2 at a wavelength of 290 nm was attributed to $B^2\Pi_u{}^+ - X^2\Sigma_g{}^+$, as shown in Figure 2 (inset) [24,25]. For carbon emissions, the atomic emission lines at 427 nm and 465 nm were attributed to a singly ionized carbon line (C II) and a doubly ionized carbon line (C III), respectively. Finally, the molecular emission band in the range of 469–474 nm was ascribed to the transition of C_2 ($d^3\Pi_g - a^3\Pi_\mu$) [16]. From this result, it could be seen that the DC-pulsed magnetron sputtering equipment produced enough carbon and nitrogen active species to form the CN-contained layers on the surface of NW-PP fabrics.

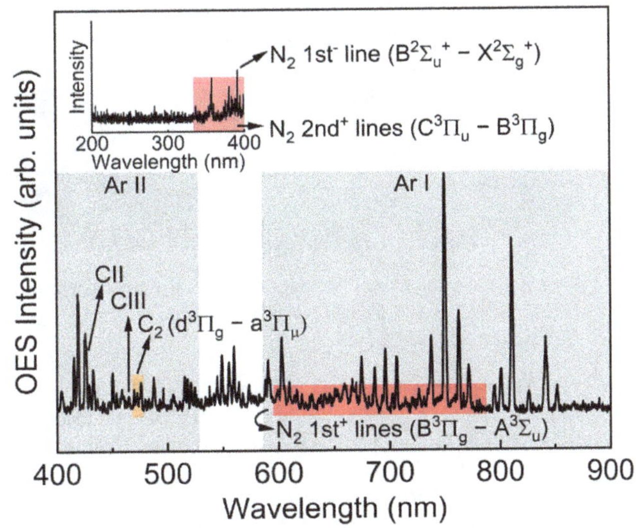

Figure 2. Optical emission spectrum of Ar/N_2-C plasma generated by the DC-pulsed sputtering (Pressure: 0.23 Pa (Ar: 180 sccm, N_2: 20 sccm), power: 450 W (voltage: 450 V, current: 1 A), square pulse frequency: 60 kHz (duty ratio of 50%)).

Figure 3 shows SEM images of the surface morphology for ordinary NW-PP and CN-formed NW-PP fabrics and the EDX image of elemental distribution for C, N, and O for CN-formed NW-PP fabric. It is shown that there were no differences in the surface morphology between the two samples, and the fabric structure remained undamaged after plasma treatment. The weight ratios of C, N, and O were 97.23, 1.51, and 1.27%, respectively. As shown in Figure 3c, the CN layer uniformly covered the surface of the NW-PP fabric. The latter could be explained by the fact that a short plasma exposure time can cause less damage, and the pulsed power could effectively reduce the charging damage of the substrates [16,17,26]. The thickness of the CN layer formed on the fabrics was measured indirectly due to the short processing time of 20 s and the low sputtering yield of the graphite target. According to the literature, the sputtering yield of carbon was approximately 0.2 for argon ions striking the carbon target with a kinetic energy of 600 eV [16,27,28]. This value is very low compared to other target materials. The growth rate was found to be approximately 0.3 Å/s by converting the thickness after 2 min of deposition on the glass substrate. In addition, no significant differences were observed by the FT-IR and Raman measurements (not shown here). The difference between ordinary and CN-formed NW-PP fabrics was only observed in XPS measurements, which was a surface-sensitive analysis with a typical penetration depth of a few nanometers. From these results, it could be estimated that the CN layer was not a thin film completely independent of the NW-PP fabric, but a composite formed on the surface of the NW-PP fabric.

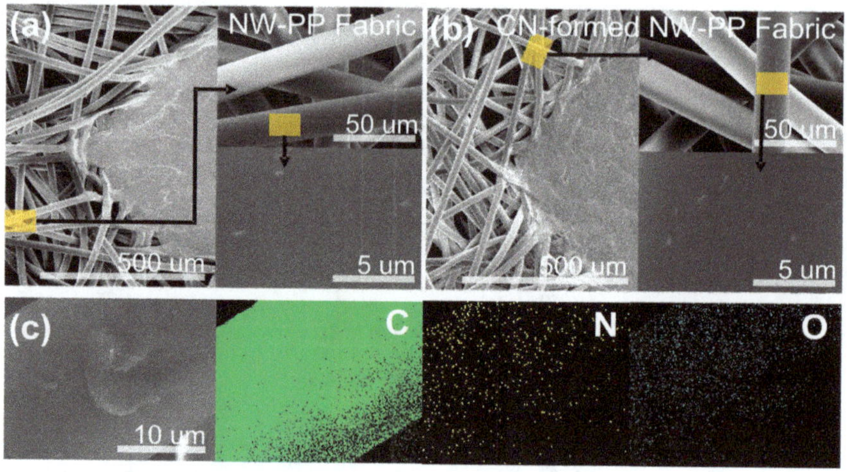

Figure 3. Morphological SEM images of the non-woven polypropylene (NW-PP) fabrics: (**a**) ordinary NW-PP fabric, (**b**) CN-formed NW-PP fabric, and (**c**) EDX elemental mapping images of C, N, and O for CN-formed NW-PP fabric.

Figure 4 shows the C 1s, N 1s, and O 1s XPS spectra of NW-PP and CN-formed NW-PP fabrics. The C 1s peak in the NW-PP fabrics was observed at 284.5 eV, corresponding to the C–C/C–H bond. For the CN-formed NW-PP fabrics, deconvoluted peaks were mainly observed in the C 1s at 284.5 eV, 285.5 eV, and 288.5 eV, which were attributed to C–C/C–H, C–N, and C=O bonds, respectively. The N 1s peak could be deconvoluted to two peaks positioned at 398.7 eV and 399.6 eV, which were attributed to sp^2 nitrogen atoms with two neighbors (=N-) and sp^2 nitrogen atoms with three neighbors (–N<), respectively. The presence of the C=O bond was due to chemical adsorption in atmospheric conditions and confirmed by the O 1s peak. Usually, the oxygen peaks range between 529 and 535 eV. The peaks related to oxygen adsorption were observed between 530 and 535 eV, while the peaks below 530 eV represented lattice oxygen [29,30]. From these results, the decrease in C–C/C–H bonds in the C 1s indicated the removal of a methyl group (CH_3 group) from the surface of the NW-PP fabrics, which was the result of easy interactions between the broken sites of the PP fabric and active nitrogen elements in the plasma. Considering the short plasma exposure time and SEM results, the C and N atoms seemed to bind to the sites broken by the high-energy plasma. This implied the formation of very thin C-N composites on the surface of the NW-PP fabrics.

Figure 5 shows contact angle images of DI water (polar liquid) and CH_2I_2 (non-polar liquid) droplets on the NW-PP and CN-formed NW-PP fabrics. The contact angles of the DI droplet on the NW-PP and CN-formed NW-PP fabrics were approximately 130° and 115°, respectively. Both samples showed a strong hydrophobicity against the polar liquid. However, the contact angle in the CN-formed NW-PP fabrics was slightly reduced compared to the NW-PP fabrics. This can be explained as the appearance of C=O bonds on the CN-formed NW-PP fabrics, which had a polar property, as shown in XPS results. In addition, the droplet of CH_2I_2 on NW-PP could only maintain its shape for a few seconds and then spread widely, while that on CN-formed NW-PP fabric was immediately spread. Thus, the CN-formed NW-PP fabric represented the full-wetting characteristic for a non-polar liquid, and this implied that as-formed CN composites on the fabrics played an important role in enhancing the wetting property for non-polar liquids. The commonly acceptable definition of hydrophobicity and hydrophilicity for water is based on 90°. It can be defined as a hydrophilic surface when the static contact angle (θ) for water is <90°, a hydrophobic surface when θ is >90°, and a superhydrophobic surface when θ is >150° [31]. Thus, the change in the contact angle was simply considered as the result of the competition between cohesion and wetting on the surface. In terms of the interaction between the water and the surface, it can be represented that the hydrophilic surface exhibited a strong affinity, while the hydrophobic surface showed little affinity for water solution. The intermolecular attraction was the source of the surface energy and was the sum of the contributions of polar and non-polar (or dispersive) components. The polar interaction between the surface and the solution was a Coulomb interaction between permanent dipoles or between permanent and induced dipoles, while the dispersive interaction was caused by temporal fluctuations in charge distribution in the surface and the solution. In addition, it is also important to identify the types of major biomolecules to determine the reactions between polymers and bacteria. Biological molecules are mainly composed of carbon and hydrogen atoms and can contain many different types of functional groups. The important functional groups in biological molecules are (-OH), methyl (-CH_3), carbonyl (C=O), carboxyl (-COOH), amino (-NH_2), sulfhydryl (-SH), and phosphate (-PO_4) groups [32]. These functional groups play a crucial role in forming biomolecules such as proteins, carbohydrates, lips, and DNA. The functional groups can be divided as having polar and non-polar characteristics, which depends on their molecular structure and compositions. The methyl group is the only non-polar functional group, while other groups are polar groups. This meant that the methyl group represented a hydrophobic property, and the remaining group exhibited hydrophilic properties. The polar (γ^p) and dispersive surface energy (γ^d) on the surface of NW-PP fabrics were determined using the contact angles for DI (γ_L^p: 51 mJ/m^2, γ_L^d: 21.9 mJ/m^2, γ_L: 72.8 mJ/m^2) and CH_2I_2 (γ_L^p: 0 mJ/m^2, γ_L^d: 50.8 mJ/m^2, γ_L: 50.8 mJ/m^2) [20,33]. The calculations were performed by the Owen–Wendt method [20]. In this case, the polar elements of ordinary and CN-formed NW-PP fabrics were 2.01 mJ/m^2 and 3.32 mJ/m^2, respectively, showing little difference. Otherwise, the dispersive surface energy of the NW-PP fabrics after plasma treatment increased from 23.2 mJ/m^2 to 59.2 mJ/m^2, which in turn increased the total surface energy from 25.2 mJ/m^2 to 62.5 mJ/m^2.

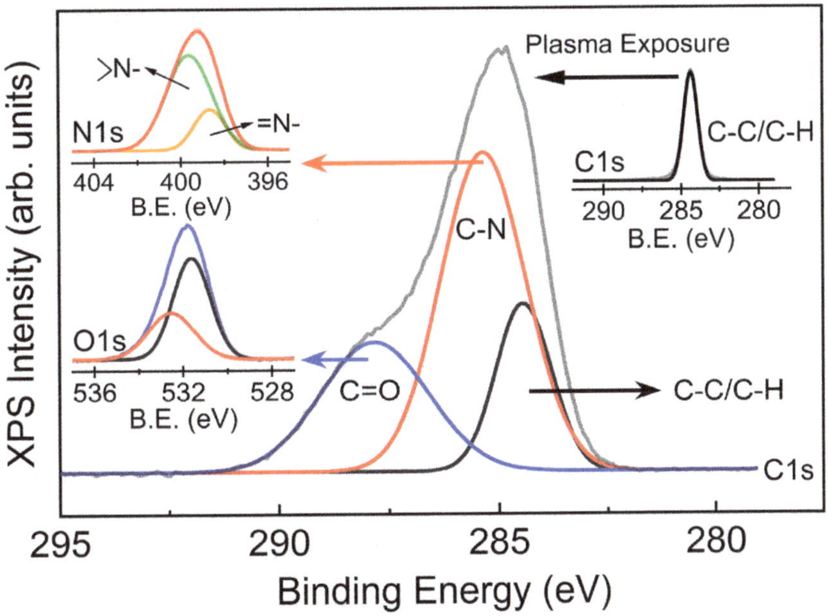

Figure 4. XPS C 1s spectra of non-woven polypropylene (NW-PP) and CN-formed NW-PP fabrics.

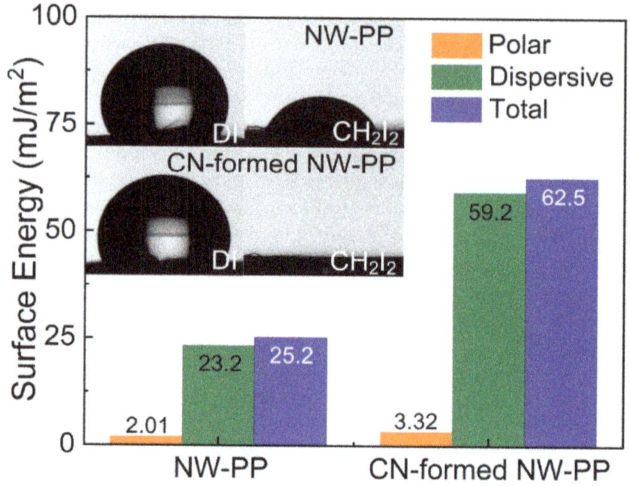

Figure 5. Analysis of the surface energy for the non-woven polypropylene (NW-PP) and CN-formed NW-PP fabric; (inset) photo-images of distilled H_2O and CH_2I_2 droplets on the NW-PP and CN-formed NW-PP fabrics.

Figure 6 shows the photographic images of the evolution of two different bacteria for the antibacterial test, and *S. aureus* and *K. pneumoniae* were used to verify the antibacterial effect of the CN-formed NW-PP fabrics. The images represented that the CN-formed NW-PP fabrics suppressed the growth of two different types of Gram-positive and Gram-negative bacteria. A brief description of the two types of bacteria is as follows. Gram-positive and Gram-negative bacteria are two different types of bacteria. Gram-positive bacteria appear blue or purple after Gram-staining in a laboratory test. They have thick cell walls (peptidoglycan cell walls). Gram-negative bacteria appear pink or red on staining and have thin walls (thinner peptidoglycan cell walls). They release different toxins and affect the body in different ways. *S. aureus* is a Gram-positive bacterium with a spherical shape and is a facultative anaerobe that can grow without the need for oxygen. *S. aureus* can become the main reason for opportunistic pathogens, skin infections, respiratory infections, and food poisoning. *K. pneumoniae* is a Gram-negative bacterium with a rod shape. This is a non-motile, encapsulated, lactose-fermenting, and facultative anaerobe bacteria. *K. pneumoniae* can cause destructive changes to human and animal lungs [34]. Table 1 summarizes the concentration changes of *S. aureus* and *K. pneumoniae* on the CN-formed NW-PP fabrics initially and after 18 h. The reduction rate (R) was calculated as follows:

$$R\ (\%) = \frac{B - A}{B} \times 100, \tag{1}$$

where A is the colony-forming unit (CFU) per mL of the control group, and B is the experimental group [21,35,36]. The reduction rate of the CN-formed NW-PP fabric was 89.0% and 91.6% for *S. aureus* (ATCC 6538) and *K. pneumoniae* (ATCC 4352), respectively, while that of the NW-PP fabric was 84.3% and 85.5%, respectively. In addition, the performance stability of the CN-formed NW-PP fabric was tested using samples left in the atmosphere for about 3 months. The reduction rate for the CN-formed NW-PP exhibited a similar value compared to previous samples. The reduction rate for the aged sample was 89.4% for *S. aureus* and 91.2% for *K. pneumoniae*.

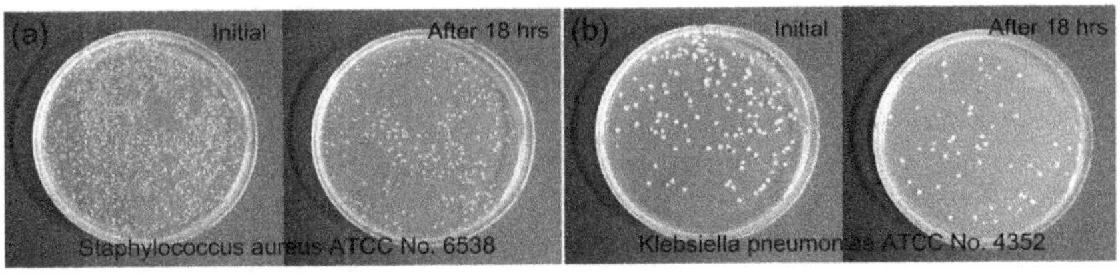

Figure 6. Photographs of bacteria evolutions for antibacterial test (KS K 0693 test): (**a**) *Staphylococcus aureus* (ATCC 6538) and (**b**) *Klebsiella pneumoniae* (ATCC 4352) on the CN-formed NW-PP fabrics. In (**a**,**b**), the left side shows the control groups and the right side shows the experimental groups after 18 h, which were incubated at a temperature of 37 °C for 18 h.

Table 1. The reduction rate (*R*) of *Staphylococcus aureus* (ATCC 6538) and *Klebsiella pneumoniae* (ATCC 4352) on the non-woven polypropylene (NW-PP) fabric and CN-formed NW-PP fabrics compared to the reference fabrics (Antibacterial test method: KS K 0693, 2016).

Bacteria Name	Bacteria Type	Testing Sample	Concentration (CFU/mL)		*R* (%)
			Initial	After 18 h.	
ATCC 6538 (*S. aureus*)	Gram-Negative	Reference NW-PP fabric	6.0×10^4	2.8×10^6 4.4×10^5	84.3
		Reference CN-formed NW-PP	8.9×10^4	4.2×10^6 4.6×10^5	89.0
ATCC 4352 (*K. pneumoniae*)	Gram-Positive	Reference NW-PP fabric	4.2×10^4	2.0×10^6 2.9×10^5	85.5
		Reference CN-formed NW-PP	3.2×10^4	1.9×10^6 1.6×10^5	91.6

The CN-formed NW-PP fabric showed an antibacterial reduction, which could be explained by the strong hydrophobicity of the fabrics. The CN-formed NW-PP fabric also demonstrated extreme wetting characteristics for non-polar liquids, as mentioned above, which could also have a significant impact on the respective antibacterial effects. Typically, the surface charge plays an important role in the interactions between materials. Bacteria can be adsorbed onto a material surface under electrostatic action, including Coulomb and van der Waals attractive forces. Usually, bacteria cells are negatively charged and coated with a peptidoglycan layer of sugars and amino acids [14]. Super-hydrophilic and -hydrophobic surfaces could yield low bacterial adhesion, affecting the antibacterial effects [9,14,15]. Recently, Yang et al. reported the antibacterial effects of a nanodiamond coating with C=O functional groups, and the appearance of C=O can also affect the antibacterial activity. They explained the antibacterial effect that the C=O bonds damaged the cell wall through the blockage and the repulsive electric force [37]. Thus, it was thought that the antibacterial activity for CN-formed NW-PP fabrics originated from the strong hydrophobicity due to the CH_3 bond of NW-PP fabric, the enhanced wetting property due to CN bonds, and the antibacterial activity due to C=O bonds. Consequently, the proposed method is a one-step process that is simple, intuitive, and inexpensive. This is an eco-friendly method not requiring harmful precursors or toxic gases. Moreover, this method can be applied to most weak substrates because there is little damage to the sample. This suits mass production and provides an eco-friendly approach to producing antibacterial fabrics enhancing the antibacterial effect [9,14,15].

4. Conclusions

The non-woven polypropylene (NW-PP) fabric was plasma-treated to modify the surface properties using a modified DC-pulsed sputtering method with a roll-to-roll system.

Compared to the commercially available sputtering gun, the configuration of the magnet was modified into alternately repeated N-polarity lines and S-polarity lines, which formed a magnetic field parallel to the substrate. After Ar/N_2–C plasma treatments, there were no structural damages in the CN-formed NW-PP fabrics. It was confirmed that the combination of the modified sputtering gun and the DC-pulsed power could minimize damage to the surface. From XPS results, the chemical bonds on the surface were converted from C–C/C–H into C–C/C–H, C–N, and C=O bonds. CN-formed NW-PP exhibited a strong hydrophobicity for H_2O (polar liquid) and full-wetting characteristics for CH_2I_2 (non-polar liquid). In addition, the CN-formed NW-PP fabric showed enhanced antibacterial properties compared to the NW-PP fabric. The reduction rate of the CN-formed NW-PP fabric was 89.0% and 91.6% for *Staphylococcus aureus* (ATCC 6538, Gram-positive) and *Klebsiella pneumoniae* (ATCC4352, Gram-negative), respectively, confirming that the CN layer represented antibacterial characteristics against both Gram-positive and Gram-negative bacteria. To sum up, it was concluded that the antibacterial activity for CN-formed NW-PP fabrics originated from the strong hydrophobicity due to CH_3 bonds of NW-PP fabric, the enhanced wetting property due to CN bonds, and the antibacterial activity due to C=O bonds. However, it is thought that more research is needed to clarify the antibacterial effect. The proposed method is a one-step process that is simple, intuitive, and inexpensive, and does not involve any harmful precursors or toxic gases. The roll-to-roll process is suitable for the mass processing of flexible substrates, including polymers, fabrics, and polymer-like materials. Moreover, this eco-friendly method can be applied to the majority of weak substrates available because it does not cause any significant damage to the sample. Consequently, we believe that our study presents an effective eco-friendly method that allows the mass production of antibacterial fabrics.

Author Contributions: Conceptualization, formal analysis, and validation, H.T.K. and Y.-S.S.; methodology, investigation, and software, S.K.J. and S.-Y.L.; writing—original draft preparation and visualization, H.T.K. and S.K.J.; writing—review and editing, S.-Y.L. and Y.-S.S.; funding acquisition, S.-Y.L. and H.T.K. All authors have read and agreed to the published version of the manuscript.

Funding: This research was supported by the Basic Science Research Program through the National Research Foundation of Korea (NRF) funded by the Ministry of Education (NRF-2018R1D1A1B07049180, NRF-2019R1I1A1A01059259).

Institutional Review Board Statement: Not applicable.

Data Availability Statement: Not applicable.

Conflicts of Interest: "Sang Kooun Jung" in this manuscript was employed by the company. The remaining authors declare that the research was conducted in the absence of any commercial or financial relationships that could be construed as a potential conflict of interest.

References

1. Mansurogu, D.; Uzun-Kaymak, U. Argon and nitrogen plasma modified polypropylene: Surface characterization along with the optical emission results. *Surf. Coat. Technol.* **2019**, *358*, 551–559. [CrossRef]
2. Zille, A.; Oliveira, F.R.; Souto, A.P. Plasma treatment in textile industry. *Plasma Process. Polym.* **2015**, *12*, 98–131. [CrossRef]
3. Gomathi, N.; Sureshkumar, A.; Neogi, S. RF plasma-treated polymers for biomedical applications. *Curr. Sci.* **2008**, *94*, 1478–1486.
4. Salah, I.; Parkin, I.P.; Allan, E. Copper as an antimicrobial agent: Recent advances. *RSC Adv.* **2021**, *11*, 18179–18186. [CrossRef]
5. Yin, I.X.; Zhang, J.; Zhao, I.S.; Mei, M.L.; Li, Q.; Chu, C.H. The antibacterial mechanism of silver nanoparticles and its application in dentistry. *Int. J. Nanomed.* **2020**, *15*, 2555–2562. [CrossRef]
6. Swathy, J.R.; Sankar, M.U.; Chaudhary, A.; Aigal, S.; Anshup; Pradeep, T. Antimicrobial silver: An unprecedented anion effect. *Sci. Rep.* **2014**, *4*, 7161. [CrossRef] [PubMed]
7. Gudkov, S.V.; Burmistrov, D.E.; Serov, D.A.; Rebezov, M.B.; Semenova, A.A.; Lisitsyn, A.B. A Mini Review of Antibacterial Properties of ZnO Nanoparticles. *Front. Phys.* **2021**, *9*, 641481. [CrossRef]
8. Kessler, A.; Hedberg, J.; Blomberg, E.; Odnevall, I. Reactive Oxygen Species Formed by Metal and Metal Oxide Nanoparticles in Physiological Media-A Review of Reactions of Importance to Nanotoxicity and Proposal for Categorization. *Nanomaterials* **2022**, *12*, 1922. [CrossRef]
9. Luan, Y.; Liu, S.; Pihl, M.; van der Mei, H.C.; Liu, J.; Hizal, F.; Choi, C.-H.; Chen, H.; Ren, Y.; Busscher, H.J. Bacterial interactions with nanostructured surfaces. *Curr. Opin. Colloid Interface Sci.* **2018**, *38*, 170–189. [CrossRef]

10. Gutiérrez, J.M.; Conceição, K.; de Andrade, V.M.; Trava-Airoldi, V.J.; Capote, G. High antibacterial properties of DLC film doped with nanodiamond. *Surf. Coat. Technol.* **2019**, *375*, 395–401. [CrossRef]
11. Antunes, J.; Matos, K.; Carvalho, S.; Cavaleiro, A.; Cruz, S.M.A.; Ferreira, F. Carbon-based coatings in medical textiles surface functionalisation: An overview. *Processes* **2021**, *9*, 1997. [CrossRef]
12. Kong, X.; Liu, X.; Zheng, Y.; Chu, P.K.; Zhang, Y.; Wu, S. Graphitic carbon nitride-based materials for photocatalytic antibacterial application. *Mater. Sci. Eng. R-Rep.* **2021**, *145*, 100610. [CrossRef]
13. Li, H.; Ding, M.; Luo, L.; Yang, G.; Shi, F.; Huo, Y. Nanomesh-structured graphitic carbon nitride polymer for effective capture and photocatalytic elimination of bacteria. *ChemCatChem* **2020**, *12*, 1334–1340. [CrossRef]
14. Li, W.; Thian, E.S.; Wang, M.; Wang, Z.; Ren, L. Surface design for antibacterial materials: From fundamentals to advanced strategies. *Adv. Sci.* **2021**, *8*, 2100368. [CrossRef]
15. Yuan, Y.; Hays, M.P.; Hardwidge, P.R.; Kim, J. Surface characteristics influencing bacterial adhesion to polymeric substrates. *RSC Adv.* **2017**, *7*, 14254–14261. [CrossRef]
16. Kim, H.T.; Jung, S.K.; Kim, D.-E.; Park, C.Y.; Lee, S.-Y. Wettability control of paper through substitution between the hydroxyl group and carbon elements using argon-carbon plasma treatment. *Vacuum* **2022**, *205*, 111398. [CrossRef]
17. Kim, H.T.; Jung, S.K.; Lee, S.-Y. Properties of ITO films deposited on paper sheets using a low-frequency (60 Hz) DC-pulsed magnetron sputtering. *Vacuum* **2021**, *187*, 110056. [CrossRef]
18. Finite Element Method Magnetics. Available online: https://www.femm.info/wiki/HomePage (accessed on 26 May 2023).
19. Rasband, W.S.; ImageJ, U.S. National Institutes of Health, Bethesda, Maryland, USA. 1997–2018. Available online: https://imagej.nih.gov/ij/ (accessed on 9 June 2023).
20. Yu, X.; Hörst, S.M.; He, C.; McGuiggan, P.; Kristiansen, K.; Zhang, X. Surface energy of the Titan aerosol analog "tholin". *Astrophys. J.* **2020**, *905*, 88. [CrossRef]
21. *KS K, 0693: 2016*; Test Method for Antibacterial Activity of Textile Materials. Korean Agency for Technology and Standards: Maengdong-myeon, Republic of Korea, 2016.
22. National Institute of Standards and Technology (NIST). Atomic Spectra Database Data. Available online: https://www.nist.gov/pml/atomic-spectra-database (accessed on 9 June 2023).
23. Kim, H.T.; Kim, C.D.; Pyo, M.S.; Park, C. Electrical and optical characteristics of Ar plasma generated by low-frequency (60Hz) power source. *Korean J. Chem. Eng.* **2014**, *31*, 1892–1897. [CrossRef]
24. Kim, H.T.; Nquyen, T.P.N.; Park, C. Damage-free treatment of ITO films using nitrogen-oxygen (N_2-O_2) molecular DC plasma. *Curr. Photovolt. Res.* **2015**, *3*, 112–115.
25. Naz, M.Y.; Shukrullah, S.; Rehman, S.U.; Khan, Y.; Al-Arainy, A.A.; Meer, R. Optical characterization of non-thermal plasma jet energy carriers for effective catalytic processing of industrial wastewaters. *Sci. Rep.* **2021**, *11*, 2896. [CrossRef]
26. Kelly, P.J.; Bradley, J.W. Pulsed magnetron sputtering—Process overview and applications. *J. Optoelectron. Adv. Mater.* **2009**, *11*, 1101–1107.
27. Greene, J.E. Review Article: Tracing the recorded history of thin-film sputter deposition: From the 1800s to 2017. *J. Vac. Sci. Technol. A* **2017**, *35*, 05C204. [CrossRef]
28. Wasa, K.; Kanno, I.; Kotera, H. *Handbook of Sputter Deposition Technology*; William Andrew: Waltham, MA, USA, 2012.
29. Kim, H.T.; Kim, C.D.; Min, B.K.; Park, C. Growth and network structured carbon-doped GaN (GaN:C) nanowires using modified metal-organic vapor deposition. *Vacuum* **2016**, *123*, 82–85. [CrossRef]
30. Kim, H.T.; Kim, C.; Kim, C.-D.; Park, C. Formation of a Mo_2N skin layer in columnar-structural Mo films using NH_3 plasma nitridation as the Se diffusion-barrier layer. *Surf. Coat. Technol.* **2016**, *302*, 463–467. [CrossRef]
31. Law, K.-Y. Definitions for Hydrophilicity, Hydrophobicity, and Superhydrophobicity: Getting the Basics Right. *J. Phys. Chem. Lett.* **2014**, *5*, 686–688. [CrossRef]
32. Reece, J.B.; Urry, L.A.; Cain, M.L.; Wasserman, S.A.; Minorsky, P.V.; Jackson, R.B. *Camp-Bell Biology*, 10th ed.; Pearson Education Inc.: London, UK, 2020.
33. Ren, Z.; Chen, G.; Wei, Z.; Sang, L.; Qi, M. Hemocompatibility evaluation of polyurethane film with surface-grafted poly(ethylene glycol) and carboxymethyl-chitosan. *J. Appl. Polym. Sci.* **2013**, *127*, 308–315. [CrossRef]
34. Lloyd, W.C., III. What is the difference between Gram-positive and Gram-negative bacteria? Medical News Today. Available online: https://www.medicalnewstoday.com/articles/gram-positive-vs-gram-negative (accessed on 26 May 2023).
35. Jung, S.; Yang, J.Y.; Jang, D.; Kim, T.; Baek, K.H.; Yoon, H.; Park, J.Y.; Kim, S.K.; Hong, J.; Ryoo, S.; et al. Sustainable antibacterial and antiviral high-performance copper-coated filter produced via ion beam treatment. *Polymers* **2022**, *14*, 1007. [CrossRef]
36. Deng, X.; Nikiforov, A.Y.; Coenye, T.; Cools, P.; Aziz, G.; Morent, R.; Geyter, N.D.; Leys, C. Antimicrobial nano-silver non-woven polyethylene terephthalate fabric via an atmospheric pressure plasma deposition process. *Sci. Rep.* **2015**, *5*, 10138. [CrossRef]
37. Yang, Y.; Zhang, Y.; Wang, L.; Miao, Z.; Zhou, K.; Yang, Q.; Yu, J.; Li, X.; Zhang, Y. Anti-bacterial Property of Oxygen-Terminated Carbon Bonds. *Adv. Func. Mater.* **2022**, *32*, 2200447. [CrossRef]

polymers

MDPI

Article

Sputtering Plasma Effect on Zinc Oxide Thin Films Produced on Photopolymer Substrates

Juan Jesus Rocha-Cuervo [1], Esmeralda Uribe-Lam [2,*], Cecilia Daniela Treviño-Quintanilla [2,3] and Dulce Viridiana Melo-Maximo [1]

[1] Tecnologico de Monterrey, School of Engineering and Sciences, Estado de Mexico Campus, Atizapán 52926, Mexico; a01752555@tec.mx (J.J.R.-C.); virimelo@tec.mx (D.V.M.-M.)
[2] Tecnologico de Monterrey, School of Engineering and Sciences, Queretaro 76130, Mexico; cdtrevino@tec.mx
[3] Tecnologico de Monterrey, Institute of Advanced Materials for Sustainable Manufacturing, Ave. Eugenio Garza Sada 2501, Monterrey 64849, Mexico
* Correspondence: euribelam@tec.mx

Abstract: This work presents a post-cured treatment alternative for photopolymer substrates considering the plasma produced via the sputtering process. The sputtering plasma effect was discussed, analyzing the properties of zinc/zinc oxide (Zn/ZnO) thin films deposited on photopolymer substrates, with and without ultraviolet (UV) treatment as a post-treatment process, after manufacturing. The polymer substrates were produced from a standard Industrial Blend resin and manufactured using stereolithography (SLA) technology. After that, the UV treatment followed the manufacturer's instructions. The influence of the sputtering plasma as an extra treatment during the deposition of the films was analyzed. Characterization was performed to determine the microstructural and adhesion properties of the films. The results showed the effect of plasma as a post-cured treatment alternative: fractures were found in thin films deposited on polymers with previous UV treatment. In the same way, the films showed a repetitive printing pattern due to the phenomenon of polymer shrinkage caused by the sputtering plasma. The plasma treatment also showed an effect on the thicknesses and roughness values of the films. Finally, according to VDI–3198 standards, coatings with acceptable adhesion failures were found. The results provide attractive properties of Zn/ZnO coatings on polymeric substrates produced by additive manufacturing.

Keywords: plasma; UV treatment; sputtering; zinc oxide (ZnO); thin films; stereolithography (SLA); photopolymerization

Citation: Rocha-Cuervo, J.J.; Uribe-Lam, E.; Treviño-Quintanilla, C.D.; Melo-Maximo, D.V. Sputtering Plasma Effect on Zinc Oxide Thin Films Produced on Photopolymer Substrates. *Polymers* **2023**, *15*, 2283. https://doi.org/10.3390/polym15102283

Academic Editor: Choon-Sang Park

Received: 20 April 2023
Revised: 9 May 2023
Accepted: 10 May 2023
Published: 12 May 2023

1. Introduction

Surface engineering produces thin films to improve one or several surface properties of solid materials [1]. Chemical vapor deposition (CVD), Physical vapor deposition (PVD), and sol–gel are examples of technologies used to produce thin films. In addition, several techniques for thin film production are based on PVD [2–4], such as Reactive magnetron sputtering (PVD-RMS). PVD-RMS allows the production of high-density coatings (e.g., metallic, ceramic, etc.) with excellent mechanical properties on several substrates (e.g., polymeric, metallic, ceramic, etc.) to provide unique surface properties in materials for various engineering applications [5,6].

PVD-RMS is an assisted plasma technology that requires the formation of a plasma environment, in which the sputtered particles will be transported to produce a thin film on a substrate. The plasma in the PVD-RMS technique is composed of ionized gas, pulverized particles of the target in constant collision, secondary electrons, UV light, visible light, and heat [7–11].

Some research discussed the influence of the substrate surface finish and the topology on the thin film growth pattern. In coating growth phenomena, the particles of the film adhere and adapt to the substrate surface where they are deposited, mimicking the substrate's surface topology [12–14]. Types of substrate materials used in PVD-RMS include metals, glass, polymers, and composites. In the case of polymeric materials, they can be used as substrates for the deposition of thin films regardless of their production method. For example, polymers developed in additive manufacturing (AM) by three-dimensional (3D) printing techniques have been used in different types of analyses to improve surface properties by adding a thin film [15–17].

Polymers could be processed via several AM techniques, such as stereolithography (SLA), digital light processing (DLP), and continuous liquid interface production (CLIP). The SLA technique is based on the photopolymerization principle to develop 3D designs into prototypes and functional objects [18]. A UV light source, usually a laser, initiates a chain reaction on a photocurable liquid resin. The laser polymerizes and solidifies one layer at a time until the desired solid part is obtained [19]. The initial polymerization process via the printer laser provides solid specimens with an excellent aesthetic finish, fine resolution (as low as 5 μm), and adequate handling strength that can be directly used for different applications [20]. The patterns produced on the final sample via SLA printers, presented in Figure 1a, are not fully polymerized from exposure to UV light in the printing process, resulting in a small volume of unpolymerized resin throughout the matrix of the pattern [21].

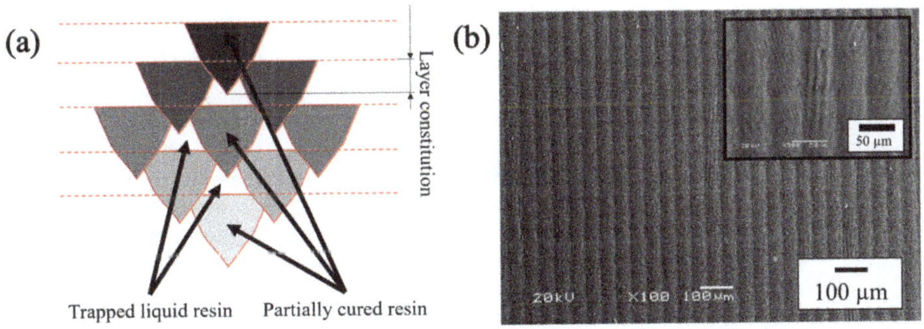

Figure 1. The final properties of the polymer substrate change between the surface and along the 3D structure. (**a**) Polymer showing the difference polymerized portions throughout the pattern; (**b**) Surface patterns from a polymer sample seen by backscattered electron imaging on scanning electron microscopy (SEM-BSE) with low and high magnification.

The pattern is known as the "green-state structure" and is characterized as being a material with different curing portions throughout the entire structure. Since the final piece is not fully polymerized, a post-cure method is often recommended to improve the final properties of the AM part [20–24].

Research analysis and experimentation determine that the plasma environment in sputtering techniques can reach temperatures between 533 and 813 K [7,9–11,25–27]. The heat produced via temperature variation can transform the surface of the polymeric substrate used for thin film production [24,28]. Considering that photosensitive polymers are susceptible to changes in properties due to high temperatures, it is important to consider these materials' limitations when used as substrates to produce thin films via plasma-assisted technologies [29]. Therefore, photosensitive resins receive a post-cure process that can be replaced during the PVD-RMS thin film production due to the nature of plasma.

This article aims to analyze the effect of plasma produced via the PVD-RMS technique, as an alternative post-curing treatment compared to the conventional UV treatment, on the microstructural and adhesion properties of a Zn/ZnO thin film on a photosensitive polymer substrate. The produced thin films were characterized by optical microscopy (OM), scanning electron microscopy (SEM) equipped with energy dispersive x-ray spectroscopy

(EDS), and atomic force microscopy (AFM). The coating substrate's adhesion was measured using a Rockwell durometer, following the VDI 3198 standard.

2. Materials and Methods

Cylindrical-shaped polymeric substrates (25 × 5 mm) were produced using an acrylic and glycol diacrylate-based photopolymer resin (Industrial Blend® developed by FunToDo®) on a stereolithography printer (Elegoo Mars UV Photocuring printer LCD MSLA, Shenzhen, China). The summary of the printed parameters is in Table 1.

Table 1. Printer parameters used for polymer substrate production.

Production Parameter	Units
Technology	LED Display Photocuring
Light source	Integrated UV light (405 nm)
XY axis resolution	0.0047 mm (2560 × 1440 px)
Z axis accuracy	0.00125 mm
Thickness per layer	0.05 mm
Exhibition time	8 s
Lower exposure time	60 s
Print speed	22.5 mm/h
Total printing time per lot	~30 min

The printed substrates were cleaned with an ultrasonic bath (Ultrasonic Cleaner Model JPS-10A, Shenzhen, Guangdong, China) in isopropanol (99.999%) for 20 min and dried with pressurized air. A UV treatment using UV light (90–260 V UV lamp, 405 nm, 60 W light effect, Shenzhen Enomaker Technology Co., Ltd.; Longhua Dist, Shenzhen, China) for 10 min was performed on substrate 1 as the manufacturer recommended [30], and substrate 2 was directly used after the dried stage in the printing process.

Zn/ZnO coatings were deposited on substrates 1 and 2 via Physical Vapor Deposition with unbalanced Reactive Magnetron Sputtering technology (PVD-RMS) in Ar and O_2 mixed atmosphere. The deposition equipment and the principal components are presented in Figure 2; details of the system have been described elsewhere [31].

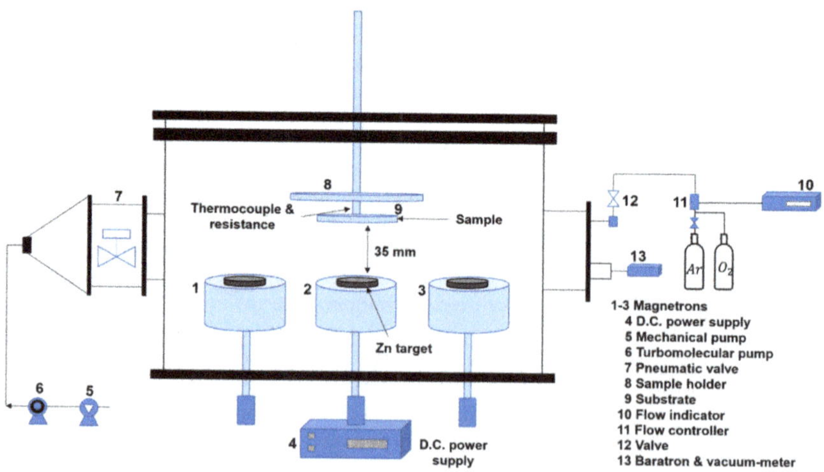

Figure 2. Diagram of the deposition setup and its main components.

Circular zinc (Zn) plate with a purity of 99.99% was selected as a target [2″ (5.08 cm) diameter × 0.25″ (0.635 cm) thick]. The polymeric substrates were placed into the reactor at ~30 mm distance from the target. After placing the substrates, the sputtering chamber was evacuated to a base pressure of ~1×10^{-4} Torr (1.3×10^{-2} Pa). Argon was introduced into the chamber at a constant flow rate of 20 SCCM (standard cubic centimeter per minute), leading to an increase in pressure around 4 Pa. In these conditions, a power of 40 W was then applied to the magnetron to initiate the process, maintaining the substrate at room temperature. The target was cleaned in these conditions for 5 min, and the pressure was adjusted to 2 Pa. A pure Zn adhesion interlayer was deposited at 40 W for 1 min in a pure Ar atmosphere. The power was reduced to 30 W to produce the ZnO layer, and the oxygen was introduced into the chamber at a constant flowrate of 7.3 SCCM. The ZnO layer was deposited for 5 min in an argon and O_2 mixed atmosphere. The total deposition time was 6 min for all the samples.

Optical microscopy (OM), Olympus Corporation, Shinjuku Tokyo, Japan, was used for the superficial inspection of samples (PGM 3 Olympus microscope) for characterization analysis. The morphology examination of the Zn/ZnO films was performed via scanning electron microscopy (SEM JEOL JSM-6360 LV) Jeol, Ltd. Akishima, Tokio, Japan, equipped with energy-dispersive spectroscopy analysis (EDS) Oxford Instruments, Santa Barbara, CA, USA, for coating chemical composition evaluation. Surface topography was characterized via atomic force microscopy (AFM—Park Systems model XE7) Park Systems Corp, Suwon-si, Gyeonggido, Korea, operating in Non-Contact Tapping mode, using a Si probe, and scanning areas of 20×20 and 5×5 μm^2. Adhesion of the films was evaluated by the VDI 3198 standard, using a Rockwell durometer (Louis Small model 8SSA) Louis Small Inc., Cincinnati, OH, USA; a load of 150 kg was applied with a diamond tip in the C scale.

3. Results and Discussion

3.1. Effect of UV Treatment on the Surface Properties of Zn/ZnO Thin Films Deposited on Photosensitive Polymeric Substrates

The samples were examined after the Zn/ZnO thin film deposition to analyze the effect of the PVD-RMS plasma on the surface properties of the films. Polymeric substrates were used, with 10 min of UV treatment post-process after manufacturing [30] and as manufactured without UV treatment, as mentioned in Section 2, "Materials and Methods". Subsequently, the substrates were subjected to a PVD-RMS process assisted with plasma to produce the Zn/ZnO thin film.

Figure 3 presents the OM images, at high magnification, of the ZnO thin films deposited on the polymeric substrates used, showing the characteristic surface topology produced via the SLA printer on the substrate surface of the samples. The results presented a difference between the analyzed thin films; the thin film deposited on the UV-treated polymeric substrate presents fracture cracks along the surface of the sample. In contrast, the thin film deposited on the not treated polymeric substrate presents a pattern without changes after deposition.

After applying 10 min of UV treatment, as mentioned in Section 2, "Materials and Methods" [30], fractures were presented along the sample's surface. The fracture cracks shown in Figure 3a were related to a deterioration of the polymeric substrates due to an abrupt change in temperature of the samples by direct contact with the PVD-RMS plasma environment, also associated with over-curing phenomena [21,32–38]. The ZnO thin film deposited on the photosensitive polymeric substrate without UV treatment is shown in Figure 3b. Due to growth phenomena, the thin film adheres and adapts to the substrate surface without modifying the characteristic printing pattern [12–14]. In the same way, the thin film looks homogeneous in all sections without the presence of apparent fractures.

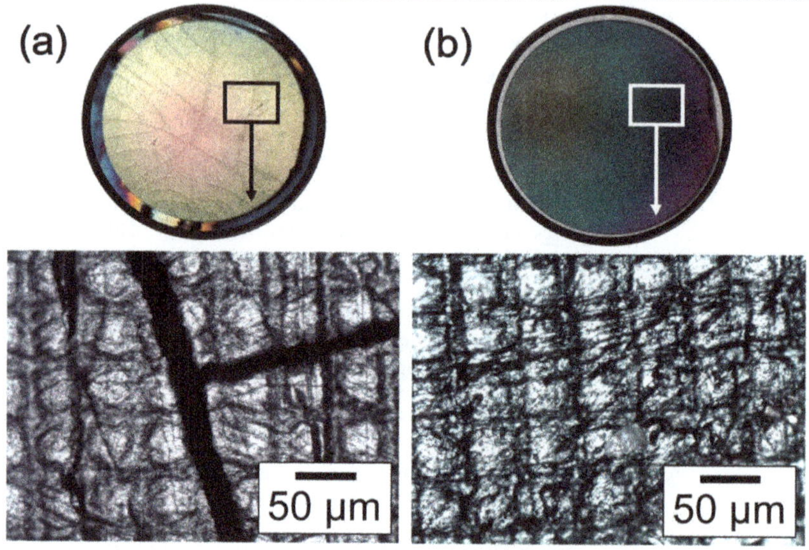

Figure 3. Macroscopic view and OM images show differences on the sample's surfaces. Polymer substrate with (**a**) 10 min of UV treatment and (**b**) without UV treatment.

Post-cure treatments can increase the mechanical properties of photosensitive polymers via polymerizing uncured resin portions trapped within the patterns and promoting the polymerization of the already cured in the "green-state structure", thus increasing the crosslinking density [23,39]. Nevertheless, restrictions may apply when UV treatments are replaced by heat treatments concerning other polymer properties such as heat deflection resistance, temperature resistance, and thermal shock resistance [21,29]. Several studies [29,40,41] have demonstrated that long exposure UV treatments do not produce remarkable changes in the final mechanical properties of photosensitive polymers, i.e., compressive strength, tensile strength, and Young's modulus reach a limit after several hours of UV treatment. However, in the case of heat treatments, if temperatures are raised above the thermal limit of the photosensitive polymers, or the photosensitive polymers are exposed to abrupt changes in temperature, fractures may appear [21]. Various studies [28,29,42] have shown that when green-state pieces are cured at high temperatures for short times, shrinkage produces fractures that weaken the final and surface properties of the desired polymer part. Therefore, it becomes necessary to increase the temperature of the post-cure treatment based on the limitations of the polymer used [22,29].

As the plasma in the PVD-RMS technique is composed mainly of ionized gas, the free electrons in its constitution move at very high velocities because of lower mass and high temperature. These energetic electrons can deliver significant power, either in the form of UV radiation or temperature, thus resulting in substrate heating during the deposition process [43,44].

According to the manufacturer, Industrial Blend® resin polymers are characterized as withstanding temperatures up to 498 K [30]. However, as some studies have shown [21,25,29,30], the plasmas in the sputtering technique can reach temperatures between 533 and 813 K. Therefore, it is suggested that the plasma–substrate interaction produced a thin film with degraded surface quality, as shown in Figure 3a, generated by changes in substrate temperature upon contact with the PVD-RMS plasma, leading to high residual stresses and originating the observed fractures.

Invariably PVD-RMS films have residual stresses that arise from the growth processes [45]. Nevertheless, it has been registered that differences in the thermal coefficients of expansion of the film and substrate in high-temperature depositions may cause plastic deformation, cracking the thin film or the substrate, or cracking at the substrate–film interface [10,45]. Although in this research thin films were not deposited at high temperatures, the energy and heat generated by the nature of the PVD-RMS plasma inevitably produced changes in the polymeric substrate, inducing constant modifications on the thin film during its deposition, allowing the high accumulation of stresses, and originating the observed fractures.

Microstructural characterization and surface analysis of as-deposited Zn/ZnO thin films were conducted via SEM technology to determine how the PVD-RMS plasma conditions change the coating topography of Zn/ZnO thin films deposited on both; a photosensitive polymeric substrate with 10 min of UV treatment before the deposition of the thin film, and a polymeric substrate directly used without UV treatment. Figure 4a,b presents backscattered electron (BSE) SEM micrographs, at high magnification, to compare the main differences found in topography. Regardless of changes in UV treatment, dense, compact, and well-adhered Zn/ZnO films were obtained; neither fractures nor detachment was found in the thin film in the analysis.

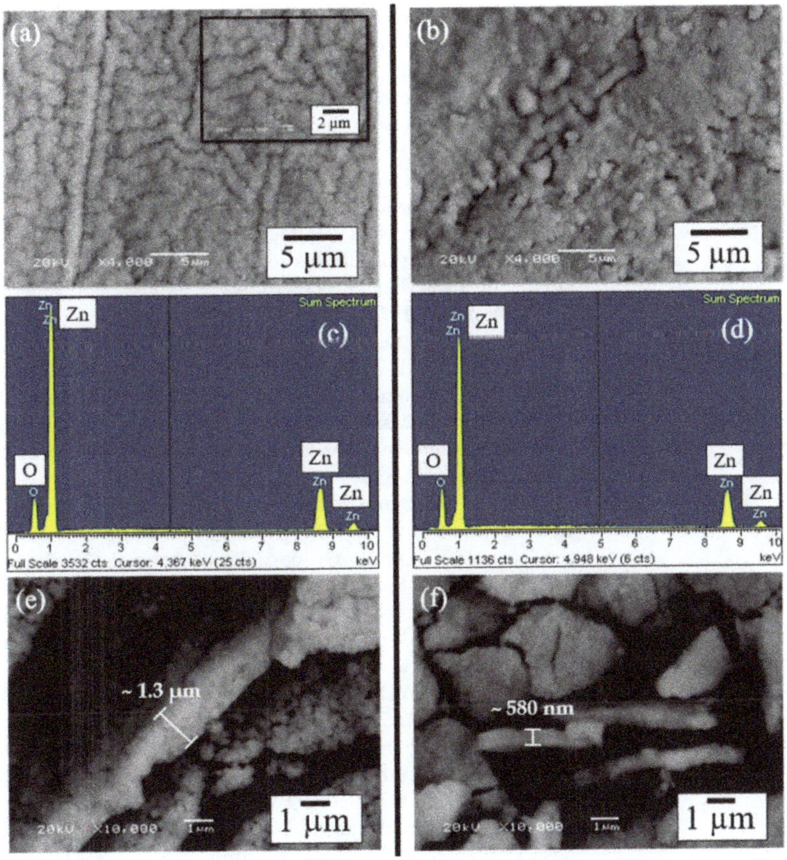

Figure 4. SEM-BSE surface views at high magnification showing differences in the film's surface topology. (**a**) Polymer with 10 min of UV treatment exhibits a homogeneous growth, and (**b**) polymer without UV treatment presents a topology with an irregular pattern. Elemental mapping via EDS in the area. (**c**) Polymer with 10 min of UV treatment and (**d**) polymer without UV treatment. SEM micrographs showing the thickness of the samples. (**e**) Polymer with 10 min of UV treatment presents a ~1.3 µm thickness, and (**f**) polymer without UV treatment presents a ~580 nm thickness.

As observed from SEM images, it was determined that the topology of the films changed depending on the substrate treatment. From Figure 4a, a homogeneous growth in the form of a labyrinth is observed, presenting a lower density since not only a more significant increase in roughness can be observed, but also a greater separation between the columns and the number of particulates on the surface becomes noticeable [46,47]. This result shows a more arranged topology when the thin film is deposited on a substrate with 10 min of UV treatment. Figure 4b presents a topography with an irregular pattern produced by the roughness of the substrate, leading to the formation of agglomerates on the surface of the film. A more homogeneous growth was presented since the surface is observed to be smooth without microdefects such as fractures or the growth of macroparticles.

The homogeneous topology growth presented in Figure 4a is related to a shrinkage phenomenon of the polymeric substrates due to the temperature changes produced via the interaction with the PVD-RMS plasma environment [24,28,29,38]. The most common post-cured treatments reported for photosensitive polymers are UV and heat treatments [21,22,28,40,42,48,49]. Different studies [22,29] have shown that any post-cure treatment is effective in the obtention of a polymer with improved final properties compared to a green-state polymer structure. However, the interaction between the green-state polymer and any post-cure treatment results in an undesired effect known as shrinkage [24,29,48].

The shrinkage phenomenon is inherent to the curing of methacrylate-based photopolymers [18,50]. Due to the polymerization process, the start liquid resin is converted into a solid, resulting in a density change that reduces the overall volume, producing volumetric shrinkage upon curing and drying [39,50]. These phenomena lead to internal stresses and distort consequences [24,29,38,48]. Afterward, with a post-cured process, a second polymerization process is generated throughout the entire structure of the as-printed part, particularly in the areas that have different curing portions, as shown in Figure 1, increasing the distortion and internal stresses produced via the first polymerization, reducing the surface finish, and magnifying the surface roughness values of the final part [18,20,28,38,42].

Compared to UV post-cured treatments [21,22,28,40,42,48,49], studies have established that particularly thermal post-curing treatments reduce heterogeneity and anisotropy, inducing undesired higher shrinkage strains and representing an accelerated aging mechanism [51], consequently, increasing the shrinkage and the risk of cracking [52]. The shrinkage of successively built layers produces residual stresses that accumulate, generating strain deformations leading to a considerable curl-distortion of the multi-layered polymer parts [41,53].

One of the main characteristics of PVD-RMS plasma is the generation of heat; the ionized gas realizes high temperatures via the high-velocity movement of free electrons, the interaction of sputtered particles with the gas results in gas heating, and the constant bombardment of the sputtered atoms during the deposition process is released in form of heat in the substrate [7–9,25,43]. Therefore, the result of the homogeneous topology was not only the result of this accumulation of stresses, but also the result of the continuous bombardment of the molecules that form the thin film, added to the working temperature of the reactive PVD-RMS plasma, and the way the film mimics the topography of the substrate [28,41,42,51,54].

Figure 4c,d presents the EDS spectrum corresponding to the samples analyzed, showing no presence of the substrate elements with the characteristic peaks attributed to the thin film, Zn and O. The effect of the PVD-RMS plasma on the thicknesses of the Zn/ZnO samples was also analyzed via SEM. Figure 4e,f shows the SEM BSE images of the thicknesses obtained in Zn/ZnO thin films. Thicknesses between ~1.3 μm and ~580 nm were obtained for the Zn/ZnO films when they were deposited on substrates with and without UV treatment, respectively. A greater thickness was obtained when the thin films were deposited on photosensitive polymeric substrates with 10 min of UV treatment, as seen in Figure 4e. A study carried out by R. Castro et al. [55] suggests that the phenomenon of higher thicknesses is related to a higher surface roughness of the substrate.

A study performed by Zhao J. et al. [41] showed that photopolymer post-cured parts tend to have higher surface roughness values compared to green-state parts due to the shrinkage phenomena [18,24,29,48,50]. Although Zn and ZnO materials are deposited via PVD-RMS under the same conditions on both substrates; differences in roughness caused via the shrinkage phenomena and post-cure treatments resulted in the sputtered atoms being deposited under different growth dynamics caused by greater availability of the substrate area presented for deposition due to the increase in roughness. The results obtained by R. Castro et al. [55] reveal a possible explanation for the phenomenon presented in the thickness differences obtained for Zn/ZnO thin films, related to an increase in roughness caused via the shrinkage phenomena accumulation of photosensitive polymeric substrates with 10 min of UV treatment.

To analyze the effect of the UV treatment on the microstructural properties of the Zn/ZnO thin films and the changes produced via the PVD-RMS plasma, an uncoated photosensitive polymeric substrate was used to examine the differences found in the rough surface in comparison with the Zn/ZnO thin films using 3D AFM. Additionally, the coated samples were analyzed to compare growth similarities, according to Thornton's diagram, and to observe if PVD-RMS plasma treatment modifies the growth of the thin films deposited on polymeric substrates with different UV treatments [56]. Figure 5 presents the 3D topology images obtained via AFM, together with the surface roughness parameters (in values of 'Rq' = root mean square roughness), made on a scan area of 400 μm^2. Approximate values of surface roughness (Rq) of 0.0241, 0.0956, and 0.205 μm were obtained for the naked substrate (a) and the thin films ((b) and (c)), respectively.

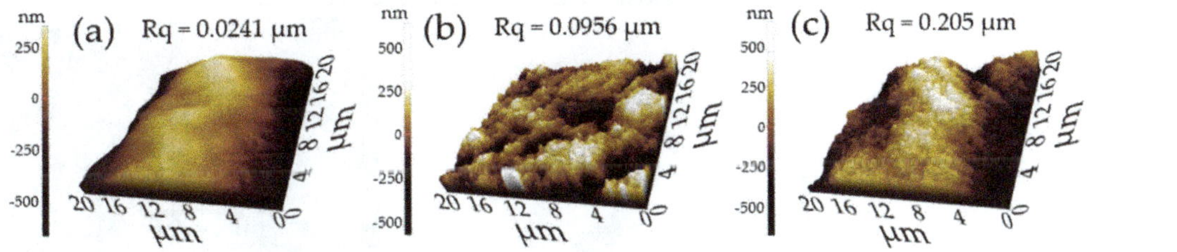

Figure 5. Three-dimensional AFM images (20 μm × 20 μm) of (**a**) uncoated polymer substrate, (**b**) Zn/ZnO thin film on polymer substrate without UV treatment, and (**c**) Zn/ZnO thin film on a polymer substrate with 10 min of UV treatment in order to compare the changes of the roughness parameters between samples.

The smoothest surface corresponds to the uncoated substrate in Figure 5a. Subsequently, due to the growth nature of the films, the surface roughness of the samples shown in Figure 5b,c increases as a result of the formation of ZnO columns with irregular boundaries and amorphous structure, related to zones 1a/1b, based on the thin film growth model presented by Thornton, characterized as being developed at low Ar pressures and low temperatures [46,47,56,57]. The same growth characteristics were observed in the Zn/ZnO films produced, showing that the substrate's post-curing treatment does not affect the type of growth of the thin films. Nevertheless, the thin film (c) produced over a polymeric substrate with UV treatment results in higher values of the surface roughness of the samples.

The increased roughness of sample (c) was related to the surface deformation caused by the shrinkage stress accumulated during the polymer substrate manufacture and post-processing. The work by D. Karalekas et al. [42] explains this correlation since their research demonstrates that residual stresses generated during polymer building and post-curing are responsible for superficial creep distortions, particularly those related to thermal post-curing treatments. In this way, it can also be considered that the plasma of the PVD-RMS technique has caused surface deformation, during the deposition of thin films, due to the plasma temperatures [21,25,29,30]. The accumulation of stresses produced by the

UV treatment of sample (c) caused a surface deformation in the polymer, which due to growth phenomena, allowed the films to adhere and adapt to the substrate surface to grow with columns of different sizes that resulted in an increase in surface roughness [12–14]. Although the change in roughness was significant via the post-cured treatment of the substrates, the interaction between the plasma environment and the substrates did not produce a difference in the observed growth of the thin films.

3.2. *Effect of UV Treatment on the Adhesion Properties of Zn/ZnO Thin Films*

Rockwell marks were observed using SEM-EDS for qualitative analysis and to identify how the effect of the PVD-RMS plasma modifies the adhesion properties of the samples. Figure 6 shows the SEM-BSE images of the Zn/ZnO thin films deposited on two photosensitive polymer substrates with different UV treatments.

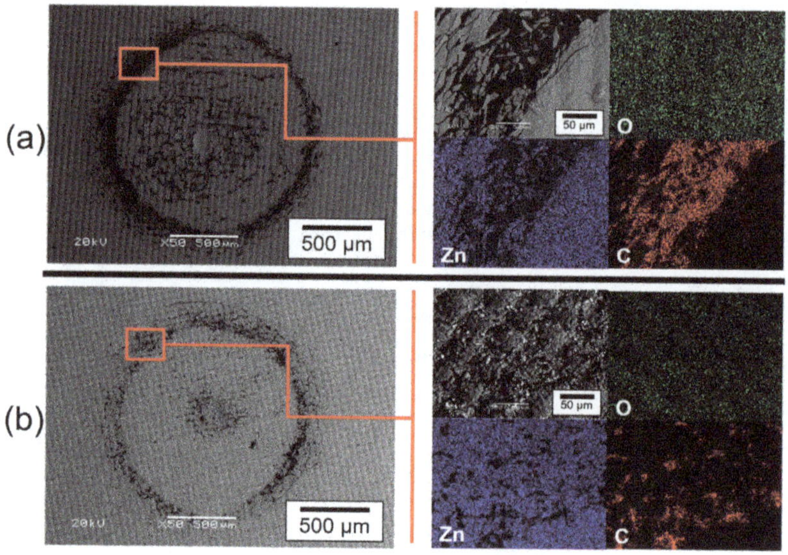

Figure 6. SEM-BSE images of the adhesion test of the Zn/ZnO thin films at low magnification, using a Rockwell durometer following the VDI 3198 standard. (**a**) Polymeric substrate with 10 min of UV treatment and (**b**) polymeric substrate without UV treatment.

Coatings exhibited acceptable failures according to the VDI 3198 standard, with the presence of cracks and delamination around the indentation marks. The same phenomenon was identified for both samples, as observed in the images obtained via SEM at low magnification. However, the degree of delamination increased slightly when the thin film was deposited on a photosensitive polymer substrate with 10 min of UV treatment.

The adhesion properties of thin films can be affected by different factors [45,58]. In the research carried out by Thouless et al. [52], it is mentioned that the adhesion of a thin film will depend on the residual stress accumulated during its production, which, in turn, is related to various deposition parameters, such as growth temperature, the temperature of the substrate, deposition rate, etc. [45]. It has been shown that those thin films that grow densely, uniformly, and with limited thickness dimensions tend to reduce the distribution of stresses during their production, thus presenting better adhesion properties [45,58,59].

However, in addition to the deposition parameters, the substrate is also a determining factor in this research. The interaction between PVD-RMS plasma with a photosensitive polymer substrate causes the substrate itself to accumulate its own stresses, affecting the interfacial bonds between substrate and film and, therefore, the final adhesion properties [59]. The origin of residual stresses is related to a discrepancy in the coefficient of thermal expansion between substrate and film. Its magnitude and size depend as well on the thickness and size of the substrate and coating [10,60–62].

Since the plasma performs a thermal post-curing treatment, its effect on the polymeric substrate reduces the superficial heterogeneity and anisotropy, inducing strains [51] and, consequently, increasing the shrinkage [52], producing residual stresses that accumulate, generating strain deformations of the multi-layered polymer parts that then were transferred to the films during the deposition. Differences in roughness caused by the shrinkage also resulted in a greater thickness obtained for sample (a), leading to high residual film stresses generated during deposition [45,58]. In addition, the columnar morphology presented with high roughness in sample (a) is generally not desirable. Local stresses can be found in films with non-homogeneous growth, allowing poor adhesion [45].

The 10 min of UV treatment implemented in only one polymeric substrate (sample a) causes them to behave as two different substrates. A study [59] has shown that two thin films of the same material, deposited under the same growth parameters, can present completely opposite adhesion properties when deposited on different substrates. Although two polymeric substrates of the same material were used in this investigation, the other thermal treatments used during their production resulted in a series of changes in the microstructural and adhesion properties of the thin films.

The image magnification was increased to carry out a deeper analysis in a specific zone of the indentation footprint [red square in Figure 6 using the elemental mapping method via EDS equipped in SEM to identify the elements present in the microfractures and delamination. Elemental mapping shows the presence of carbon in the fracture zones and a decrease in the elements zinc and oxygen. Since the element carbon comes from the substrate, the detachment of the film and the adhesion layer was confirmed. These results have demonstrated that the changes in curing treatment parameters of the sample in Figure 6a produced a fracturing effect in the adhesion tests, showing that the adhesion properties of the films decrease when photosensitive polymers with 10 min of UV treatment are used as substrates since more catastrophic failures were found, presenting extended delamination and fractures at the vicinity of the imprint, indicating a poor interfacial adhesion between the film and the substrate.

The evaluation criteria of the Rockwell test, following the VDI-3198 standard, says that the deposited Zn/ZnO coatings presented in Figure 6 belong to the category of acceptable failures; the sample in Figure 6b showed stronger interfacial bonds with fewer cracks and delamination and, therefore, a higher degree of adhesion [59,63].

The study by Vidakis et al. [59] could adequately explain the differences observed between the two polymeric substrates used with different curing treatments. Throughout this research, it has been observed that PVD-RMS plasma modifies the surface properties of a photosensitive polymeric material used as a substrate. So, because of the final thin film properties, such as increased roughness, thickness and weakening of the structure of the film, a substrate with 10 min of UV treatment is not favorable for the adhesion properties of the thin film [46].

The degree of adhesion of the films is not related to a single microstructural property but a combination of them. Therefore, the sum of the properties found on the sample (a) (greater thickness, greater roughness, and less homogeneous surface with irregular growth and fractures of the substrate) has caused the decrement in the adhesion properties of the film when it is deposited on a substrate with 10 min of UV and plasma treatment.

One fact to highlight is that, with fewer defects, the accumulation of residual stresses in the films decreases, and this promotes better adhesion. If each of the indicated microdefects has caused residual stresses not only in the substrate but also in the film, it is normal to find that sample (a) has presented the lowest degree of adhesion [38,63,64].

4. Conclusions

The effect of the plasma produced via the PVD-RMS technique as an alternative post-curing treatment on photosensitive polymeric substrates was demonstrated. Changes in

Zn/ZnO thin films varied according to the post-cured treatments of the substrates. Superficial fractures were observed in Zn/ZnO thin films deposited on a polymeric substrate with 10 min of UV treatment. The phenomenon was related to a deterioration of the substrate properties, as a consequence of an abrupt change in temperature by the interaction with the PVD-RMS plasma environment, inducing higher shrinkage strains and cracking. The plasma treatment showed an effect on the topology changes of the films. A particular homogeneous topology pattern was observed in the sample with 10 min of UV treatment. The singular topology was not only the result of residual stresses accumulated on the successively built layers but also the way the film mimics the topology of the substrate. Thicknesses between 1.3 μm and 580 nm were obtained for the Zn/ZnO films when they were deposited on substrates with and without UV treatment, respectively. The thickness differences were related to an increase in roughness caused by the shrinkage phenomena of photosensitive polymeric substrates with 10 min of UV treatment. The plasma treatment showed an effect on the roughness changes of the films. Surface roughness (Rq) of 0.205 and 0.0956 μm were obtained for the thin films deposited on substrates with and without UV treatment, respectively. The accumulation of stresses produced via the UV treatment caused a surface deformation in the polymer, allowing the films to grow with columns of different sizes that resulted in an increase in surface roughness. The columnar growth of the thin films was not altered from the plasma environment. The adhesion tests showed detachment and microfractures in the analyzed samples, thus obtaining acceptable failures according to the VDI 3198 standard for all samples. The plasma effect showed a slight decrease in the adhesion properties of the films. The sputtering technology allows a high degree of freedom regarding the deposition materials and substrates employed [65]. With the suitable changes, an alternative methodology can be applied to allow the use of polymeric substrates obtained via stereolithography, without altering the general properties of the substrate, for the deposition of thin films via the sputtering technique and its use in different applications that require the direct creation of functional prototypes [66]. Since the thin films produced present good properties in general and accepted adhesion failures, we expect that they could be suitable for applications that require high surface roughness.

Author Contributions: Conceptualization, J.J.R.-C. and C.D.T.-Q.; methodology, E.U.-L. and C.D.T.-Q.; software, D.V.M.-M.; validation, J.J.R.-C., E.U.-L., C.D.T.-Q. and D.V.M.-M.; formal analysis, J.J.R.-C., E.U.-L., C.D.T.-Q. and D.V.M.-M.; investigation, J.J.R.-C. and D.V.M.-M.; resources, E.U.-L. and D.V.M.-M.; writing—original draft preparation, J.J.R.-C.; writing—review and editing, J.J.R.-C., E.U.-L., C.D.T.-Q. and D.V.M.-M.; visualization, J.J.R.-C. and D.V.M.-M.; supervision, E.U.-L. and D.V.M.-M.; project administration, E.U.-L.; funding acquisition, J.J.R.-C. All authors have read and agreed to the published version of the manuscript.

Funding: This research was supported by the Consejo Nacional de Ciencia y Tecnología (CONACYT)—México under Grant [911597]; and Tecnologico de Monterrey under Grant [A01752555].

Institutional Review Board Statement: Not applicable.

Data Availability Statement: The data presented in this study are available on request from the corresponding author.

Acknowledgments: All the financial support provided by Tecnologico de Monterrey and CONACYT-Mexico is thankfully acknowledged.

Conflicts of Interest: The authors declare no conflict of interest.

References

1. Hovsepian, P.E.; Lewis, D.B.; Luo, Q.; Münz, W.D.; Mayrhofer, P.H.; Mitterer, C.; Zhou, Z.; Rainforth, W.M. TiAlN Based Nanoscale Multilayer Coatings Designed to Adapt Their Tribological Properties at Elevated Temperatures. *Thin Solid Films* **2005**, *485*, 160–168. [CrossRef]
2. Prabu, R.; Ramesh, S.; Savitha, M.; Balachandar, M. Review of Physical Vapour Deposition (Pvd) Techniques. In Proceedings of the International Conference on "Sustainable Manufacturing"; Coinbatore Institute of Technology: Coinbatore, India, 2013; pp. 427–434. [CrossRef]

3. Abegunde, O.O.; Akinlabi, E.T.; Oladijo, O.P.; Akinlabi, S.; Ude, A.U. Overview of Thin Film Deposition Techniques. *AIMS Mater. Sci.* **2019**, *6*, 174–199. [CrossRef]
4. Yang, P.F.; Wen, H.C.; Jian, S.R.; Lai, Y.S.; Wu, S.; Chen, R.S. Characteristics of ZnO Thin Films Prepared by Radio Frequency Magnetron Sputtering. *Microelectron. Reliab.* **2008**, *48*, 389–394. [CrossRef]
5. Constantin, D.G.; Apreutesei, M.; Arvinte, R.; Marin, A.; Andrei, O.C.; Munteanu, D. Magnetron Sputtering Technique Used for Coatings Deposition; Technologies and Applications. In Proceedings of the 7th International Conference on Materials Science and Engineering, Citeseer, Brasov, Romania, 24–26 February 2011; Volume 12, pp. 29–33.
6. Mubarak, A.M.A.; Hamzah, E.H.E.; Tofr, M.R.M.T.M.R.M. Review of Physical Vapour Deposition (PVD) Techniques for Hard Coating. *J. Mek.* **2005**, *20*, 42–51.
7. Thornton, J.A. Substrate Heating in Cylindrical Magnetron Sputtering Sources. *Thin Solid Films* **1978**, *54*, 23–31. [CrossRef]
8. Braun, M. Magnetron Sputtering Technique. In *Handbook of Manufacturing Engineering and Technology*; Springer: London, UK, 2015; pp. 2929–2957. [CrossRef]
9. Shul, R.J.; Pearton, S.J. (Eds.) *Handbook of Advanced Plasma Processing Techniques*; Springer Science & Business Media: Berlin/Heidelberg, Germany, 2001; Volume 43, p. 372. [CrossRef]
10. Shah, S.I.; Jaffari, G.H.; Yassitepe, E.; Ali, B. Chapter 4—Evaporation: Processes, Bulk Microstructures, and Mechanical Properties. In *Handbook of Deposition Technologies for Films and Coatings*, 3rd ed.; Martin, P.M., Ed.; William Andrew Publishing: Boston, MA, USA, 2010; pp. 135–252. [CrossRef]
11. Martin, P.M. (Ed.) Chapter 6—Ion Plating. In *Handbook of Deposition Technologies for Films and Coatings*, 3rd ed.; William Andrew Publishing: Boston, MA, USA, 2010; pp. 297–313. [CrossRef]
12. Holmberg, K.; Matthews, A. *Coatings Tribology: Properties, Mechanisms, Techniques and Applications in Surface Engineering*; Elsevier: Amsterdam, The Netherlands, 2009.
13. Elmas, S.; Korkmaz, Ş. Deposition of Al Doped ZnO Thin Films on the Different Substrates with Radio Frequency Magnetron Sputtering. *J. Non Cryst. Solids* **2013**, *359*, 69–72. [CrossRef]
14. Jian, S.R.; Chen, H.G.; Chen, G.J.; Jang, J.S.C.; Juang, J.Y. Structural and Nanomechanical Properties of A-Plane ZnO Thin Films Deposited under Different Oxygen Partial Pressures. *Curr. Appl. Phys.* **2012**, *12*, 849–853. [CrossRef]
15. Sukwisute, P.; Sakdanuphab, R.; Sakulkalavek, A. Hardness and Wear Resistance Improvement of ABS Surface by CrN Thin Film. In *Materials Today: Proceedings*; Elsevier: Amsterdam, The Netherlands, 2017; Volume 4, pp. 6553–6561. [CrossRef]
16. Jiang, P.; Ji, Z.; Wang, X.; Zhou, F. Surface Functionalization—A New Functional Dimension Added to 3D Printing. *J. Mater. Chem. C* **2020**, *8*, 12380–12411. [CrossRef]
17. Cheng, C.; Gupta, M. Surface Functionalization of 3D-Printed Plastics via Initiated Chemical Vapor Deposition. *Beilstein J. Nanotechnol.* **2017**, *8*, 1629–1636. [CrossRef]
18. Bagheri, A.; Jin, J. Photopolymerization in 3D Printing. *ACS Appl. Polym. Mater.* **2019**, *1*, 593–611. [CrossRef]
19. Melchels, F.P.W.; Feijen, J.; Grijpma, D.W. A Review on Stereolithography and Its Applications in Biomedical Engineering. *Biomaterials* **2010**, *31*, 6121–6130. [CrossRef]
20. Voet, V.S.D.; Strating, T.; Schnelting, G.H.M.; Dijkstra, P.; Tietema, M.; Xu, J.; Woortman, A.J.J.; Loos, K.; Jager, J.; Folkersma, R. Biobased Acrylate Photocurable Resin Formulation for Stereolithography 3D Printing. *ACS Omega* **2018**, *3*, 1403–1408. [CrossRef]
21. Binnion, J. A New Method for Preparing 3D Acrylic Photopolymer Patterns for Investment Casting. In Proceedings of the Santa Fe Symposium on Jewelry Manufacturing Technology, Albuquerque, NM, USA, 15–18 May 2016; pp. 103–122.
22. Salmoria, G.V.; Ahrens, C.H.; Beal, V.E.; Pires, A.T.N.; Soldi, V. Evaluation of Post-Curing and Laser Manufacturing Parameters on the Properties of SOMOS 7110 Photosensitive Resin Used in Stereolithography. *Mater. Des.* **2009**, *30*, 758–763. [CrossRef]
23. Cheah, C.M.; Fuh, J.Y.H.; Nee, A.Y.C.; Lu, L.; Choo, Y.S.; Miyazawa, T. Characteristics of Photopolymeric Material Used in Rapid Prototypes: Part II. Mechanical Properties at Post-Cured State. *J. Mater. Process. Technol.* **1997**, *67*, 46–49. [CrossRef]
24. Fuh, J.Y.H.; Lu, L.; Tan, C.C.; Shen, Z.X.; Chew, S. Curing Characteristics of Acrylic Photopolymer Used in Stereolithography Process. *Rapid Prototyp. J.* **1999**, *5*, 27–34. [CrossRef]
25. Peter, M.M. *Handbook of Deposition Technologies for Films and Coatings: Science, Applications and Technology*; Elsevier: Berkeley, CA, USA, 2010; pp. 32–92,253–296.
26. Brodie, I.; Lamont, L.T., Jr.; Myers, D.O. Substrate Bombardment during RF Sputtering. *Shinku* **1969**, *12*, 259–263. [CrossRef]
27. Lamont, L.T.; Lang, A. Reduction of Substrate Heating during Rf Sputtering. *J. Vac. Sci. Technol.* **1970**, *7*, 198–200. [CrossRef]
28. Karalekas, D.; Aggelopoulos, A. Study of Shrinkage Strains in a Stereolithography Cured Acrylic Photopolymer Resin. *J. Mater. Process. Technol.* **2003**, *136*, 146–150. [CrossRef]
29. Watters, M.P.; Bernhardt, M.L. Curing Parameters to Improve the Mechanical Properties of Stereolithographic Printed Specimens. *Rapid Prototyp. J.* **2018**, *24*, 46–51. [CrossRef]
30. Fun To Do®. Industrial Blend—Resina FunToDo. Available online: https://funtodo.es/producto/industrial-blend-resina-funtodo/ (accessed on 8 June 2022).
31. Acosta, J.; Rojo, A.; Salas, O.; Oseguera, J. Process Monitoring during AlN Deposition by Reactive Magnetron Sputtering. *Surf. Coat. Technol.* **2007**, *201*, 7992–7999. [CrossRef]
32. Billmeyer, F.W., Jr. *Ciencia De Los Polimeros*; Editorial Reverte: Barcelona, Spain, 1975.
33. Ding, R.; Leonov, A.I. A Kinetic Model for Sulfur Accelerated Vulcanization of a Natural Rubber Compound. *J. Appl. Polym. Sci.* **1996**, *61*, 455–463. [CrossRef]

34. Coran, A.Y. Vulcanization: Conventional and Dynamic. *Rubber Chem. Technol.* **1995**, *68*, 351–375. [CrossRef]
35. Ghosh, P.; Katare, S.; Patkar, P. Sulfur Vulcanization of Natural Rubber for Benzothiazole Accelerated Formulations. *Rubber Chem. Technol.* **2003**, *76*, 592–693. [CrossRef]
36. Mukhopadhyay, R.; De, S.K.; Chakraborty, S.N. Effect of Vulcanization Temperature and Vulcanization Systems on the Structure and Properties of Natural Rubber Vulcanizates. *Polymer* **1977**, *18*, 1243–1249. [CrossRef]
37. Morrison, N.J.; Porter, M. Temperature Effects on the Stability of Intermediates and Crosslinks in Sulfur Vulcanization. *Rubber Chem. Technol.* **1984**, *57*, 63–85. [CrossRef]
38. Maji, D.; Das, S. Analysis of Plasma-Induced Morphological Changes in Sputtered Thin Films over Compliant Elastomer. *J. Phys. D Appl. Phys.* **2014**, *47*, 105401. [CrossRef]
39. Oakdale, J.S.; Ye, J.; Smith, W.L.; Biener, J. Post-Print UV Curing Method for Improving the Mechanical Properties of Prototypes Derived from Two-Photon Lithography. *Opt. Express* **2016**, *24*, 27077. [CrossRef]
40. Chantarapanich, N.; Puttawibul, P.; Sitthiseripratip, K.; Sucharitpwatskul, S.; Chantaweroad, S. Study of the Mechanical Properties of Photo-Cured Epoxy Resin Fabricated by Stereolithography Process. *Songklanakarin J. Sci. Technol.* **2013**, *35*, 91–98.
41. Zhao, J.; Yang, Y.; Li, L. A Comprehensive Evaluation for Different Post-Curing Methods Used in Stereolithography Additive Manufacturing. *J. Manuf. Process.* **2020**, *56*, 867–877. [CrossRef]
42. Karalekas, D.; Rapti, D.; Gdoutos, E.E.; Aggelopoulos, A. Investigation of Shrinkage-Induced Stresses in Stereolithography Photo-Curable Resins. *Exp. Mech.* **2002**, *42*, 439–444. [CrossRef]
43. Depla, D.; Mahieu, S.; Greene, J.E. Chapter 5—Sputter Deposition Processes. In *Handbook of Deposition Technologies for Films and Coatings*, 3rd ed.; Martin, P.M., Ed.; William Andrew Publishing: Boston, MA, USA, 2010; pp. 253–296. [CrossRef]
44. Iglesias, E.J.; Hecimovic, A.; Mitschker, F.; Fiebrandt, M.; Bibinov, N.; Awakowicz, P. Ultraviolet/Vacuum-Ultraviolet Emission from a High Power Magnetron Sputtering Plasma with an Aluminum Target. *J. Phys. D Appl. Phys.* **2020**, *53*, 55202. [CrossRef]
45. Mattox, D.M. Chapter 12—Adhesion and Deadhesion. In *Handbook of Physical Vapor Deposition (PVD) Processing*, 2nd ed.; Mattox, D.M., Ed.; William Andrew Publishing: Boston, MA, USA, 2010; pp. 439–474. [CrossRef]
46. Gao, W.; Li, Z. ZnO Thin Films Produced by Magnetron Sputtering. *Ceram. Int.* **2004**, *30*, 1155–1159. [CrossRef]
47. Gonçalves, R.S.; Barrozo, P.; Brito, G.L.; Viana, B.C.; Cunha, F. The Effect of Thickness on Optical, Structural and Growth Mechanism of ZnO Thin Film Prepared by Magnetron Sputtering. *Thin Solid Films* **2018**, *661*, 40–45. [CrossRef]
48. Zguris, Z. How Mechanical Properties of Stereolithography 3D Prints Are Affected by UV Curing. *Formlabs White Pap.* **2016**, 1–11.
49. Oskui, S.M.; Diamante, G.; Liao, C.; Shi, W.; Gan, J.; Schlenk, D.; Grover, W.H. Assessing and Reducing the Toxicity of 3D-Printed Parts. *Environ. Sci. Technol. Lett.* **2016**, *3*, 1–6. [CrossRef]
50. Schricker, S.R. Composite Resin Polymerization and Relevant Parameters. In *Orthodontic Applications of Biomaterials*; Woodhead Publishing: Soston, UK, 2017; pp. 153–170. [CrossRef]
51. Bertana, V.; De Pasquale, G.; Ferrero, S.; Scaltrito, L.; Catania, F.; Nicosia, C.; Marasso, S.L.; Cocuzza, M.; Perrucci, F. 3D Printing with the Commercial UV-Curable Standard Blend Resin: Optimized Process Parameters towards the Fabrication of Tiny Functional Parts. *Polymers* **2019**, *11*, 292. [CrossRef]
52. Ngo, T.D.; Kashani, A.; Imbalzano, G.; Nguyen, K.T.Q.; Hui, D. Additive Manufacturing (3D Printing): A Review of Materials, Methods, Applications and Challenges. *Compos. Part B Eng.* **2018**, *143*, 172–196. [CrossRef]
53. Xu, K.; Chen, Y. Photocuring Temperature Study for Curl Distortion Control in Projection-Based Stereolithography. *J. Manuf. Sci. Eng.* **2016**, *139*, 021002. [CrossRef]
54. Wu, D.; Zhao, Z.; Zhang, Q.; Qi, H.J.; Fang, D. Mechanics of Shape Distortion of DLP 3D Printed Structures during UV Post-Curing. *Soft Matter* **2019**, *15*, 6151–6159. [CrossRef]
55. Castro-Rodríguez, R.; Oliva, A.I.; Sosa, V.; Caballero-Briones, F.; Peña, J.L. Effect of Indium Tin Oxide Substrate Roughness on the Morphology, Structural and Optical Properties of CdS Thin Films. *Appl. Surf. Sci.* **2000**, *161*, 340–346. [CrossRef]
56. Thornton, J.A.; Hoffman, D.W. Stress-Related Effects in Thin Films. *Thin Solid Films* **1989**, *171*, 5–31. [CrossRef]
57. Tvarozek, V.; Novotny, I.; Sutta, P.; Flickyngerova, S.; Schtereva, K.; Vavrinsky, E. Influence of Sputtering Parameters on Crystalline Structure of ZnO Thin Films. *Thin Solid Films* **2007**, *515*, 8756–8760. [CrossRef]
58. Jankowski, A.F.; Bionta, R.M.; Gabriele, P.C. Internal Stress Minimization in the Fabrication of Transmissive Multilayer X-ray Optics. *J. Vac. Sci. Technol. A Vac. Surf. Film.* **1989**, *7*, 210–213. [CrossRef]
59. Vidakis, N.; Antoniadis, A.; Bilalis, N. The VDI 3198 Indentation Test Evaluation of a Reliable Qualitative Control for Layered Compounds. *J. Mater. Process. Technol.* **2003**, *143*, 481–485. [CrossRef]
60. Raghuram, A.C.; Bunshah, R.F. The Effect of Substrate Temperature on the Structure of Titanium Carbide Deposited by Activated Reactive Evaporation. *J. Vac. Sci. Technol.* **1972**, *9*, 1389–1394. [CrossRef]
61. Karlsson, L.; Hultman, L.; Sundgren, J.-E. Influence of Residual Stresses on the Mechanical Properties of TiCxN1−x (X= 0, 0.15, 0.45) Thin Films Deposited by Arc Evaporation. *Thin Solid Films* **2000**, *371*, 167–177. [CrossRef]
62. Thornton, J.A. New Industries and Applications for Advanced Materials Technology. *SAMPE* **1994**, *19*, 443.
63. Mishra, S.K.; Bhattacharyya, A.S. Effect of Substrate Temperature on the Adhesion Properties of Magnetron Sputtered Nano-Composite Si-C-N Hard Thin Films. *Mater. Lett.* **2008**, *62*, 398–402. [CrossRef]
64. Wu, B.; Yu, Y.; Wu, J.; Shchelkanov, I.; Ruzic, D.N.; Huang, N.; Leng, Y.X. Tailoring of Titanium Thin Film Properties in High Power Pulsed Magnetron Sputtering. *Vacuum* **2018**, *150*, 144–154. [CrossRef]

65. Park, S.-Y.; Rho, S.-H.; Lee, H.-S.; Kim, K.-M.; Lee, H.-C. Fabrication of Highly Porous and Pure Zinc Oxide Films Using Modified DC Magnetron Sputtering and Post-Oxidation. *Materials* **2021**, *14*, 6112. [CrossRef]
66. Mueller, B. Additive Manufacturing Technologies—Rapid Prototyping to Direct Digital Manufacturing. *Assem. Autom.* **2012**, *32*, 415–435. [CrossRef]

polymers

MDPI

Article

Simultaneous Treatment of Both Sides of the Polymer with a Conical-Shaped Atmospheric Pressure Plasma Jet

Felipe Vicente de Paula Kodaira [1,*], Bruno Henrique Silva Leal [1], Thayna Fernandes Tavares [1], Antje Quade [2], Luis Rogerio de Oliveira Hein [3], William Chiappim [1,*] and Konstantin Georgiev Kostov [1]

1 Laboratory of Plasmas and Applications, Department of Physics, Faculty of Engineering and Sciences, São Paulo State University (UNESP), Guaratinguetá, São Paulo 12516-410, Brazil
2 Leibniz Institute for Plasma Science and Technology—INP, 17489 Greifswald, Germany
3 Department of Materials and Technology, São Paulo State University, UNESP, Guaratinguetá, São Paulo 12516-410, Brazil
* Correspondence: kodaira.felipe@gmail.com (F.V.d.P.K.); william.chiappim@unesp.br (W.C.); Tel.: +55-12-3123-2165 (W.C.)

Abstract: A conical-shaped atmospheric pressure plasma jet (CS-APPJ) was developed to overcome a standard limitation of APPJs, which is their small treatment area. The CS-APPJs increase the treatment area but use the same gas flow. In the present work, polypropylene samples were treated by CS-APPJ and characterized by scanning electron microscope (SEM), the contact angle, Fourier-transformed infrared spectroscopy (FTIR), and X-ray photoelectron spectroscopy (XPS). It was observed that the treatment co-occurs on the face directly in contact with the plasma and on the opposite face (OF) of the samples, i.e., no contact. However, the treatment changed the chemical composition on each side; the OF is rougher than the direct contact face (DCF), probably due to the oxygen groups in excess at the DCF and nitrogen in quantity at the OF. Although simultaneous treatment of both sides of the sample occurs for most atmospheric plasma treatments, this phenomenon is not explored in the literature.

Keywords: conical-shaped; atmospheric pressure plasma jet; polymer; polypropylene; SEM; FTIR; XPS

Citation: Kodaira, F.V.d.P.; Leal, B.H.S.; Tavares, T.F.; Quade, A.; Hein, L.R.d.O.; Chiappim, W.; Kostov, K.G. Simultaneous Treatment of Both Sides of the Polymer with a Conical-Shaped Atmospheric Pressure Plasma Jet. *Polymers* **2023**, *15*, 461. https://doi.org/10.3390/polym15020461

Academic Editor: Choon-Sang Park

Received: 11 December 2022
Revised: 4 January 2023
Accepted: 5 January 2023
Published: 16 January 2023

1. Introduction

Although surface modification of polymers by non-thermal atmospheric pressure plasmas (NTAPPs) has become a well-established technology [1–3], there is still much to be improved in this device development and optimization. The processing of materials by NTAPP has several advantages over other industrial processes, highlighting the advantage of being eco-friendly, i.e., it does not involve potentially dangerous solvents to the environment or workers [4,5]. Its versatility in modifying the surface of polymers without changing the bulk properties is worth mentioning, in addition to being suitable for most polymers that are sensitive to high temperatures [6–8]. Furthermore, plasma processes can take place in an open or controlled environment, with the treatment time varying from a few seconds to several minutes, depending on the configuration of the device used [8–12]. It is worth mentioning that this process can be carried out continuously in a line of production or individually [13,14]. Among the NTAPP devices, dielectric barrier discharge (DBD) is the most widely used [11–20]. A significant advantage of DBD over other devices that generate electrical discharges is the higher electron density induced by micro-discharges caused by a large number of tiny current filaments that pass through the dielectric material covering one or both electrodes [21,22]. This high electron density improves the functionalization of the polymers due to the uniform incidence of discharge on the treated surface [6,8]. However, the maximum distance between the electrodes is limited, affecting the device's operation in open systems [23].

In the late 1990s, atmospheric pressure plasma jets (APPJs) appeared to mitigate, reduce or suppress the deficiency mentioned above. APPJs generate plasma plumes in open spaces, enabling the direct treatment of samples of different shapes and sizes [24–34]. Over the years, APPJs have been successfully tested in many applications, such as material treatment, sterilization, cancer and wound treatment, aesthetic applications, and dentistry [35–49]. Many different configurations of APPJs are reported in the literature. In his review article Lu, X et al. [32] presented APPJ with a single electrode, APPJ without dielectrics, DBD jets with ring and pin electrodes, and combinations between them.

However, plasma jets have diameters of a few millimeters, making them suitable for biomedical applications [24,29–31,50]. It is known that the most intense effect of APPJ is in the central region, with a gradual decrease as it moves away from the plasma plume region. Thus, its minor diameter limits the treatment area or surface activation, which makes its application time-consuming and expensive for industrial applications. The exposure time to APPJ can be increased to overcome the technological limitation, which increases the treatment area or surface activation [51,52]. The problem with increasing exposure time to APPJ is that there may be damage to the treated surface. Therefore, there is a need to improve the APPJs for use in larger areas. Recently, Abdelaziz et al. [53] investigated a configuration of a wide tube APPJ with a diameter of up to 30.0 mm. However, for this configuration, the gas flow used is high, which generates a non-uniform movement of the gas inside wide tubes [53]. Some works have tried to enlarge the diameter of the APPJs using complex and 2D jet arrays [54,55]. However, these configurations bring challenges, such as interactions between neighboring jets [56,57] and increased gas consumption, which increases with the number of jets in the array. Another exciting configuration is a gas-flowless pin-to-ring geometry presented by Khun et al. [58], where they reported achieving a more extensive treatment area than most APPJs.

In order to improve and explore novel APPJ configurations, this work used a conical-shaped APPJ (CS-APPJ) with a 75.0 mm diameter outlet nozzle. Discharge was initiated between a high voltage pin electrode inserted into the thin part of the conical-shaped (CS). In contrast, a grounded flat electrode covered by a 4.0 mm thick glass insulator was inserted under the CS. The glass plate also served as a sample holder. Compared to most APPJs, this device allows treatment over a significantly larger area, allowing for greater sample thickness and shape variety. The surface modification effect extends over the entire area covered by the CS and exhibits a considerable degree of uniformity.

Furthermore, our results prove that plasma treatments co-occur on both sides of the flat samples. Although the discharge geometry resembles a point-to-plane corona configuration, it also has a dielectric barrier covering the ground electrode, which makes it a DBD-type device. Finally, due to the gas flow required for the discharge propagation, the system also has some similarities with the APPJs. However, in the case of the CS-APPJ, there is no plasma plume but rather a filament discharge composed of multiple filaments that start from the top electrode pin and travel along the inner wall of the funnel to end in the glass substrate. The polymer chosen for this work was polypropylene (PP) because it is a common material widely used in several industrial applications, such as packaging, decoration, electronics, and medicine [59]. The extensive use of PP comes from its versatility, stability, and good mechanical properties [59]. However, as with most polymeric materials, PP has low surface energy, making it difficult to paint or glue [60,61]. Plasma processing of polymers conveniently increases their surface energy while keeping the beneficial bulk properties unchanged [60,61].

Therefore, the present work contributes to developing plasma jets, mainly of the DBD type. One can also highlight the contribution in the application of the device on polymeric surfaces, which are of great applicability in industry and the daily life of the entire world population.

2. Materials and Methods

2.1. Assembly Setup and Electrical Measurements

The device employs a 3.0 mm thick commercial glass funnel (conical-shaped) comprised of an 8.0 mm narrow tube section ending in a 75.0 mm conical horn. It is placed vertically with its wide part facing downwards (see Figure 1a), while the top of the funnel is closed by a dielectric support through which a 1.0 mm diameter pin electrode is introduced. The pin-to-plate electrode geometry is formed by placing a grounded plate electrode under the 4.0 mm thick glass-covered funnel, which also serves as a sample holder. Working gas (99.2% Ar from AirLiquid, São José dos Campos, Brazil) was injected into the system through an orifice in the dielectric support at a flow rate of 2.0 SLM. The sharp pin electrode was connected to a commercial high-voltage power supply (model Minipuls 6, GBS Elektronik GmbH, Radeberg, Germany). It was operated in burst mode, i.e., generating 12 consecutive high voltage oscillations at a frequency of 25.0 kHz followed by a period of voltage off. The burst repetition period was set to 2.0 ms. Electrical characterization was performed by obtaining the applied voltage (directly from a voltage divider in the power supply) and calculating the transferred charge and discharge current by measuring the voltage drop across a serial capacitor (10.0 nF) or resistor (120.0 Ω), respectively, coupled as shown in Figure 1a. The discharge power was calculated from the electrical energy contained in a burst (calculated by the area of the Q-V Lissajous figure) divided by the repetition period [62]. A photo of the filament discharge generated inside the funnel with 2.0 SLM Ar is shown in Figure 1b. As can be seen, the discharge filaments start at the tip of the high voltage electrodes, and after reaching the glass funnel, the inner wall propagates downstream along the conical horn until it reaches the glass substrate. The discharge power was calculated as 8.0 ± 0.2 W, using the area of the $Q \times V$ Lissajous figure of an entire burst period. Figure 1c shows the Lissajous figure of a cutout of two oscillations within the burst. Figure 1d depicts the typical voltage signal showing two consecutive (high-voltage) HV bursts of 12.0 cycles with a 2.0 ms repetition period. In contrast, a detailed view of the voltage and current waveforms in a burst is shown in Figure 1e. For this, a magnification interval between 0.0 and 1.0 ms was used. Finally, Figure 1f shows a clearer view of the signal indicating only two periods of oscillation (magnification between 0.40 and 0.50 s). The current signal in Figure 1f exhibits the typical form of a DBD discharge, i.e., multiple current spikes superimposed on the capacitive current. Thus, although the geometry of the discharge resembles a point-to-plane corona configuration, the dielectric barrier that covers the ground electrode makes it a DBD-type device.

Therefore, this new CS-APPJ configuration used in this work can be characterized as a new DBD-type APPJ and is an essential contribution to the field of plasma surface treatment. It is important to note that in the case of CS-APPJ, there is no plasma plume but a filament discharge composed of several filaments that start at the pin of the upper electrode and run along the inner wall of the funnel, and end at the substrate (as can be seen in Figure 1b).

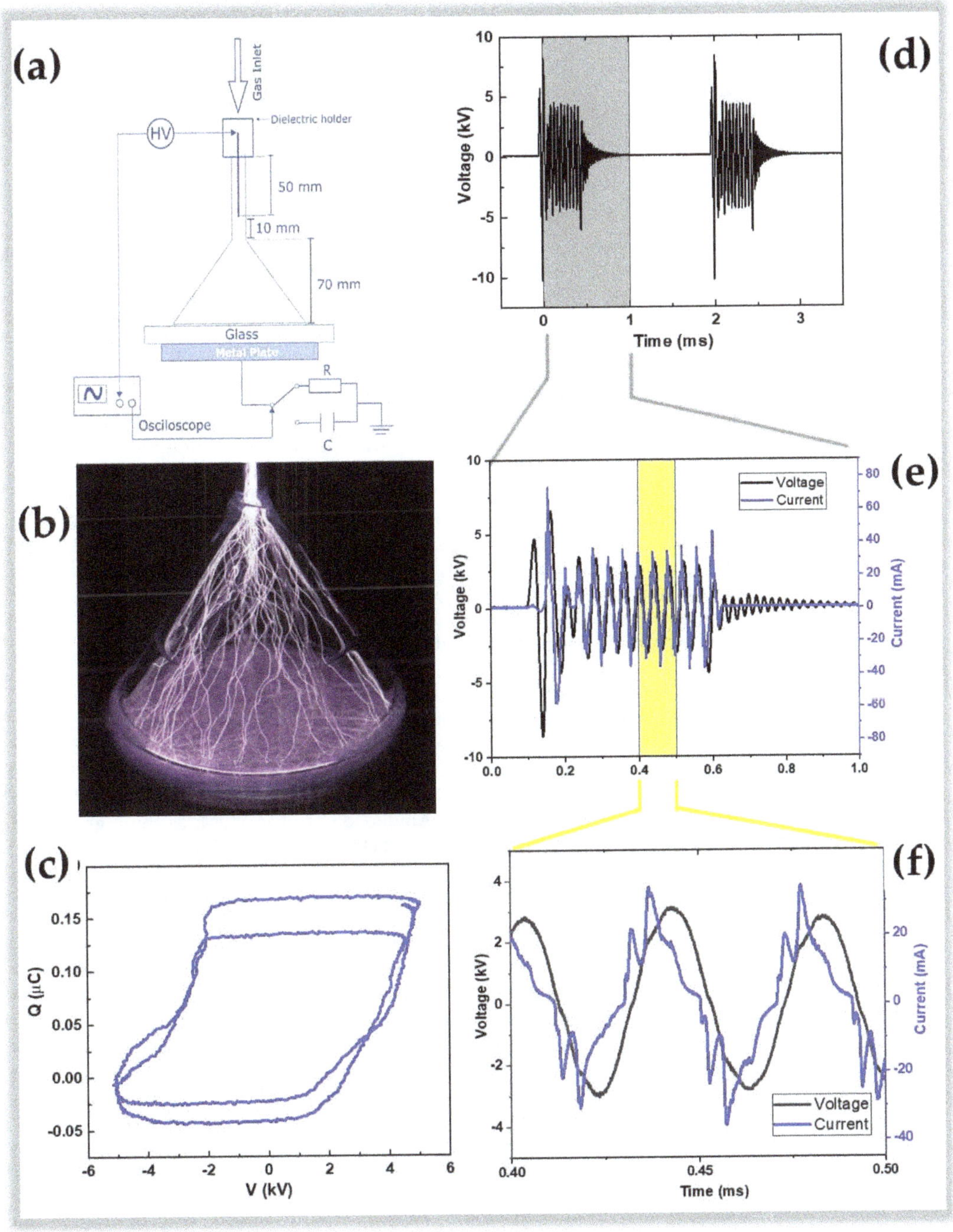

Figure 1. (**a**) Schematic layout of the conical-shaped atmospheric pressure plasma jet (CS-APPJ) system; (**b**) Photo of the filamentary discharge inside the conical horn; (**c**) Lissajous Figure (Q × V). Overview of typical electrical parameters of (**d**) applied voltage; (**e**) voltage and current waveforms in a full burst of 12 high-voltage oscillations; (**f**) in two wave periods from the middle of the burst signal.

2.2. Polypropylene Samples Preparation

The polypropylene flat samples (0.95 g/cm^3; 1.0 mm thickness) were cut into rectangular shapes (20.0 × 10.0 mm^2), then cleaned for 20.0 min in an ultrasonic bath with distilled water and finally rinsed in isopropanol for another 20.0 min. Finally, the samples were dried at room temperature and treated up to 5.0 min.

2.3. Characterization of Polypropylene Samples

Scanning electron microscope (SEM) images were performed by a Carl Zeiss EVO LS 15 tool at low pressure (~10^{-3} Pa) and high electron tension of 5.0 kV. Before the measurements, 20.0 × 10.0 mm samples were covered by a 6.0 nm thick gold layer deposited by magnetron sputtering. Topography images were carried out from both the direct contact face (DCF) and opposite face (OF) of samples. In addition, the roughness values were obtained from the SEM images. The wettability was carried out on a Ramé-Hart 300 F1 goniometer by depositing deionized water droplets (1.0 µL). The droplets were placed along the longer sample axis spaced 5.0 mm from each other, and nine samples were measured for each treatment time. Both faces were measured. The mean water contact angle (WCA) value of the droplets on all samples was then calculated and plotted to evaluate the dependence on the treatment time and the difference of the treatment on both faces of the PP samples. The changes in the molecular structure of PP samples were carried out by Fourier-transform infrared spectroscopy (FTIR) coupled with attenuated total reflectance (ATR), which was obtained by a Perkin Elmer Spectrum 100 FTIR spectrometer. X-ray photoelectron spectroscopy (XPS) analysis was performed in a Kratos AXIS Ultra. The composition of the surface was scanned along its longest symmetry axis, looking for carbon, oxygen, and nitrogen. Again, both faces of the samples were analyzed. XPS data were also used to ascertain the homogeneity of the treatment by inspecting the chemical composition along the surface of the samples.

3. Results and Discussion

The results obtained were discussed separately for each characterization technique in this section. In the SEM images, it is possible to observe differences in both faces of the sample, with the formation of oligomers of low molecular weight oxidized materials (LMWOM) in the OF. The constant water angle measurements corroborate the hypothesis that the structures formed on the underside are LMWOM; as these groups are polar, lower contact angle values are expected [18], as shown in Section 3.2. The spectroscopic characterization techniques (FTIR and XPS) confirm different structures and chemical compositions for both sides of the samples treated with oxygen groups in DFC and nitrogen groups in OF.

3.1. Scanning Electron Microscopy (SEM)

Figure 2 shows SEM images at 5000× *g* magnification. Figure 2a shows the PP without treatment, and Figure 2b,c show the images of the PP with the face directly in contact with the discharge and the opposite face, respectively. Figure 2d shows the roughness of the respective samples, evaluated from the images in Figure 2a–c. As can be seen, the untreated surface has less roughness compared to the PP treated for 5.0 min. Another interesting point concerns DCF (Figure 2b), which presents a slightly higher surface roughness than untreated samples. In contrast, Figure 2c shows the OF treated under the same conditions; however, it has a greater surface roughness with some small granular structures that spread over the entire surface. The granular forms are probably oligomers, known as oxidized low molecular weight materials (LMWOM). The formation of this type of structure on the surface of the sample is typically observed in the case of polymers treated with DBD and directly applied APPJs [6,18,35]. It is possible to follow in Figure 1b that the discharge filaments run along the wall of the conical structure of the reactor and on the surface of the sample holder during the treatment. However, in polymer treatment, the filaments also pass under the sample, which makes the bottom side of the sample closer to the discharge.

In addition, due to the loading of charges on the sample surface due to the sample/plasma contact, there is the generation of an electric field between the sample surface and the sample holder, which can lead to the formation of a DBD-planar type under the bottom side of the sample. This generation of a DBD discharge in the OF may be the cause of the greater roughness in this face compared to the DFC.

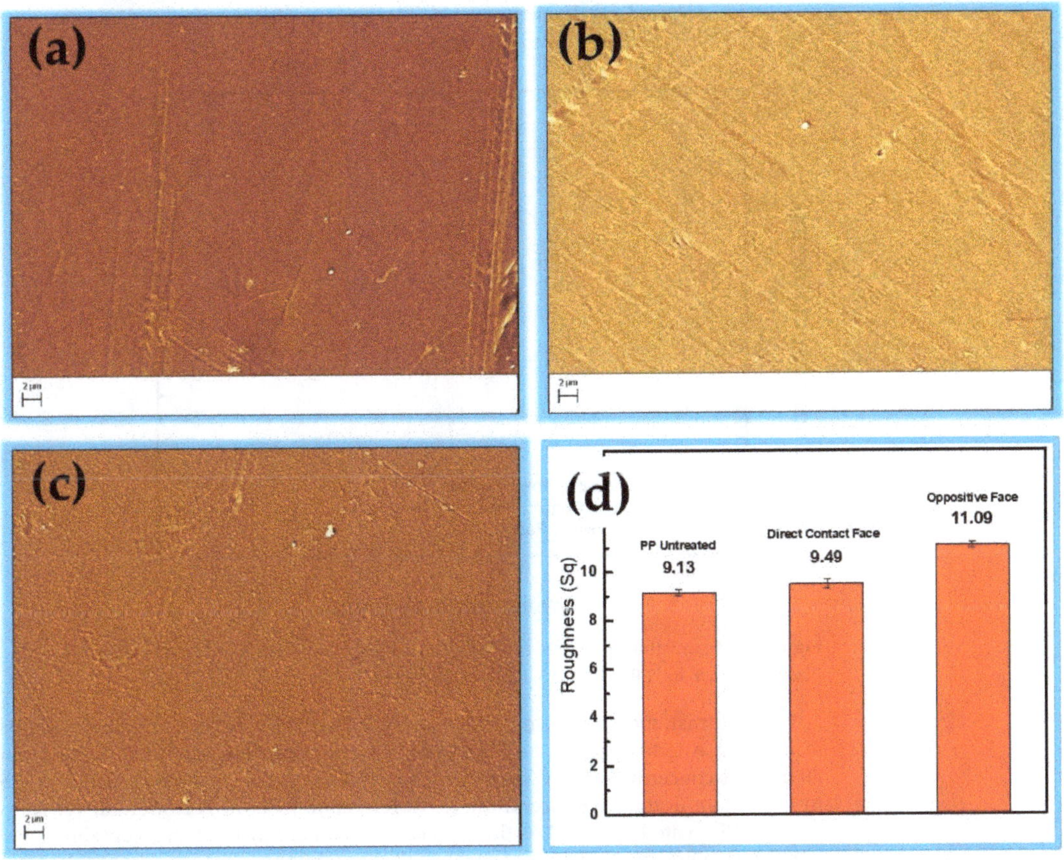

Figure 2. Scanning electron microscopy (SEM) images of polypoprylene samples: (**a**) untreated; (**b**) treated by plasma with direct face contact (DFC); (**c**) treated by plasma through the opposite face (OF); and (**d**) roughness values for all samples. All samples were treated for 3 min.

Another point that may contribute to the different roughness of the OF is the lower pressure of the working gas (Ar) that reaches the surface of the polymer. In contrast, there is a more significant interaction with free radicals from the formation of the planar type DBD generated between the sample surface and the sample holder, which increases the energy of electrons and free radicals, increasing ablation and interactions with reactive molecules such as N_2^+, N_4^+, N+, O_2^+, H_2O^+, O_2^- and O^- [8,63]. However, due to the high complexity of the interaction between the plasma and the polymer surface, it is not possible to differentiate by SEM images whether the attack is related to the reduction of the chemical structure of the surface caused by plasma or is associated with the material removed through the impact of surface plasma species [8,63].

3.2. Water Contact Angle (WCA)

On both sides, the contact angle with water (WCA) was measured on 20.0 × 10.0 mm PP samples. The WCA of the untreated substrate was 97.1°, according to [64,65]. Mean values are shown in Figure 3, these values were measured along the longest axis of the sample, spaced 5.0 mm apart, and nine samples were measured for each treatment time, thus obtaining an average value with an error bar (as seen in Figure 3). On both sides, the treatment promoted a reduction in the WCA. It can be observed that the treatment of DFC presented WCA between 70 and 80 degrees.

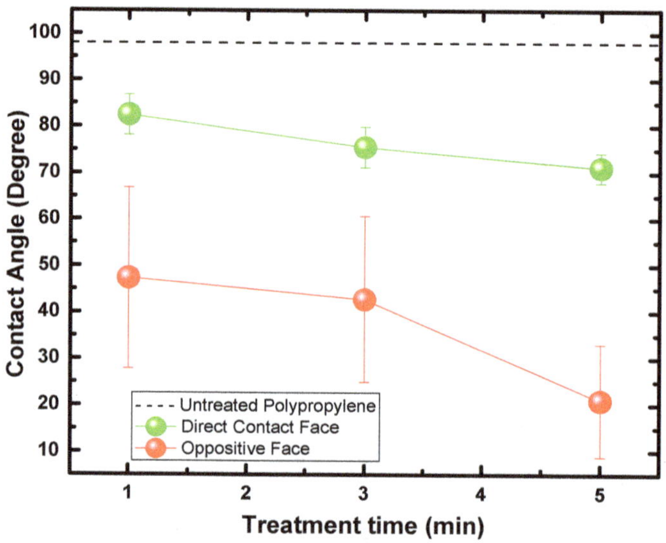

Figure 3. Mean values of water contact angle of polypropylene samples measured on DCF with plasma and OF for 1.0, 3.0, and 5.0 min of treatment.

In contrast, the OF showed a more pronounced WCA reduction, with values below 50 degrees. Another point to be observed is the error bars that present variations of up to 20 degrees concerning the average value, which indicates a less homogeneous treatment in the OF about the DCF. According to Morent et al. [66], the considerable reduction in WCA of PP treated with DBD is due to the formation of functionalities containing oxygen, with a prevalence of C–O, O–C=O, and C=O. Other authors show similar results for other polymeric substrates, such as PA6 and PA66 [67]. Therefore, this behavior corroborates the hypothesis suggested by the SEM images, which assumes the formation of LMWOM in the substrate treated with predominance in the OF, which favors the reduction of WCA.

3.3. Chemical Analysis of the Treated and Untreated Polypropylene

Figure 4 shows the FTIR spectra for PP samples treated with CS-APPJ for periods of 1.0, 3.0, and 5.0 min, in addition to the spectrum of the untreated PP sample. It is essential to highlight that due to the reach of the technique, which varies between 0.5–5.0 μm (4000–400 cm^{-1}) in depth, FTIR becomes an excellent tool for understanding and verifying possible chemical changes in the inner layers of the polypropylene. Figure 4a shows the bands related to the DCF internal connections. In contrast, Figure 4b shows the OF-related bands.

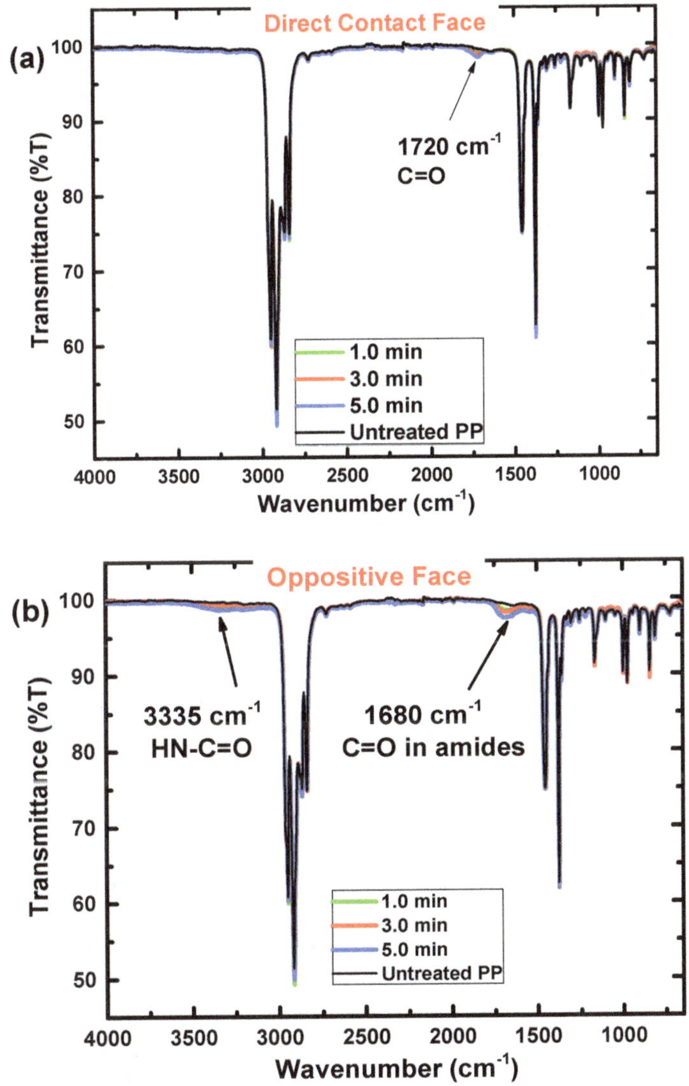

Figure 4. FTIR full spectra of polypropylene samples untreated, and treated by 1.0, 3.0, and 5.0 min. (**a**) DCF; and (**b**) OF.

The absorption bands associated with PP demonstrated in Figure 4a,b are (i) bands related to the asymmetric elongation vibration $-CH_3$ at 2952 cm^{-1}; (ii) bands related to symmetric flexion $-CH_2-$, symmetric elongation $-CH_2-$ and asymmetric elongation $-CH_2-$, respectively, at 1455, 2838 and 2917 cm^{-1}; (iii) symmetric bending vibration bands of the $-CH_3$ group is detected at 1375 cm^{-1}; (iv) the bands attributed to the oscillating vibration $-CH_3$ are at 972, 997 and 1165 cm^{-1}; (v) the band located at 840 cm^{-1} is attributed to the $C-CH_3$ stretching vibration. It is essential to highlight that the bands mentioned above appear due to the presence of the methyl group in polypropylene [66–70]. After CS-APPJ treatment for periods of 1.0, 3.0, and 5.0 min, new bands were detected in the FTIR spectra. First, in DCF (Figure 4a), a peak appears at 1720 cm^{-1} related to C=O groups [23]. Its intensity increases with treatment time. In contrast, Figure 4b shows that in the OF of the sample, a new band at approximately 1680 cm^{-1} is observed. This new band is only detected in the OF of the treated samples and can be attributed to C=O stretching groups in amides. Figure 4b also shows a band at approximately 3335 cm^{-1}, which may be associated with elongation in the HN-C=O group [71], confirming the presence of nitrogen detected in the 1680 cm^{-1} band.

Figure 5 shows the growth in intensity of bands associated with CS-APPJ treatment (1720, 1680, and 3335 cm^{-1}). In DCF (Figure 5a), it is observed that the C=O groups grows in intensity with the increase in the period of exposure to CS-APPJ. The same behavior is kept for the C=O stretching groups in amides and elongation in the HN-C=O group, referring to OF (Figure 5b).

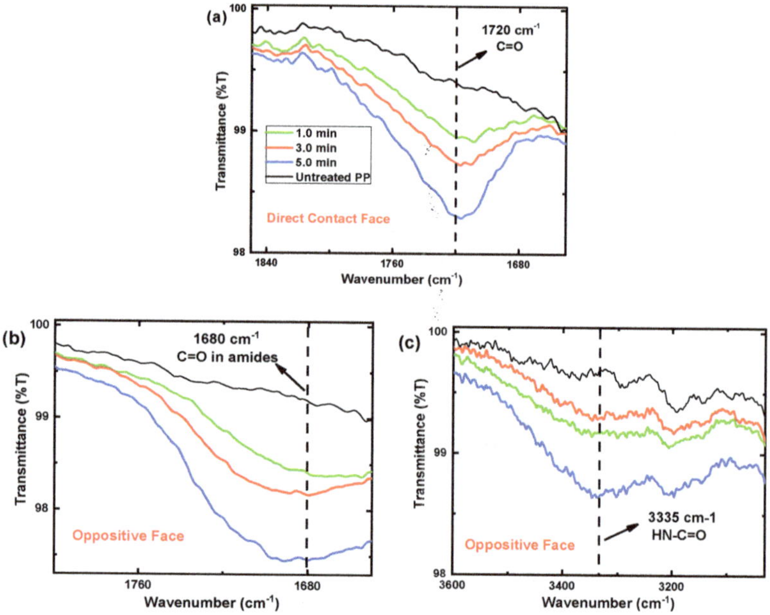

Figure 5. FTIR spectra of polypropylene samples untreated, and treated by 1.0, 3.0, and 5.0 min. (**a**) DCF in the region between 1850 to 1650 cm^{-1}; (**b**) OF in the region between 1800 to 1650 cm^{-1}; and (**c**) OF in the region between 3600 to 3000 cm^{-1}.

FTIR spectra show that plasma treatment on polymers often results in incorporating oxygen and nitrogen atoms into the material's surface [72]. Therefore, to study the degree of chemical modifications on the surface of substrates with a depth range between 5.0 and 10.0 nm, XPS analysis was used (Figure 6). With this technique, it is possible to evaluate the surface oxidation and the percentage of atomic species on the surface of the PP. For this, the sum of C, O, N, and Si was considered 100%, i.e., the proportion of H in the calculations was neglected. Figure 6a shows the different groups of carbon bonds detected in the untreated and the samples treated for 5.0 min, and it can be observed that the formation of other groups of carbon bonds occurs in the DCF and the OF, compared with the PP untreatment.

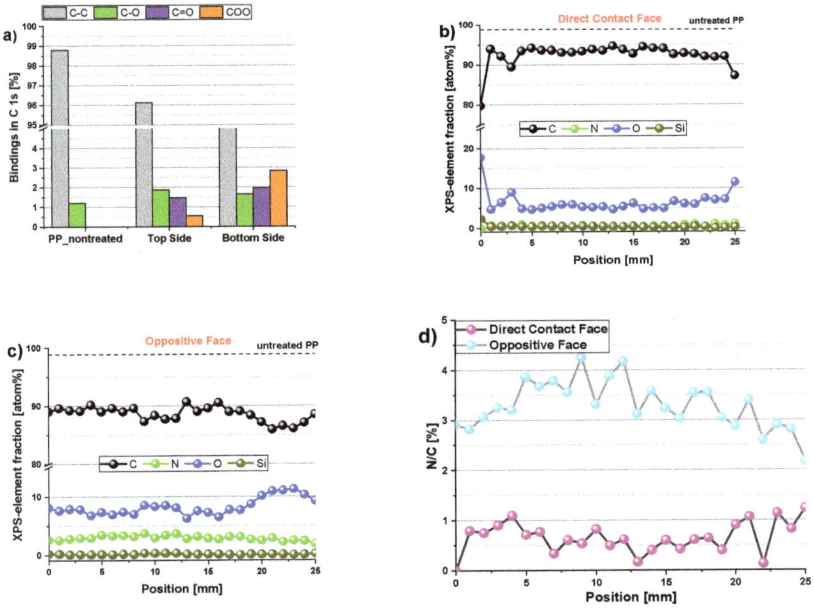

Figure 6. Distribution of the elemental composition of the plasma-treated PP samples; (**a**) Comparison of different carbon groups detected on the untreated PP and both sides of the plasma-treated sample; (**b**) on the DCF; (**c**) on the OF; (**d**) longitudinal distribution of the N/C ratio on both sample sides.

This corroborates the data found in the FTIR spectra. The untreated sample was observed to contain carbon primarily with only traces of O (less than 2%) due to surface oxidation. After treatment with CS-APPJ, the amount of O on both sides of the sample increased. The OF presented a higher oxygen content with improved COO groups (Figure 6a). This result corroborates the WCA data (Figure 3), which shows a marked reduction of the WCA in the OF. To assess the distribution of chemical elements on both sides of the PP sample treated for 5.0 min, a piece of PP of area 25.0 $\times$ 20.0 mm^2 was used for high-resolution XPS scan measurements along the central axis (25.0 mm). The results are shown in Figure 6b,c, indicating that the CS-APPJ treatment added O atoms and a small amount of N with only traces of Si atoms on the PP surface. Figure 6b,c show that the species are homogeneously distributed along both sides of the sample. This shows the uniformity in the treatment with CS-APPJ, which corroborates the SEM images (Figure 2). Figure 6d shows the N/C percentage ratio of both faces, where a balance between 0.0 and 1.0% (DCF) and between 2.5 and 4.0% is observed in the phase opposite to direct contact with the plasma. It is essential to highlight that the XPS results corroborate the FTIR spectra, showing the formation of different groups on opposite sides of the treated samples.

4. Conclusions

Despite resembling the geometry of a point-to-plane corona configuration, the CS-APPJ device behaves similar to a DBD-type device, as observed in the typical electric current signal shown in Figure 1. In the CS-APPJ, the filaments of discharge run along the wall of the conical structure until they reach the surface of the sample holder during the treatment. The formation of a DBD-planar under the OF of the sample was observed due to the loading of charges on the sample surface caused by the sample/plasma contact. The generation of a planar-DBD under the OF was responsible for simultaneously treating both sides of the PP, which led to different surface changes in the OF compared to the DCF. Treatment with the CS-APPJ device induced physical and chemical changes on the surface of commercial PP. The SEM images show slight surface alterations, with a more significant modification occurring in the OF samples that present greater roughness. Corroborating the SEM and roughness images, the WCA measurements show the formation of LMWOM, which favors the reduction of WCA with increasing treatment time with CS-APPJ. The FTIR analysis showed that the treatment with CS-APPJ affected the internal bonds in the deep layer of the polymer in different ways, with the detection of the presence of amide groups in the OF and C=O in the DCF. Furthermore, FTIR spectra indicated that the amount of these functional groups tends to increase with treatment time. The XPS measurements showed uniformity in the surface treatment on both sides of the sample, highlighting the difference in functional groups incorporated on both sides of the material, with a more significant presence of nitrogen groups in the OF, corroborating the FTIR spectra.

Author Contributions: Conceptualization, F.V.d.P.K. and K.G.K.; methodology, F.V.d.P.K., B.H.S.L., T.F.T. and A.Q.; formal analysis, F.V.d.P.K. and L.R.d.O.H.; investigation, F.V.d.P.K., B.H.S.L., T.F.T., L.R.d.O.H. and A.Q.; resources, K.G.K.; data curation, F.V.d.P.K. and T.F.T.; writing—original draft preparation, F.V.d.P.K., W.C. and K.G.K.; writing—review and editing, F.V.d.P.K., W.C. and K.G.K.; supervision, F.V.d.P.K. and K.G.K.; project administration, K.G.K.; funding acquisition, W.C. and K.G.K. All authors have read and agreed to the published version of the manuscript.

Funding: This work was supported by the São Paulo State Research Foundation (FAPESP) [2019/05856-7; 2020/10450-7] and by Coordination for the Improvement of Higher Education (CAPES).

Data Availability Statement: Not applicable.

Conflicts of Interest: The authors declare no conflict of interest.

References

1. Cullen, P.J.; Lalor, J.; Scally, L.; Boehm, D.; Milosavljević, V.; Bourke, P.; Keener, K. Translation of plasma technology from the lab to the food industry. *Plasma Process. Polym.* **2017**, *15*, 1700085. [CrossRef]
2. Sundriyal, P.; Pandey, M.; Bhattacharya, S. Plasma-assisted surface alteration of industrial polymers for improved adhesive bonding. *Int. J. Adhes. Adhes.* **2020**, *101*, 102626. [CrossRef]
3. Nishime, T.M.C.; Wagner, R.; Kostov, K.G. Study of Modified Area of Polymer Samples Exposed to a He Atmospheric Pressure Plasma Jet Using Different Treatment Conditions. *Polymers* **2020**, *12*, 1028. [CrossRef] [PubMed]
4. Singh, M.; Vajpayee, M.; Ledwani, L. Eco-friendly surface modification of natural fibres to improve dye uptake using natural dyes and application of natural dyes in fabric finishing: A review. *Mater. Today Proc.* **2021**, *43*, 2868–2871. [CrossRef]
5. Moridi Mahdieh, Z.; Shekarriz, S.; Afshar Taromi, F. Fabrication of Antibacterial and Self-Cleaning Polyester/Cellulose Fabric by Corona Air Plasma via an Eco-Friendly Approach. *Clean Technol. Environ. Policy* **2022**, *24*, 2143–2159. [CrossRef]
6. Károly, Z.; Kalácska, G.; Zsidai, L.; Mohai, M.; Klébert, S. Improvement of Adhesion Properties of Polyamide 6 and Polyoxymethylene-Copolymer by Atmospheric Cold Plasma Treatment. *Polymers* **2018**, *10*, 1380. [CrossRef]
7. Dufay, M.; Jimenez, M.; Degoutin, S. Effect of Cold Plasma Treatment on Electrospun Nanofibers Properties: A Review. *ACS Appl. Bio. Mater.* **2020**, *3*, 4696–4716. [CrossRef]
8. Nascimento, L.; Gasi, F.; Landers, R.; Sobrinho, A.D.S.; Aragão, E.; Fraga, M.; Petraconi, G.; Chiappim, W.; Pessoa, R. Physicochemical Studies on the Surface of Polyamide 6.6 Fabrics Functionalized by DBD Plasmas Operated at Atmospheric and Sub-Atmospheric Pressures. *Polymers* **2020**, *12*, 2128. [CrossRef]
9. Lee, C.; Yang, S.; Choi, D.; Kim, W.; Kim, J.; Hong, J. Chemically surface-engineered polydimethylsiloxane layer via plasma treatment for advancing textile-based triboelectric nanogenerators. *Nano Energy* **2019**, *57*, 353–362. [CrossRef]
10. Yang, J.; Pu, Y.; Miao, D.; Ning, X. Fabrication of Durably Superhydrophobic Cotton Fabrics by Atmospheric Pressure Plasma Treatment with a Siloxane Precursor. *Polymers* **2018**, *10*, 460. [CrossRef]
11. Nguyen Thi Kim, T.; Vu Thi Hong, K.; Vu Thi, N.; Vu Manh, H. The Effect of DBD Plasma Activation Time on the Dyeability of Woven Polyester Fabric with Disperse Dye. *Polymers* **2021**, *13*, 1434. [CrossRef]
12. Ngo, H.-T.; Hong, K.V.T.; Nguyen, T.-B. Surface Modification by the DBD Plasma to Improve the Flame-Retardant Treatment for Dyed Polyester Fabric. *Polymers* **2021**, *13*, 3011. [CrossRef] [PubMed]
13. Lee, Y.-H.; Yeom, G.-Y. Properties and Applications of a Modified Dielectric Barrier Discharge Generated at Atmospheric Pressure. *Jpn. J. Appl. Phys.* **2005**, *44*, 1076–1080. [CrossRef]
14. Dell'Orto, E.C.; Vaccaro, A.; Riccardi, C. Morphological and Chemical Analysis of PP Film Treated by Dielectric Barrier Discharge. *J. Phys. Conf. Ser.* **2014**, *550*, 012032. [CrossRef]
15. Ma, Y.; Ha, C.S.; Hwang, S.W.; Lee, H.J.; Kim, G.C.; Lee, K.-W.; Song, K. Non-Thermal Atmospheric Pressure Plasma Preferentially Induces Apoptosis in p53-Mutated Cancer Cells by Activating ROS Stress-Response Pathways. *PLoS ONE* **2014**, *9*, e91947. [CrossRef] [PubMed]

16. Lim, M.T.; Zulkifli, A.Z.S.; Jayapalan, K.K.; Chin, O. Development of a dimensionless parameter for characterization of dielectric barrier discharge devices with respect to geometrical features. *Plasma Sci. Technol.* **2017**, *19*, 095402. [CrossRef]
17. Motyka-Pomagruk, A.; Dzimitrowicz, A.; Orlowski, J.; Babinska, W.; Terefinko, D.; Rychlowski, M.; Prusinski, M.; Pohl, P.; Lojkowska, E.; Jamroz, P.; et al. Implementation of a Non-Thermal Atmospheric Pressure Plasma for Eradication of Plant Pathogens from a Surface of Economically Important Seeds. *Int. J. Mol. Sci.* **2021**, *22*, 9256. [CrossRef]
18. Mastanaiah, N.; Banerjee, P.; Johnson, J.A.; Roy, S. Examining the Role of Ozone in Surface Plasma Sterilization Using Dielectric Barrier Discharge (DBD) Plasma. *Plasma Process. Polym.* **2013**, *10*, 1120–1133. [CrossRef]
19. Park, J.-H.; Park, J.-S.; Lee, J.-H.; Jeong, B.-H. Space Sterilization Effect through High-Density Plasma Ozone Using DBD Device. *J. Electr. Eng. Technol.* **2022**, *17*, 2771–2778. [CrossRef]
20. Cui, Y.; Cheng, J.; Chen, Q.; Yin, Z. The Types of Plasma Reactors in Wastewater Treatment. *Proc. IOP Conf. Ser. Earth Environ. Sci.* **2018**, *208*, 012002. [CrossRef]
21. Das, S.P.; Dalei, G.; Barik, A. A Dielectric Barrier Discharge (DBD) Plasma Reactor: An Efficient Tool to Measure the Sustainability of Non-Thermal Plasmas through the Electrical Breakdown of Gases. *Proc. IOP Conf. Ser. Mater. Sci. Eng.* **2018**, *410*, 012004. [CrossRef]
22. Kostov, K.; Nishime, T.; Hein, L.; Toth, A. Study of polypropylene surface modification by air dielectric barrier discharge operated at two different frequencies. *Surf. Coat. Technol.* **2013**, *234*, 60–66. [CrossRef]
23. de Souza, I.A.; Neto, A.B.D.N.; de Queiroz, J.C.A.; Matamoros, E.P.; Costa, T.H.D.C.; Feitor, M.C.; de Souza, J.M.L.; Camara, N.T.; Sobrinho, V.D.S.S. Study of the Influence of Variation in Distances Between Electrodes in Spectral DBD Plasma Excitation. *Mater. Res.* **2016**, *19*, 202–206. [CrossRef]
24. Kondeti, V.S.K.; Phan, C.Q.; Wende, K.; Jablonowski, H.; Gangal, U.; Granick, J.L.; Hunter, R.C.; Bruggeman, P.J. Long-lived and short-lived reactive species produced by a cold atmospheric pressure plasma jet for the inactivation of Pseudomonas aeruginosa and Staphylococcus aureus. *Free. Radic. Biol. Med.* **2018**, *124*, 275–287. [CrossRef] [PubMed]
25. Nascimento, F.; Petroski, K.; Kostov, K. Effects of O_2 Addition on the Discharge Parameters and Production of Reactive Species of a Transferred Atmospheric Pressure Plasma Jet. *Appl. Sci.* **2021**, *11*, 6311. [CrossRef]
26. Van Gils, C.; Hofmann, S.; Boekema, B.K.H.L.; Brandenburg, R.; Bruggeman, P.J. Mechanisms of bacterial inactivation in the liquid phase induced by a remote RF cold atmospheric pressure plasma jet. *J. Phys. D Appl. Phys.* **2013**, *46*, 175203. [CrossRef]
27. Jirásek, V.; Lukeš, P. Formation of reactive chlorine species in saline solution treated by non-equilibrium atmospheric pressure He/O_2 plasma jet. *Plasma Sources Sci. Technol.* **2019**, *28*, 035015. [CrossRef]
28. Morabit, Y.; Hasan, M.I.; Whalley, R.D.; Robert, E.; Modic, M.; Walsh, J.L. A review of the gas and liquid phase interactions in low-temperature plasma jets used for biomedical applications. *Eur. Phys. J.* **2021**, *75*, 32. [CrossRef]
29. Leite, L.D.P.; de Oliveira, M.A.C.; Vegian, M.R.D.C.; Sampaio, A.D.G.; Nishime, T.M.C.; Kostov, K.G.; Koga-Ito, C.Y. Effect of Cold Atmospheric Plasma Jet Associated to Polyene Antifungals on *Candida albicans* Biofilms. *Molecules* **2021**, *26*, 5815. [CrossRef]
30. Lima, G.D.M.G.; Borges, A.C.; Nishime, T.M.C.; Santana-Melo, G.d.F.; Kostov, K.G.; Mayer, M.P.A.; Koga-Ito, C.Y. Cold Atmospheric Plasma Jet as a Possible Adjuvant Therapy for Periodontal Disease. *Molecules* **2021**, *26*, 5590. [CrossRef]
31. Doria, A.C.O.C.; Figueira, F.R.; de Lima, J.S.B.; Figueira, J.A.N.; Castro, A.H.R.; Sismanoglu, B.N.; Petraconi, G.; Maciel, H.S.; Khouri, S.; Pessoa, R.S. Inactivation of Candida albicans biofilms by atmospheric gliding arc plasma jet: Effect of gas chemistry/flow and plasma pulsing. *Plasma Res. Express* **2018**, *1*, 015001. [CrossRef]
32. Lu, X.; Laroussi, M.; Puech, V. On atmospheric-pressure non-equilibrium plasma jets and plasma bullets. *Plasma Sources Sci. Technol.* **2012**, *21*, 034005. [CrossRef]
33. Kutasi, K.; Popović, D.; Krstulović, N.; Milošević, S. Tuning the composition of plasma-activated water by a surface-wave microwave discharge and a kHz plasma jet. *Plasma Sources Sci. Technol.* **2019**, *28*, 095010. [CrossRef]
34. Laroussi, M.; Akan, T. Arc-Free Atmospheric Pressure Cold Plasma Jets: A Review. *Plasma Process. Polym.* **2007**, *4*, 777–788. [CrossRef]
35. Oh, J.-S.; Szili, E.J.; Gaur, N.; Hong, S.-H.; Furuta, H.; Kurita, H.; Mizuno, A.; Hatta, A.; Short, R.D. How to assess the plasma delivery of RONS into tissue fluid and tissue. *J. Phys. D Appl. Phys.* **2016**, *49*, 304005. [CrossRef]
36. Liu, Z.; Zhou, C.; Liu, D.; He, T.; Guo, L.; Xu, D.; Kong, M. Quantifying the concentration and penetration depth of long-lived RONS in plasma-activated water by UV absorption spectroscopy. *AIP Adv.* **2019**, *9*, 015014. [CrossRef]
37. Milhan, N.V.M.; Chiappim, W.; Sampaio, A.D.G.; da Cruz Vegian, M.R.; Pessoa, R.S.; Koga-Ito, C.Y. Applications of Plasma-Activated Water in Dentistry: A Review. *Int. J. Mol. Sci.* **2022**, *23*, 4131. [CrossRef]
38. Ranjan, R.; Krishnamraju, P.V.; Shankar, T.; Gowd, S. Nonthermal Plasma in Dentistry: An Update. *J. Int. Soc. Prev. Community Dent.* **2017**, *7*, 71–75.
39. Gherardi, M.; Tonini, R.; Colombo, V. Plasma in Dentistry: Brief History and Current Status. *Trends Biotechnol.* **2017**, *36*, 583–585. [CrossRef]
40. Lata, S.; Chakravorty, S.; Mitra, T.; Pradhan, P.K.; Mohanty, S.; Patel, P.; Jha, E.; Panda, P.K.; Verma, S.K.; Suar, M. Aurora Borealis in dentistry: The applications of cold plasma in biomedicine. *Mater. Today Bio.* **2022**, *13*, 100200. [CrossRef]
41. Thirumdas, R.; Kothakota, A.; Annapure, U.; Siliveru, K.; Blundell, R.; Gatt, R.; Valdramidis, V.P. Plasma activated water (PAW): Chemistry, physico-chemical properties, applications in food and agriculture. *Trends Food Sci. Technol.* **2018**, *77*, 21–31. [CrossRef]
42. Subramanian, P.G.; Rao, H.; Shivapuji, A.M.; Girard-Lauriault, P.L.; Rao, L. Plasma-activated water from DBD as a source of nitrogen for agriculture: Specific energy and stability studies. *J. Appl. Phys.* **2021**, *129*, 093303. [CrossRef]

43. Zhou, R.; Zhou, R.; Wang, P.; Xian, Y.; Mai-Prochnow, A.; Lu, X.P.; Cullen, P.J.; Ostrikov, K.; Bazaka, K. Plasma-activated water: Generation, origin of reactive species and biological applications. *J. Phys. D Appl. Phys.* **2020**, *53*, 303001. [CrossRef]
44. Pan, J.; Li, Y.L.; Liu, C.M.; Tian, Y.; Yu, S.; Wang, K.L.; Zhang, J.; Fang, J. Investigation of Cold Atmospheric Plasma-Activated Water for the Dental Unit Waterline System Contamination and Safety Evaluation in Vitro. *Plasma Chem. Plasma Process.* **2017**, *37*, 1091–1103. [CrossRef]
45. Guo, J.-J.; Li, M.-D.; Sun, Q.-Q.; Yang, W.; Zhou, P.; Ding, S.-J.; Zhang, D.W. A Water-free Low Temperature Process for Atomic Layer Deposition of Al2O3Films. *Chem. Vap. Depos.* **2013**, *19*, 156–160. [CrossRef]
46. De Melo, T.F.; Rocha, L.C.; Silva, R.P.; Pessoa, R.S.; Negreiros, A.M.P.; Sales Júnior, R.; Tavares, M.B.; Alves Junior, C. Plasma–Saline Water Interaction: A Systematic Review. *Materials* **2022**, *15*, 4854. [CrossRef]
47. Chiappim, W.; Sampaio, A.; Miranda, F.; Fraga, M.; Petraconi, G.; Sobrinho, A.D.S.; Kostov, K.; Koga-Ito, C.; Pessoa, R. Antimicrobial Effect of Plasma-Activated Tap Water on *Staphylococcus aureus*, *Escherichia coli*, and *Candida albicans*. *Water* **2021**, *13*, 1480. [CrossRef]
48. Chiappim, W.; Sampaio, A.; Miranda, F.; Petraconi, G.; Sobrinho, A.S.; Cardoso, P.; Kostov, K.; Koga-Ito, C.; Pessoa, R. Nebulized plasma-activated water has an effective antimicrobial effect on medically relevant microbial species and maintains its physicochemical properties in tube lengths from 0.1 up to 1.0 m. *Plasma Process. Polym.* **2021**, *18*, 2100010. [CrossRef]
49. Sampaio, A.D.G.; Chiappim, W.; Milhan, N.V.M.; Neto, B.B.; Pessoa, R.; Koga-Ito, C.Y. Effect of the pH on the Antibacterial Potential and Cytotoxicity of Different Plasma-Activated Liquids. *Int. J. Mol. Sci.* **2022**, *23*, 13893. [CrossRef]
50. Lima, G.D.M.G.; Carta, C.F.L.; Borges, A.C.; Nishime, T.M.C.; da Silva, C.A.V.; Caliari, M.V.; Mayer, M.P.A.; Kostov, K.G.; Koga-Ito, C.Y. Cold Atmospheric Pressure Plasma Is Effective against *P. gingivalis* (HW24D-1) Mature Biofilms and Non-Genotoxic to Oral Cells. *Appl. Sci.* **2022**, *12*, 7247. [CrossRef]
51. Vesel, A.; Primc, G. Investigation of Surface Modification of Polystyrene by a Direct and Remote Atmospheric-Pressure Plasma Jet Treatment. *Materials* **2020**, *13*, 2435. [CrossRef] [PubMed]
52. Baniya, H.B.; Shrestha, R.; Guragain, R.P.; Kshetri, M.B.; Pandey, B.P.; Subedi, D.P. Generation and Characterization of an Atmospheric-Pressure Plasma Jet (APPJ) and Its Application in the Surface Modification of Polyethylene Terephthalate. *Int. J. Polym. Sci.* **2020**, *2020*, 9247642. [CrossRef]
53. Abdelaziz, A.; Kim, H.-H.; Teramoto, Y.; Takeuchi, N. Towards launching a stable wide plasma jet from a single tube: I. The importance of controlling the gas dynamics. *J. Phys. D Appl. Phys.* **2021**, *54*, 395203. [CrossRef]
54. Kim, J.Y.; Ballato, J.; Kim, S.-O. Intense and Energetic Atmospheric Pressure Plasma Jet Arrays. *Plasma Process. Polym.* **2012**, *9*, 253–260. [CrossRef]
55. Cao, Z.; Walsh, J.L.; Kong, M.G. Atmospheric plasma jet array in parallel electric and gas flow fields for three-dimensional surface treatment. *Appl. Phys. Lett.* **2009**, *94*, 021501. [CrossRef]
56. Liu, F.; Zhang, B.; Fang, Z.; Wan, M.; Wan, H.; Ostrikov, K.K. Jet-to-jet interactions in atmospheric-pressure plasma jet arrays for surface processing. *Plasma Process. Polym.* **2017**, *15*, 1700114. [CrossRef]
57. Ghasemi, M.; Olszewski, P.; Bradley, J.; Walsh, J. Interaction of multiple plasma plumes in an atmospheric pressure plasma jet array. *J. Phys. D Appl. Phys.* **2013**, *46*, 052001. [CrossRef]
58. Khun, J.; Machková, A.; Kašparová, P.; Klenivskyi, M.; Vaňková, E.; Galář, P.; Julák, J.; Scholtz, V. Non-Thermal Plasma Sources Based on Cometary and Point-to-Ring Discharges. *Molecules* **2022**, *27*, 238. [CrossRef]
59. Maddah, H.; Maddah, H.A. Polypropylene as a Promising Plastic: A Review. *Am. J. Polym. Sci.* **2016**, *6*, 1–11. [CrossRef]
60. Mandolfino, C. Polypropylene surface modification by low pressure plasma to increase adhesive bonding: Effect of process parameters. *Surf. Coat. Technol.* **2019**, *366*, 331–337. [CrossRef]
61. Pandiyaraj, K.N.; Selvarajan, V.; Deshmukh, R.; Gao, C. Adhesive properties of polypropylene (PP) and polyethylene terephthalate (PET) film surfaces treated by DC glow discharge plasma. *Vacuum* **2008**, *83*, 332–339. [CrossRef]
62. Mui, T.S.M.; Mota, R.P.; Quade, A.; Hein, L.R.D.O.; Kostov, K.G. Uniform surface modification of polyethylene terephthalate (PET) by atmospheric pressure plasma jet with a horn-like nozzle. *Surf. Coat. Technol.* **2018**, *352*, 338–347. [CrossRef]
63. Karahan, H.A.; Özdo, E. Improvements of Surface Functionality of Cotton Fibers by Atmospheric Plasma Treatment. *Fibers Polym.* **2008**, *9*, 21–26. [CrossRef]
64. Chen, W.; Yu, J.; Hu, W.; Chen, G. Partial hydrophilic modification of biaxially oriented polypropylene film by an atmospheric pressure plasma jet with the allylamine monomer. *Appl. Surf. Sci.* **2016**, *387*, 957–964. [CrossRef]
65. Mandolfino, C.; Lertora, E.; Gambaro, C.; Pizzorni, M. Functionalization of Neutral Polypropylene by Using Low Pressure Plasma Treatment: Effects on Surface Characteristics and Adhesion Properties. *Polymers* **2019**, *11*, 202. [CrossRef]
66. Morent, R.; De Geyter, N.; Leys, C.; Gengembre, L.; Payen, E. Comparison between XPS- and FTIR-analysis of plasma-treated polypropylene film surfaces. *Surf. Interface Anal.* **2008**, *40*, 597–600. [CrossRef]
67. Mandolfino, C.; Lertora, E.; Gambaro, C. Influence of cold plasma treatment parameters on the mechanical properties of polyamide homogeneous bonded joints. *Surf. Coat. Technol.* **2017**, *313*, 222–229. [CrossRef]
68. Mylläri, V.; Ruoko, T.-P.; Syrjälä, S. A comparison of rheology and FTIR in the study of polypropylene and polystyrene photodegradation. *J. Appl. Polym. Sci.* **2015**, *132*. [CrossRef]
69. Gopanna, A.; Mandapati, R.N.; Thomas, S.P.; Rajan, K.; Chavali, M. Fourier transform infrared spectroscopy (FTIR), Raman spectroscopy and wide-angle X-ray scattering (WAXS) of polypropylene (PP)/cyclic olefin copolymer (COC) blends for qualitative and quantitative analysis. *Polym. Bull.* **2018**, *76*, 4259–4274. [CrossRef]

70. Fang, J.; Zhang, L.; Sutton, D.; Wang, X.; Lin, T. Needleless Melt-Electrospinning of Polypropylene Nanofibres. *J. Nanomater.* **2012**, *2012*, 382639. [CrossRef]
71. Mahindrakar, J.; Patil, Y.; Salunkhe, P.; Ankushrao, S.; Kadam, V.; Ubale, V.; Ghanwat, A. Optically transparent, organosoluble poly(ether-amide)s bearing triptycene unit; synthesis and characterization. *J. Macromol. Sci. Part A* **2018**, *55*, 658–667. [CrossRef]
72. Booth, J.-P.; Mozetic, M.; Nikiforov, A.Y.; Oehr, C. Foundations of plasma surface functionalization of polymers for industrial and biological applications. *Plasma Sources Sci. Technol.* **2022**, *31*, 103001. [CrossRef]

polymers

Article

Influence of Dielectric Barrier Discharge Cold Plasma Treatment on Starch, Gelatin, and Bacterial Cellulose Biodegradable Polymeric Films

Mayara Lima Goiana [1], Adriano Lincoln Albuquerque Mattos [2], Henriette Monteiro Cordeiro de Azeredo [3], Morsyleide de Freitas Rosa [2] and Fabiano André Narciso Fernandes [1,*]

1 Departamento de Engenharia Química, Universidade Federal do Ceará, Fortaleza 60440-900, CE, Brazil
2 Embrapa Agroindustria Tropical, R. Dra. Sara Mesquita, 2270, Fortaleza 60511-110, CE, Brazil
3 Embrapa Instrumentação, R. 15 de Novembro, 1452, São Carlos 13560-970, SP, Brazil
* Correspondence: fabiano@ufc.br

Abstract: The environmental damage caused by plastic packaging and the need to reduce pollution requires actions to substitute plastic materials for more sustainable and biodegradable materials. Starch, gelatin, and bacterial cellulose films are three potential biodegradable polymeric films for use in packaging. However, these materials need improvements in their physical, chemical, and mechanical properties to be used in packaging. In this work, these films were treated with cold plasma to evaluate the effects of treatment conditions on several physical, chemical, and mechanical properties. The dielectric barrier discharge plasma technology was applied with varying treatment times (0 to 20 min) and excitation frequencies (50 to 900 Hz) at 20 kV. The optimal excitation frequency for starch films (50 Hz) was different from the optimal frequency for gelatin and bacterial cellulose films (900 Hz), indicating a high dependency on the treatment in this variable that is often neglected. Plasma treatment improved the hydrophobicity, surface morphology, water resistance, and mechanical properties of all three films, with the advantage of not recurring to chemical or biological additives.

Keywords: biopolymer; edible films; starch; gelatin; bacterial cellulose

Citation: Goiana, M.L.; Mattos, A.L.A.; de Azeredo, H.M.C.; de Freitas Rosa, M.; Fernandes, F.A.N. Influence of Dielectric Barrier Discharge Cold Plasma Treatment on Starch, Gelatin, and Bacterial Cellulose Biodegradable Polymeric Films. *Polymers* **2022**, *14*, 5215. https://doi.org/10.3390/polym14235215

Academic Editors: Choon-Sang Park and Nadege Follain

Received: 17 October 2022
Accepted: 26 November 2022
Published: 30 November 2022

1. Introduction

Environmental damage caused by the accumulation of non-biodegradable plastic packaging waste has generated high interest in developing possible biodegradable substitutes. Natural biodegradable polymers-based packaging is especially demanded by the food industry, which is one of the largest consumers of plastic packaging [1].

Biomolecules such as starch, bacterial cellulose, and gelatin have been used to develop packaging materials [2]. These packaging materials are considered biodegradable, sustainable, and environmentally friendly [3,4]. However, some disadvantages must be overcome, such as their high hydrophilicity, low water vapor, gas, and light barrier, and low mechanical resistance [1,5].

Starch can be obtained from several vegetable sources, such as maize, seed kernels, cereal grains, roots, and fruits. Starches are biodegradable and usually have low costs. It consists of two fractions: amylose (straight chain) and amylopectin (branched chain). Starches have good film-forming properties but present some disadvantages, including low mechanical performance and high hydrophilicity, which result in rapid moisture absorption during use and storage [6].

Bacterial cellulose is a natural nanostructured polymer produced by several bacteria. It has good thermal stability, high porosity, good mechanical properties, amphiphilic character, and a high liquid sorption capacity [7]. However, it has a high permeability to water vapor, which limits its application [8].

Gelatin is a linear protein resulting from collagen denaturation of bovine, swine, or fish origin. It has a good ability to form films; however, it has a limited moisture barrier property [9].

Improvements on polymeric films can be carried out by several techniques, such as cold plasma [10], ultrasonics [11], doping with additives [12], production of composites [13], and other procedures. In this study, we have applied cold plasma technology to induce changes to starch, gelatin, and bacterial cellulose films.

Cold plasma is a technology that can be applied for surface treatment, including modification or functionalization of polymers. Recently, it has emerged as an innovative, effective, and sustainable alternative to traditional chemical treatments. One of the plasma generation methods is the dielectric barrier discharge (DBD), which operates at room temperature and atmospheric pressure and does not involve using petroleum-based solvents. Cold plasma generates ultraviolet light, high-energy electrons, and reactive particles, which can initiate several physical-chemical reactions on the surface of polymers [14,15].

Several studies have investigated the effect of DBD plasma on polymers, varying the exposure time, voltage, and excitation frequency. Wu et al. [15] showed that cold plasma improved casein-based films' barrier and mechanical properties by increasing the plasma generation voltage. Films from corn starch with high amylose content and treated with high-voltage cold plasma (70 and 80 kV) increased the film roughness and hydrophilicity [16]. New functional groups containing oxygen were introduced to the sodium caseinate film surface, improving its hydrophilicity by applying short plasma exposure times (1 to 5 min, at 70 kV). At the same time, no changes occurred in the helical structure of the protein [17].

Studies on plasma treatment of bacterial cellulose films are still scarce and more work is needed to understand the changes induced by plasma species on the properties of bacterial cellulose films. Studies have been carried out on the degradation of toxic compounds from the raw material used as a carbon source for bacterial cellulose production [18,19] and for the deposition of oxides and other materials on the surface of bacterial cellulose [20]. However, there are no studies on plasma treatment of bacterial cellulose films aiming toward the improvement of its physical, chemical, and mechanical properties.

Modification of starch using plasma technology has been successfully addressed. Plasma increases carbonyl and carboxyl groups, resulting in improved paste viscosity, hydration properties, and digestibility [21,22]. These studies showed that the changes in starch properties were greater at excitation frequencies between 100 and 300 Hz and not at 50 Hz as commonly applied [10,21]. Plasma treatment in starch films also indicated that the optimal excitation frequency range was between 200 and 300 Hz [10,23].

Most studies on cold plasma treatment on polymers were carried out at a single plasma excitation frequency, usually at 50 Hz. However, the excitation frequency has been proven to be an essential factor in cold plasma applications [24–26]. The excitation frequency influences the concentration and type of ions and free radicals generated in the cold plasma, inducing different chemical transformations in the treated product. Plasma treatments on bacterial cellulose and gelatin films at 50 Hz have not resulted in good improvements in film properties. Due to the importance of the excitation frequency in the generation of plasma in many applications, such as starch modification [10,21], aroma modulation [24], phenolics content [27], and mitigation of off-flavors [28], this work has addressed the significance of the excitation frequency in three different biodegradable polymer films.

This study applied DBD plasma at three different plasma excitation frequencies to three films produced from natural polymers (gelatin, starch, and bacterial cellulose). The effects of cold plasma treatment on the film water solubility, hydrophobicity, chemical group composition, mechanical properties, and surface morphology were evaluated.

2. Materials and Methods

2.1. Materials

Commercial corn starch (Unilever, brand Maizena™, Garanhuns, Brazil) and commercial fish gelatin (Fitoway Laboratório Nutricional Ltd. a, brand São Pedro™, Tarumã, Brazil) were purchased in the local supermarket. Bacterial cellulose membranes were supplied by Seven Indústria de Produtos Biotecnológicas Ltd. (Ibiporã, Brazil). Glycerol was purchased from Dinâmica (São Paulo, Brazil).

2.2. Elaboration of the Biopolymer Films

The casting technique was used in the production of the films. Glycerol acted as a plasticizer. The degassing step of the filmogenic dispersions to remove the air bubbles formed during the process, common to the three films, was carried out before the final drying. The process took place in a vacuum system with a Vacuubrand-1C pump.

The starch film was produced following the methodology proposed by Oliveira et al. [29]. In brief, 5 g of corn starch was dissolved in 100 mL of distilled water (5% w/v) and heated to 95 °C under magnetic stirring for 30 min for complete starch gelatinization. Glycerol was added to the starch solution in a 25% w/w proportion and maintained at 60 °C for 15 min under magnetic stirring. Afterward, the dispersion was homogenized in an Ultra-Turrax (IKA model T25) at 10,000 rpm for 15 min. The degassed filmogenic dispersion was poured onto glass plates coated with a polyester film (Mylar™) and allowed to dry under ambient conditions (25 °C) for 24 h.

Bacterial cellulose membranes were oven-dried for 48 h at 50 °C and ground in an analytical mill (IKA model A11). The dried bacterial cellulose underwent an oxidation process following the method presented by Saito et al. [30]. After oxidation, a mechanical treatment took place in a colloid mill (Meteor model Rex Inox I-V-N) for 10 min, obtaining nano-fibrillated bacterial cellulose (NFBC). The NFBC was freeze-dried (Liotop model LP510). The methodology proposed by Nascimento et al. [31] was used to obtain the filmogenic dispersion of bacterial cellulose. For this purpose, an aqueous dispersion was prepared to contain 1% (w/v) of lyophilized NFBC and 50% (w/w) of glycerol on a dry basis of the matrix. The dispersion was homogenized in an Ultra-Turrax (IKA model T25) at 13,000 rpm for 15 min, sonicated for 2 min at 19 kHz (Unique model DES500) for degassing, and dried in stainless-steel trays coated with Mylar™ in an oven at 50 °C for 48 h.

Gelatin films were prepared following the procedure described by Santos et al. [32]. An aqueous solution was prepared with 9.6% w/v gelatin and 25% w/v glycerol (on a dry basis). The gelatin was hydrated and heated to 50 °C for 15 min, adding the glycerol in the last 5 min, under constant agitation. The solution was homogenized in an Ultra-Turrax (IKA model T25) at 10,000 rpm for 10 min. The degassed film-forming solution was poured onto glass plates covered with Mylar™ and allowed to dry at 25 °C for 24 h.

2.3. Dielectric Barrier Discharge Plasma Treatment

The films were treated with cold plasma in dielectric barrier discharge equipment consisting of a power source (Inergiae model PLS0130) and two 8 cm-diameter aluminum electrodes separated by 2 mm acrylic plates. The films were placed within a 3 cm gap between the dielectric barriers centered with the electrodes.

Cold plasma was generated at 20 kV (maximal voltage achieved with the equipment) and at excitation frequencies of 50, 400, and 900 Hz. The films were treated for 5 min. The films were treated adhered to the polyester support (Mylar™) and without the polyester support. All assays were carried out in triplicates.

The optimal excitation frequency for plasma treatment was defined as the frequency at which greater changes were observed in the chemical composition, hydrophobicity, and film solubility. The best films regarding hydrophilicity and solubility were plasma-treated for 10, 20, and 30 min under the optimal excitation frequency for further analysis of the films' water solubility, hydrophobicity, and chemical group composition. The films with the best results for hydrophobicity, water solubility, and chemical composition were further tested for changes in surface morphology and mechanical properties.

2.4. *Physical and Chemical Analysis*

2.4.1. Thickness and Humidity

The thickness of the films was measured using a digital micrometer (Mitutoyo model IP65). Ten replicates were taken.

The film humidity was measured in an infrared moisture balance (Marte model ID50) operating at 100 °C. The humidity was based on 1 g of sample. Three replicates were taken.

2.4.2. Hydrophobicity

Changes in the hydrophobicity of the films were determined through the water contact angle (GBX Instrumentation Specifique), attained according to ASTM D-5725-99 [33]. Film samples (2 $\times$ 2 cm) were fixed to a glass holder, and an image was captured (Pixe Link Nikon camera) when the drop touched the film surface. The test was carried out at room temperature (23 $\pm$ 2 °C). All assays were performed in quadruplicate.

2.4.3. Water Solubility

The water solubility was determined following the method described by Pena-Serna and Lopes-Filho [34]. The film samples (2 cm-diameter discs) were dried in an oven at 105 °C for 24 h, weighed (W_0), and immersed in 50 mL of distilled water at 25 $\pm$ 2 °C for 24 h under agitation in an orbital shaker (Tecnal model TE-142) at 100 rpm. After this period, the samples were removed and dried in an oven (105 °C for 24 h) to determine the mass of the material that was not solubilized (W). The solubility was calculated using Equation (1):

$$S(\%) = \frac{W - W_0}{W_0} 100 \tag{1}$$

2.4.4. FTIR Analysis

The absence or formation of chemical groups on the surfaces of the films was evaluated using Fourier transform infrared (FTIR) spectra. A Perkin-Elmer Spectrum Two equipment (Perkin-Elmer, Waltham, MA, USA) was used in attenuated total reflection mode (ATR). The spectra were collected in wavelengths between 4000 and 650 cm^{-1}, using 32 scans and a 4 cm^{-1} resolution.

2.5. *Surface Morphology*

The surface morphology was characterized using a Quanta FEG SEM at an accelerating voltage of 10 kV. The films were cut into 5 $\times$ 5 mm squares, fixed on the sample table with conductive tape, and covered with a thin layer of gold.

2.6. *Mechanical Properties*

The mechanical properties of the films were determined using an EMIC DL-3000 equipment, following the standard method D882-18 [35]. The tensile strength (TS), elongation at break (EB), and elasticity module (EM) were determined. The films were cut into 4 $\times$ 75 mm dumbbell shapes. The mechanical tests were performed with an initial clamping distance of 50 mm and a tensile rate of 50 mm/s. All measurements were carried out in quintuplicate.

2.7. *Selection of the Best Films and Cold Plasma Operating Conditions*

The selection of the best biodegradable polymeric films and their respective cold plasma operating conditions was based on a desirability function that considered the highest change in functional groups measured by FTIR, highest hydrophobicity, and lowest water solubility (Equation (1)). These three factors were chosen because the goal of cold plasma processing is to induce physical and chemical changes in the film surface and chemical structure, and because films with high hydrophobicity and low water solubility are desired for food applications. The weight of the hydrophobicity in the desirability function was higher than the other two factors due to the importance of this property in food packaging. The contact angle divided by 90° was used to address the hydrophobicity, measuring how close the contact angle was from the value for which a film is considered hydrophobic. For the changes in the chemical structure, the absorbance that presented the largest change was used in the function.

$$Score = \frac{(100 - S(\%))}{100} + \frac{(Abs^{mas} - Abs)}{Abs^{max}} + 2\frac{CA}{90} \quad (2)$$

The starch, gelatin, and bacterial cellulose with the highest scores were considered the best biodegradable films. Surface analysis and mechanical properties' analysis were carried out only with the best films.

2.8. Statistical Analysis

The results were presented as mean ± standard deviation. The statistical analysis was performed using analysis of variance (ANOVA), followed by the Tukey's test, with the Statistica version 7 software.

3. Results

The films were produced by the casting method, which enabled attaining films with low thickness and humidity (Table 1). The starch and gelatin films were thinner than the bacterial cellulose film, while their humidity was not significantly different.

Table 1. Width and humidity (mass basis) of the starch, bacterial cellulose, and gelatin films. For each property, values followed by at least one common letter are not different from each other ($p > 0.05$).

Film	Thickness (mm)	Humidity (%)
Starch	0.054 ± 0.002 [b]	10.2 ± 0.5 [a]
Bacterial cellulose	0.126 ± 0.011 [a]	9.6 ± 2.4 [a]
Gelatin	0.061 ± 0.003 [b]	12.5 ± 1.2 [a]

3.1. Hydrophobicity

Table 2 presents the effects of the excitation frequency on the hydrophobic character and water solubility of the films subjected to plasma treatment.

Table 2. The angle of contact and water solubility of the starch, bacterial cellulose, and gelatin films treated with dielectric barrier discharge plasma: with and without Mylar™ support. For each property, values followed by at least one common letter are not different from each other ($p > 0.05$).

Plasma Excitation Frequency (Hz)	Contact Angle (°)			Water Solubility (%)		
	Starch	Bacterial Cellulose	Gelatin	Starch	Bacterial Cellulose	Gelatin
With Mylar™ support						
Control	55.7 ± 0.8 [a]	80.3 ± 0.3 [a]	69.9 ± 0.4 [a]	96.9 ± 1.0 [a]	16.7 ± 1.0 [b]	34.3 ± 0.4 [c]
50	55.1 ± 1.0 [a]	77.9 ± 0.6 [b]	68.9 ± 0.9 [a]	96.6 ± 0.1 [a]	18.5 ± 0.4 [a]	37.2 ± 0.8 [a]
400	53.8 ± 0.6 [b]	78.4 ± 0.2 [b]	69.7 ± 0.7 [a]	97.4 ± 0.5 [a]	17.3 ± 0.2 [a]	36.3 ± 0.5 [b]
900	54.2 ± 0.5 [b]	79.6 ± 1.1 [a]	70.3 ± 0.5 [a]	96.2 ± 0.3 [a]	17.1 ± 0.7 [b]	35.4 ± 0.8 [c]
Without Mylar™ support						
Control	55.7 ± 0.8 [a]	80.3 ± 0.3 [c]	69.9 ± 0.4 [c]	96.9 ± 1.0 [a]	16.7 ± 1.0 [a]	34.3 ± 0.4 [b]
50	56.3 ± 1.0 [a]	79.6 ± 0.3 [c]	70.6 ± 0.2 [c]	94.8 ± 0.3 [b]	15.5 ± 0.3 [a]	37.9 ± 0.6 [a]
400	55.1 ± 0.7 [a]	81.2 ± 0.3 [b]	72.5 ± 0.8 [b]	95.4 ± 0.4 [b]	13.9 ± 0.6 [b]	34.8 ± 0.5 [b]
900	55.1 ± 0.4 [a]	83.0 ± 0.4 [a]	77.4 ± 0.8 [a]	97.3 ± 0.2 [a]	11.3 ± 0.4 [c]	32.4 ± 0.4 [c]

The hydrophobicity of the film is an essential feature in deciding the packaging application of any polymeric film [16]. A totally hydrophobic film is characterized by contact angles equal to or greater than 90° [14]. Among the films tested, the bacterial cellulose film had the highest hydrophobicity, while the starch film had the lowest hydrophobicity.

The Mylar support acted as an extra barrier between the electrodes and the films, reducing the effects of plasma treatment. The hydrophobicity of the films did not change significantly or decrease when the Mylar support was used. Thus, its use is not recommended when applying cold plasma.

Plasma treatment increased the hydrophobicity of the bacterial cellulose and gelatin films, while no significant changes were observed in starch films when the Mylar support was not used. The highest increase in hydrophobicity was observed at a plasma excitation frequency of 900 Hz. At 50 Hz, no significant change was observed. Thus, proper setting of the excitation frequency is critical to increasing the hydrophobicity of the films.

The films were subjected to plasma treatment for an extended period at the optimal excitation frequency of each film (Table 3). The increase in processing time increased the contact angle of all films, indicating the production of films with higher hydrophobicity. Thus, the processing time was a significant variable for the process ($p < 0.05$). Plasma application increased the hydrophobicity of the films by 15%, 6%, and 20%, respectively, for the starch, bacterial cellulose, and gelatin films.

Table 3. The contact angle of the starch, bacterial cellulose, and gelatin films treated with dielectric barrier discharge plasma at an optimal excitation frequency (50 Hz for the starch film and 900 Hz for the bacterial cellulose and gelatin films) without Mylar™ support. For each property, values followed by at least one common letter are not different from each other ($p > 0.05$).

Processing Time (min)	Starch	Bacterial Cellulose	Gelatin
0	55.7 ± 0.8^{d}	80.3 ± 0.3^{c}	69.9 ± 0.4^{c}
10	57.0 ± 0.6^{c}	83.0 ± 0.6^{c}	79.0 ± 0.3^{b}
15	61.0 ± 0.3^{b}	84.0 ± 0.3^{b}	81.0 ± 0.7^{b}
20	64.0 ± 0.6^{a}	85.0 ± 0.5^{a}	84.0 ± 0.3^{a}

The increase in hydrophobicity has been attributed to changes in the surface morphology of the films due to an increase in surface roughness. This will be discussed in Section 3.4.

Table 4 presents the contact angles of several films reported in the literature. Starch films have a low contact angle (between 20 and 56°) and are distant from the recommended 90° contact angle for hydrophobic surfaces. Plasma processing and high-pressure processing increase the hydrophobicity of starch films, but these physical changes have limited effects on the films' hydrophobicity. Chemical additives such as cinnamaldehyde have a higher impact on the contact angle and the hydrophilic nature of starch films.

Table 4. The contact angles of several starches, bacterial cellulose, and gelatin films reported in the literature.

Film	Contact Angle (°)	Reference
Corn starch	55.7	This work
Corn starch (plasma-treated)	64	This work
Corn starch (high amylose)	55	[36]
Corn starch + cinnamaldehyde	107.4	[36]
Potato starch	24.5	[37]
Potato starch (high-pressure-treated)	30.2	[37]
Banana starch	50.3	[10]
Banana starch (plasma-treated)	65.1	[10]

Table 4. *Cont.*

Film	Contact Angle (°)	Reference
Gelatin	69.9	This work
Gelatin (plasma-treated)	84	This work
Gelatin	68	[38]
Gelatin + chitosan	56	[38]
Bacterial cellulose	80.3	This work
Bacterial cellulose (plasma-treated)	85	This work
Bacterial cellulose	36	[39]
Bacterial cellulose (plasma-treated)	21	[39]
Bacterial cellulose	14.9	[40]
Bacterial cellulose + curcumin extract	38.6	[40]

Plasma treatment increased the contact angle of gelatin films by 17%, while the production of a gelatin + chitosan composite reduced the contact angle [38].

The use of plasma treatment was shown to increase and decrease the contact angle. Proper setting of the excitation frequency (900 Hz) improved the contact angle, while low excitation frequencies (50 Hz) decreased the contact angle, worsening the wettability properties of gelatin films [39]. This comparison evidences the importance of the excitation frequency in plasma treatment of bacterial cellulose films.

3.2. Solubility in Water

The solubility of the films is an important property that can indicate the presence of hydrophilic groups, in addition to assessing the resistance in aqueous media [41]. Overall, the starch film is the most soluble film, and the bacterial cellulose film is the least soluble film (Table 2).

The Mylar support has also attenuated the effects of plasma treatment regarding the solubility of the films. The solubility of the bacterial cellulose and gelatin films increased when the support was used, which is not a desirable result.

On the other hand, the solubility of all films decreased when the Mylar support was not used. The highest decrease in solubility was observed for the bacterial cellulose film, which was already the least soluble film.

The excitation frequency affected each film differently. A lower excitation frequency (50 Hz) decreased the solubility of the starch film, while a higher excitation frequency (1000 Hz) decreased the bacterial cellulose and the gelatin film solubilities.

The films were subjected to plasma treatment for an extended period at the optimal excitation frequency of each film (Table 5). The increase in processing time decreased the solubility, improving its potential application in packaging. Processing time was also a significant variable for solubility ($p < 0.05$). Plasma application reduced the solubility of the films by 6%, 40%, and 18%, respectively, for the starch, bacterial cellulose, and gelatin films.

Table 5. Solubility of the starch, bacterial cellulose, and gelatin films treated with dielectric barrier discharge plasma at an optimal excitation frequency (50 Hz for the starch film and 900 Hz for the bacterial cellulose and gelatin films) without Mylar™ support. For each property, values followed by at least one common letter are not different from each other ($p > 0.05$).

Processing Time (min)	Starch	Bacterial Cellulose	Gelatin
0	96.9 ± 1.0 [a]	16.7 ± 1.1 [a]	34.3 ± 0.4 [a]
10	94.0 ± 0.5 [b]	11.0 ± 0.8 [b]	31.0 ± 0.1 [bc]
15	92.0 ± 0.4 [c]	10.0 ± 0.3 [b]	30.0 ± 0.3 [bc]
20	91.0 ± 0.1 [c]	10.0 ± 0.2 [b]	28.0 ± 0.3 [c]

Although an 80% increase was observed in the insoluble matter of the starch film after plasma treatment, this film remained with an insoluble matter below that recommended for food packaging applications [42]. The potential application of a film in the food industry requires water-insoluble packaging to increase product integrity and water resistance [42].

Table 6 presents the water solubility of several films reported in the literature. Starch films have an extensive range of solubility (between 15% and 98%), which is usually related to the amylose–amylopectin ratio in the starch. Plasma processing decreased the water solubility of starch films more than adding chemical additives such as cinnamaldehyde and orange peel, while having a similar effect as the starch/date palm pits composite.

Table 6. Solubility of several starches, bacterial cellulose, and gelatin films reported in the literature.

Film	Water Solubility (%)	Reference
Corn starch	96.6	This work
Corn starch (plasma-treated)	91	This work
Corn starch (high amylose)	17	[36]
Corn starch + cinnamaldehyde	16	[36]
Corn starch	34	[43]
Corn starch + orange peel	30	[43]
Corn starch	19.4	[44]
Corn starch + date palm pits	10.2	[44]
Potato starch	28.1	[37]
Potato starch (high-pressure-treated)	29.6	[37]
Banana starch	37.4	[10]
Banana starch (plasma-treated)	30.7	[10]
Bacterial cellulose	16.7	This work
Bacterial cellulose (plasma-treated)	10	This work
Bacterial cellulose	52.8	[45]
Bacterial cellulose (nanocellulose)	27.3	[45]
Bacterial cellulose (nanocrystalline)	32.5	[45]
Whey protein	43	[46]
Whey protein + bacterial cellulose	38	[46]
Gelatin	34.3	This work
Gelatin (plasma-treated)	28	This work
Gelatin	64	[47]
Gelatin + carboxymethylcellulose	21	[47]
Gelatin + polyvinyl alcohol	8	[47]
Gelatin	35.4	[48]
Gelatin (transglutaminase-treated)	26.6	[48]

Plasma treatment decreased the solubility of bacterial cellulose by 40%, which was a similar result attained by chemical treatment to convert bacterial cellulose into nanocellulose (48%) and nanocrystalline cellulose (38%). However, the decrease achieved with plasma treatment reduced the bacterial cellulose to the lowest level reported in the literature. The physical and chemical treatments of bacterial cellulose reduced the solubility more than the formation of composites, for example, the whey protein + bacterial cellulose composite [46].

Gelatin films were less affected by plasma treatment than by enzymatic treatment and incorporation of chemical additives. Enzymatic treatment of gelatin films with transglutaminase [48] showed a slightly superior effect in decreasing the film solubility (25%) compared to the plasma treatment (18%). The addition of polyvinyl alcohol was reported as a good alternative to reduce the gelatin film solubility [47].

3.3. FTIR Analysis

The FTIR spectra of the plasma-treated films are shown in Figure 1. The spectra of the starch film presented the characteristic peaks at 3284 cm^{-1} (-OH vibration), 2900 cm^{-1}

(-CH vibration), 1150 cm^{-1} (C-O stretching), 1076 cm^{-1} (C-H bending), and 929 cm^{-1} (C-H out-of-plane bending).

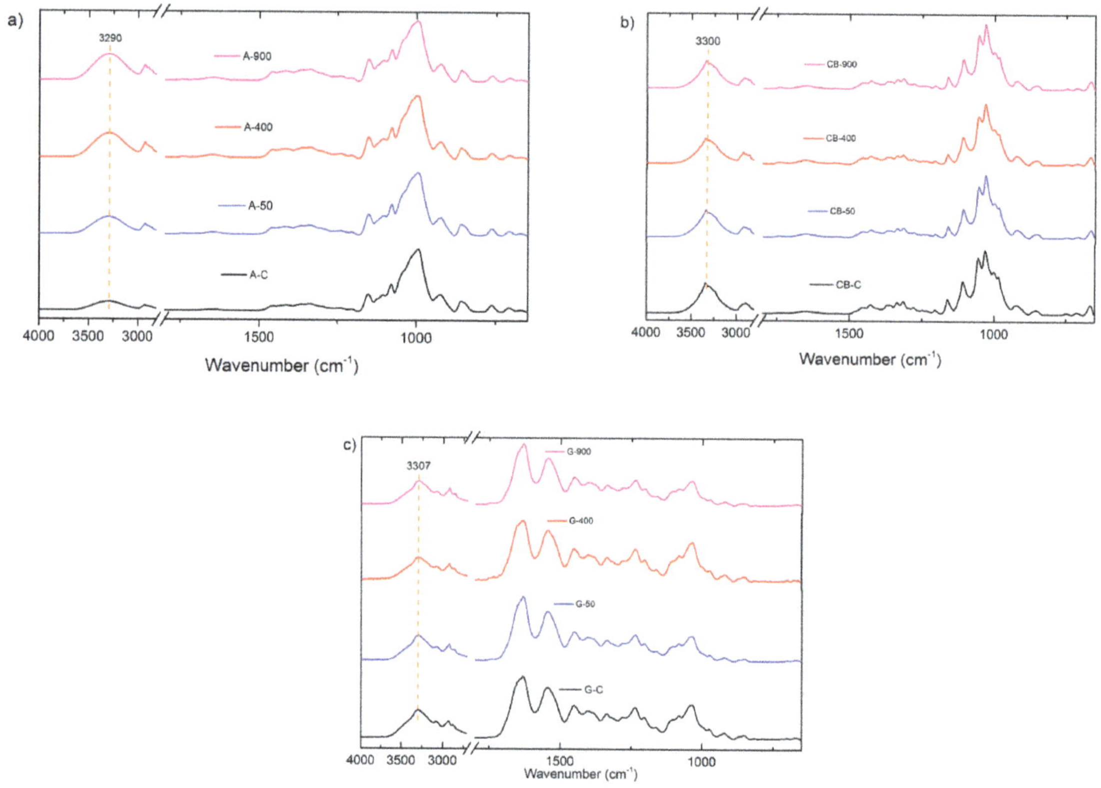

Figure 1. FTIR spectra of (**a**) starch films, (**b**) bacterial cellulose films, and (**c**) gelatin films subjected to plasma treatment.

The spectra of the bacterial cellulose film presented the characteristic peaks at 3300 cm^{-1} (-OH vibration), 2900 cm^{-1} (-CH vibration), 1160 cm^{-1} (C-O-C asymmetric stretching vibration), and 1030 cm^{-1} (C-O stretching). The spectra of the gelatin film presented the characteristic peaks at 3300 cm^{-1} (-OH vibration), 2900 cm^{-1} (-CH vibration), 1635 cm^{-1} (C = O stretching vibration), 1550 cm^{-1} (N-H bending vibration), 1160 cm^{-1} (C-O-C asymmetric stretching and NH_2 oscillation vibrations), and 1030 cm^{-1} (C-O stretching).

No new bands corresponding to the formation of chemical clusters were observed in the plasma-treated starch, bacterial cellulose, and gelatin films. Thus, plasma treatment promoted the interaction between the reactive plasma species and the functional groups already in the films [49].

The hydroxyl band (3284 cm^{-1}) showed a slight reduction in intensity in all three films. The most significant reductions occurred at 50 Hz for the starch film and 900 Hz for the bacterial cellulose and gelatin films. All other characteristic bands of the films were preserved. The hydroxyl band further reduced its intensity when treated for an extended period (20 min). This reduction can be correlated to the higher hydrophobicity of the material plasma-treated for 15 and 20 min.

3.4. Best Biodegradable Films and Plasma Operating Conditions

The results obtained for hydrophobicity, water solubility, and FTIR analysis for the films processed at different cold plasma frequencies were evaluated using the desirability function to determine the best plasma conditions for each biodegradable film. Table 7 presents the scores obtained for the desirability function.

Table 7. Scores obtained for the desirability function used to select the best starches, bacterial cellulose, and gelatin films' processing conditions.

Plasma Frequency (Hz)	Scores		
	Starch	Bacterial Cellulose	Gelatin
50	1.36	2.71	2.22
600	1.28	2.73	2.25
900	1.3	2.88	2.71

The scores of the desirability function have pointed out that the best starch film was produced at 50 Hz, while the best bacterial cellulose and gelatin films were produced at 900 Hz. These scores corroborate to point out that the cold plasma excitation frequency is an important variable when applying plasma to polymeric films.

The highest scores for each film agreed with the highest changes observed in hydrophobicity, water solubility, and chemical changes. Among the three films tested herein, the bacterial cellulose film presented the highest scores, especially due to its highest hydrophobicity, which is an important property of biodegradable films for food packaging.

3.5. Surface Morphology

The starch film showed a rough surface with no cracks (Figure 2). Plasma processing significantly changed the surface of the film. The surface became smoother than the untreated film, but many microscopic cracks appeared on the surface. Tiny white dots appeared scattered throughout the film surface, which could be related to the higher amount of insoluble particles in the film [44].

The bacterial cellulose film presented a reasonable roughness, which increased with plasma treatment. The treated film showed a more rugged structure.

The gelatin film presented a flat structure with several small dots attributed to insoluble matter. The plasma-treated film presented a different aspect, with a smoother surface. The small dots were drastically reduced, which could be attributed to a better homogenization and interaction of the soluble and insoluble matter.

The films' hydrophobicity increased more in the films that presented a higher smoothness (starch and gelatin) than in the film (bacterial cellulose) where plasma treatment increased the roughness. The hydrophobicity is linked with the availability and exposure of hydroxyl groups, which are reduced during plasma treatment. Thus, the reduction in hydrophobicity may be associated with the decrease in hydroxyl groups and the changes in surface morphology, which may have reduced the exposure of the remaining hydroxyl groups.

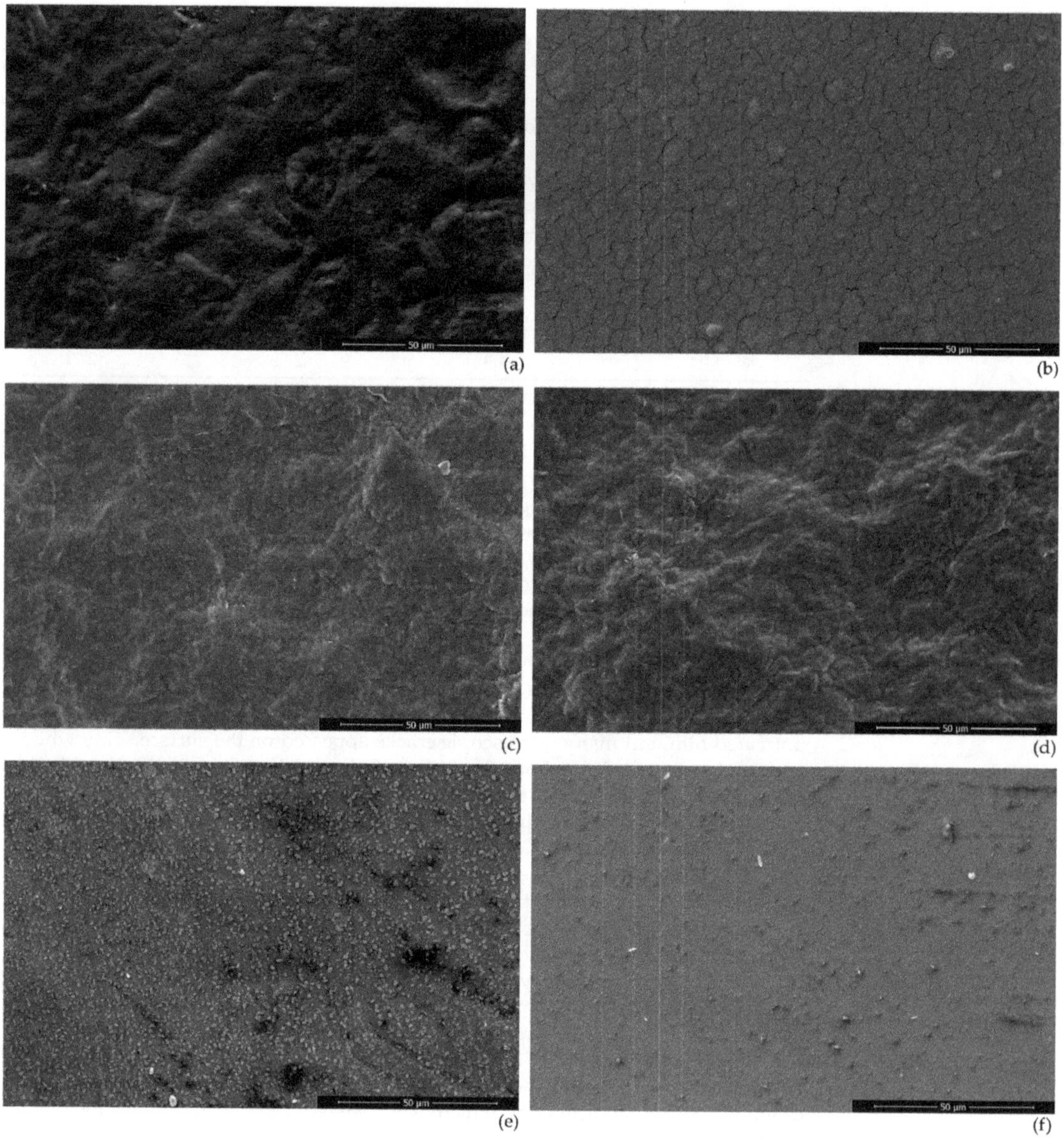

(a) (b) (c) (d) (e) (f)

Figure 2. Scanning electronic micrographs of: (**a**) untreated starch film, (**b**) plasma-treated starch film (20 min at 50 Hz), (**c**) untreated bacterial cellulose film, (**d**) plasma-treated bacterial cellulose film (20 min at 900 Hz), (**e**) untreated gelatin film, and (**f**) plasma-treated gelatin film (20 min at 900 Hz).

3.6. Mechanical Properties

The mechanical properties of the films were evaluated at the best operating conditions for each film (20 min, 50 Hz for starch, 900 Hz for bacterial cellulose and gelatin) and compared to the control (untreated films). Table 8 presents the values obtained for the tensile strength (TS), elongation at break (ER), and elasticity module (EM) for all films.

Table 8. Mechanical properties of the starch, bacterial cellulose, and gelatin films treated for 20 min with dielectric barrier discharge plasma at an optimal excitation frequency (50 Hz for the starch film and 900 Hz for the bacterial cellulose and gelatin films) without Mylar™ support. For each property, values followed by at least one common letter are not different from each other ($p > 0.05$).

	Starch	Bacterial Cellulose	Gelatin
Tensile strength (MPa)			
Control	11.25 ± 3.59 [a]	60.64 ± 5.16 [a]	28.72 ± 2.41 [a]
Plasma-treated	15.82 ± 7.79 [a]	61.32 ± 11.07 [a]	35.48 ± 6.43 [a]
Elongation at break (%)			
Control	22.50 ± 3.10 [a]	7.71 ± 1.39 [a]	16.10 ± 3.22 [a]
Plasma-treated	16.04 ± 4.17 [a]	9.67 ± 2.03 [a]	7.18 ± 5.94 [a]
Elasticity module (MPa)			
Control	531 ± 43 [a]	784 ± 91 [a]	781 ± 47 [a]
Plasma-treated	664 ± 61 [b]	809 ± 62 [a]	852 ± 101 [a]

Plasma treatment increased the mean values of the tensile strength of the films. However, due to the high standard deviation of this kind of test, it is not possible to affirm that plasma treatment has statistically increased the films' tensile strength. The same trend was observed with the elasticity module, where the mean value increased after plasma treatment. The elongation at break decreased for the starch and gelatin films, while it increased for the bacterial cellulose film. Overall, it is possible to notice that plasma treatment tended to make the films more rigid.

The tensile strength is related to the mechanical resistance of the film. Plasma processing increased the average tensile strength of the films by 41%, 1%, and 23%, respectively, for starch, bacterial cellulose, and gelatin films. Although the mechanical resistances of the films were statistically similar, there was a positive trend in the average value, leading to more resistant starch and gelatin films.

The elongation at break is related to film flexibility. Plasma treatment only increased the flexibility of the bacterial cellulose (by 25%), while a reduction of 29% and 56% was observed in the starch and gelatin films, respectively.

Table 9 presents the mechanical properties of several films reported in the literature. Compared to other techniques, plasma treatment considerably increased the tensile strength of corn starch films. The increase in mechanical resistance attained by plasma treatment was higher than the increase gained with adding natural additives and similar to the increase achieved by high-pressure processing in potato starch films. Thus, mechanical resistance seems to increase more with physical treatments than chemical treatments.

Table 9. Mechanical properties of several starches, bacterial cellulose, and gelatin films reported in the literature.

Film	TS (MPa)	EB (%)	EM (MPa)	Reference
Corn starch	11.2	22.5	531	This work
Corn starch (plasma-treated)	15.8	16	664	This work
Corn starch (high amylose)	13	43	–	[36]
Corn starch + cinnamaldehyde	14.8	45	–	[36]
Corn starch	0.4	23.5	–	[43]
Corn starch + orange peel	1.4	11.5	–	[43]
Corn starch	3.8	60.1	95	[50]
Corn starch + fireweed extract	4.5	63.6	114	[50]
Corn starch	1.2	53.1	–	[44]
Corn starch + data palm pits	2.9	18.6	–	[44]
Sorghum starch	8.3	1.7	–	[51]
Cassava starch	5	55	–	[50]

Table 9. *Cont.*

Film	TS (MPa)	EB (%)	EM (MPa)	Reference
Potato starch	4.7	65.2	–	[52]
Potato starch	6.6	19	–	[37]
Potato starch (high-pressure)	11	19	–	[37]
Banana starch	6.2	4.3	144	[10]
Banana starch (plasma-treated)	9.1	7.6	268	[10]
Bacterial cellulose	60.6	7.7	784	This work
Bacterial cellulose (plasma-treated)	61.3	9.6	809	This work
Bacterial cellulose	1.8	33.3	–	[45]
Bacterial cellulose (nanocellulose)	1.7	64.9	–	[45]
Bacterial cellulose (nanocrystalline)	3	77	–	[45]
Bacterial cellulose	125	6.8	–	[40]
Bacterial cellulose + curcumin	45	5	–	[40]
Whey protein	4	30	–	[46]
Whey protein + bacterial cellulose	8	20	–	[46]
Gelatin	28.7	16.1	781	This work
Gelatin (plasma-treated)	35.4	7.2	852	This work
Gelatin	12.2	16.9	–	[53]
Gelatin + *B. gymnorhiza* extract	11.9	17.2	–	[53]
Gelatin	10.4	17.5	–	[53]
Gelatin + *S. alba* extract	13.1	19.4	–	[53]
Gelatin	4	115	–	[47]
Gelatin + carboxymethylcellulose	36	105	–	[47]
Gelatin + polyvinyl alcohol	14	350	–	[47]
Gelatin	65.5	20.4	–	[48]
Gelatin (transglutaminase-treated)	50.6	32.2	–	[48]

However, the flexibility of starch films decreased with the physical treatment and improved when specific natural products, such as fireweed extract, were incorporated into the film. In most cases, chemical treatments decreased the flexibility of starch films more than physical treatments.

The tensile strength and elongation at break of bacterial cellulose and its composites reported in the literature have a broad range of values (from 1 to 70 MPa for the tensile strength and from 7% to 80% for the elongation at break), which are affected by the formation methodology of the film and other parameters. The methods used herein produced a film with high tensile strength and low elongation at break, while the methods used by Efthymiou et al. [45] and Papadaki et al. [46] produced a film with low tensile strength and high elongation at break. Plasma treatment increased both tensile strength and elongation at break, as occurred with the chemical treatment to convert bacterial cellulose to nanocrystalline cellulose. The amplitude of change of the tensile strength was similar between the plasma and chemical treatments. However, the chemical treatment appears to better affect the elongation at break. The production of composites such as whey protein + bacterial cellulose increases the tensile strength but decreases the elongation at break.

Plasma treatment showed a superior effect in increasing the tensile strength of gelatin films compared to the addition of natural extracts, such as extracts from *B. gymnorhiza* and *S. alba* [53]. However, the effect was lower than that attained by producing a gelatin and polyvinyl alcohol composite [47]. Regarding the elongation at break, plasma treatment reduced this property, making a more rigid film. On the other hand, the addition of natural compounds [53], enzymatic treatment [48], and the composition of polyvinyl alcohol [47] increased the flexibility of gelatin films.

4. Conclusions

Plasma treatment improved the hydrophobicity, surface morphology, tensile strength, and elasticity module and reduced the water solubility of starch, bacterial cellulose, and gelatin films. The applied excitation frequency during plasma generation significantly affected the films' properties and requires optimization for better results. A low excitation frequency (50 Hz) resulted in greater changes in starch films, while a high excitation frequency (900 Hz) induced greater changes in gelatin and bacterial cellulose films.

Films produced on Mylar™ support should have the support removed before subjecting the film to plasma treatment. Treating the films with the support resulted in less properties' modifications because the changes occurred on a single surface instead of both surfaces.

Plasma treatment is an alternative to improve the hydrophobicity, surface morphology, water resistance, and mechanical properties of biodegradable films. Its performance can be superior, similar, or lower than other options, such as chemical treatment, enzymatic treatment, the addition of chemical compounds, and composites. It has the advantage of not recurring to chemical or biological additives, and no harmful compound remains in the film after plasma application. The treatment's disadvantage is its limited effect on the mechanical properties of the films since bigger changes could be achieved by producing composites or by chemical addition.

Author Contributions: Conceptualization, M.d.F.R. and F.A.N.F.; methodology, H.M.C.d.A., M.d.F.R., and F.A.N.F.; formal analysis, M.L.G. and F.A.N.F.; investigation, M.L.G. and A.L.A.M.; writing—original draft preparation, M.L.G. and F.A.N.F.; writing—review and editing, F.A.N.F.; funding acquisition, M.d.F.R. and F.A.N.F. All authors have read and agreed to the published version of the manuscript.

Funding: This research was funded by Fundação Cearense de Apoio ao Desenvolvimento Científico e Tecnológico (FUNCAP), grant number PR2-0101-00023.01.00/15, and Coordenação de Aperfeiçoamento de Pessoal de Nível Superior (CAPES), a scholarship awarded to M.L.G.

Institutional Review Board Statement: Not applicable.

Data Availability Statement: Data are contained within the article.

Conflicts of Interest: The authors declare no conflict of interest.

References

1. Romani, V.P.; Olsen, B.; Pinto Collares, M.; Meireles Oliveira, J.R.; Prentice, C.; Martins, V.G. Cold Plasma and Carnauba Wax as Strategies to Produce Improved Bi-Layer Films for Sustainable Food Packaging. *Food Hydrocoll.* **2020**, *108*, 106087. [CrossRef]
2. Gandhi, N.; Khurana, S.; Mathur, R.; Bansal, U.; Nair, R. Trends in Food Packaging: A Comprehensive Review. *Asian J. Chem.* **2022**, *34*, 2499–2510. [CrossRef]
3. Choi, S.M.; Rao, K.M.; Zo, S.M.; Shin, E.J.; Han, S.S. Bacterial Cellulose and Its Applications. *Polymers* **2022**, *14*, 1080. [CrossRef] [PubMed]
4. Onyeaka, H.; Obileke, K.; Makaka, G.; Nwokolo, N. Current Research and Applications of Starch-Based Biodegradable Films for Food Packaging. *Polymers* **2022**, *14*, 1126. [CrossRef]
5. Moosavi, M.H.; Khani, M.R.; Shokri, B.; Hosseini, S.M.; Shojaee-Aliabadi, S.; Mirmoghtadaie, L. Modifications of Protein-Based Films Using Cold Plasma. *Int. J. Biol. Macromol.* **2020**, *142*, 769–777. [CrossRef]
6. Sifuentes-Nieves, I.; Hernández-Hernández, E.; Neira-Velázquez, G.; Morales-Sánchez, E.; Mendez-Montealvo, G.; Velazquez, G. Hexamethyldisiloxane Cold Plasma Treatment and Amylose Content Determine the Structural, Barrier and Mechanical Properties of Starch-Based Films. *Int. J. Biol. Macromol.* **2019**, *124*, 651–658. [CrossRef]
7. Azeredo, H.M.C.; Barud, H.; Farinas, C.S.; Vasconcellos, V.M.; Claro, A.M. Bacterial Cellulose as a Raw Material for Food and Food Packaging Applications. *Front. Sustain. Food Syst.* **2019**, *3*, 7. [CrossRef]
8. Díaz-Calderón, P.; MacNaughtan, B.; Hill, S.; Foster, T.; Enrione, J.; Mitchell, J. Changes in Gelatinisation and Pasting Properties of Various Starches (Wheat, Maize and Waxy Maize) by the Addition of Bacterial Cellulose Fibrils. *Food Hydrocoll.* **2018**, *80*, 274–280. [CrossRef]
9. Romani, V.P.; Olsen, B.; Pinto Collares, M.; Meireles Oliveira, J.R.; Prentice-Hernández, C.; Guimarães Martins, V. Improvement of Fish Protein Films Properties for Food Packaging through Glow Discharge Plasma Application. *Food Hydrocoll.* **2019**, *87*, 970–976. [CrossRef]
10. Marenco-Orozco, G.A.; Rosa, M.F.; Fernandes, F.A.N. Effects of Multiple-step Cold Plasma Processing on Banana (Musa Sapientum) Starch-based Films. *Packag. Technol. Sci.* **2022**, *35*, 589–601. [CrossRef]
11. Wang, Z.; Wang, K.; Yao, X.; Jiang, J.; Wang, M.; Yuan, S. Ultrasound-Assisted Preparation of $Fe(OH)_3$@bacterial Cellulose Aerogel for Efficient Removal of Organic Contamination in Water. *Appl. Surf. Sci.* **2023**, *607*, 154959. [CrossRef]

12. Šafranko, S.; Stanković, A.; Hajra, S.; Kim, H.-J.; Strelec, I.; Dutour-Sikirić, M.; Weber, I.; Bosnar, M.H.; Grbčić, P.; Pavelić, S.K.; et al. Preparation of Multifunctional N-Doped Carbon Quantum Dots from Citrus Clementina Peel: Investigating Targeted Pharmacological Activities and the Potential Application for Fe3+ Sensing. *Pharmaceuticals* **2021**, *14*, 857. [CrossRef] [PubMed]
13. Panda, S.; Hajra, S.; Jeong, H.; Panigrahi, B.K.; Pakawanit, P.; Dubal, D.; Hong, S.; Kim, H.J. Biocompatible $CaTiO_3$-PVDF Composite-Based Piezoelectric Nanogenerator for Exercise Evaluation and Energy Harvesting. *Nano Energy* **2022**, *102*, 107682. [CrossRef]
14. Chen, G.; Chen, Y.; Jin, N.; Li, J.; Dong, S.; Li, S.; Zhang, Z.; Chen, Y. Zein Films with Porous Polylactic Acid Coatings via Cold Plasma Pre-Treatment. *Ind. Crops Prod.* **2020**, *150*, 112382. [CrossRef]
15. Wu, X.; Liu, Q.; Luo, Y.; Murad, M.S.; Zhu, L.; Mu, G. Improved Packing Performance and Structure-Stability of Casein Edible Films by Dielectric Barrier Discharges (DBD) Cold Plasma. *Food Packag. Shelf Life* **2020**, *24*, 100471. [CrossRef]
16. Pankaj, S.K.K.; Bueno-Ferrer, C.; Misra, N.N.N.; O'Neill, L.; Tiwari, B.K.K.; Bourke, P.; Cullen, P.J.J. Dielectric Barrier Discharge Atmospheric Air Plasma Treatment of High Amylose Corn Starch Films. *LWT-Food Sci. Technol.* **2015**, *63*, 1076–1082. [CrossRef]
17. Pankaj, S.K.; Bueno-Ferrer, C.; Misra, N.N.; O'Neill, L.; Tiwari, B.K.; Bourke, P.; Cullen, P.J. Physicochemical Characterization of Plasma-Treated Sodium Caseinate Film. *Food Res. Int.* **2014**, *66*, 438–444. [CrossRef]
18. Lin, S.-P.; Huang, S.-H.; Ting, Y.; Hsu, H.-Y.; Cheng, K.-C. Evaluation of Detoxified Sugarcane Bagasse Hydrolysate by Atmospheric Cold Plasma for Bacterial Cellulose Production. *Int. J. Biol. Macromol.* **2022**, *204*, 136–143. [CrossRef]
19. Santoso, S.P.; Lin, S.-P.; Wang, T.-Y.; Ting, Y.; Hsieh, C.-W.; Yu, R.-C.; Angkawijaya, A.E.; Soetaredjo, F.E.; Hsu, H.-Y.; Cheng, K.-C. Atmospheric Cold Plasma-Assisted Pineapple Peel Waste Hydrolysate Detoxification for the Production of Bacterial Cellulose. *Int. J. Biol. Macromol.* **2021**, *175*, 526–534. [CrossRef]
20. Janpetch, N.; Saito, N.; Rujiravanit, R. Fabrication of Bacterial Cellulose-ZnO Composite via Solution Plasma Process for Antibacterial Applications. *Carbohydr. Polym.* **2016**, *148*, 335–344. [CrossRef]
21. Carvalho, A.P.M.G.; Barros, D.R.; da Silva, L.S.; Sanches, E.A.; da Costa Pinto, C.; de Souza, S.M.; Clerici, M.T.P.S.; Rodrigues, S.; Fernandes, F.A.N.; Campelo, P.H. Dielectric Barrier Atmospheric Cold Plasma Applied to the Modification of Ariá (Goeppertia Allouia) Starch: Effect of Plasma Generation Voltage. *Int. J. Biol. Macromol.* **2021**, *182*, 1618–1627. [CrossRef] [PubMed]
22. Guo, Z.; Gou, Q.; Yang, L.; Yu, Q.; Han, L. Dielectric Barrier Discharge Plasma: A Green Method to Change Structure of Potato Starch and Improve Physicochemical Properties of Potato Starch Films. *Food Chem.* **2022**, *370*, 130992. [CrossRef] [PubMed]
23. Goiana, M.L.; de Brito, E.S.; Alves Filho, E.G.; de Castro Miguel, E.; Fernandes, F.A.N.; de Azeredo, H.M.C.; de Freitas Rosa, M. Corn Starch Based Films Treated by Dielectric Barrier Discharge Plasma. *Int. J. Biol. Macromol.* **2021**, *183*, 2009–2016. [CrossRef] [PubMed]
24. Campelo, P.H.; Alves Filho, E.G.; Silva, L.M.A.; de Brito, E.S.; Rodrigues, S.; Fernandes, F.A.N. Modulation of Aroma and Flavor Using Glow Discharge Plasma Technology. *Innov. Food Sci. Emerg. Technol.* **2020**, *62*, 102363. [CrossRef]
25. Alves Filho, E.G.; Silva, L.M.A.; Oiram Filho, F.; Rodrigues, S.; Fernandes, F.A.N.; Gallão, M.I.; Mattison, C.P.; de Brito, E.S. Cold Plasma Processing Effect on Cashew Nuts Composition and Allergenicity. *Food Res. Int.* **2019**, *125*, 108621. [CrossRef]
26. Farias, T.R.B.; Rodrigues, S.; Fernandes, F.A.N. Effect of Dielectric Barrier Discharge Plasma Excitation Frequency on the Enzymatic Activity, Antioxidant Capacity and Phenolic Content of Apple Cubes and Apple Juice. *Food Res. Int.* **2020**, *136*, 109617. [CrossRef]
27. Almeida, F.D.L.; Cavalcante, R.S.; Cullen, P.J.; Frias, J.M.; Bourke, P.; Fernandes, F.A.N.N.; Rodrigues, S. Effects of Atmospheric Cold Plasma and Ozone on Prebiotic Orange Juice. *Innov. Food Sci. Emerg. Technol.* **2015**, *32*, 127–135. [CrossRef]
28. Rodrigues, S.; Fernandes, F.A.N. Glow Discharge Plasma Processing for the Improvement of Pasteurized Orange Juice's Aroma and Off-Flavor. *Processes* **2022**, *10*, 1812. [CrossRef]
29. Oliveira, A.V.; da Silva, A.P.M.; Barros, M.O.; de sá M. Souza Filho, M.; Rosa, M.F.; Azeredo, H.M.C. Nanocomposite Films from Mango Kernel or Corn Starch with Starch Nanocrystals. *Starch-Stärke* **2018**, *70*, 1800028. [CrossRef]
30. Saito, T.; Kimura, S.; Nishiyama, Y.; Isogai, A. Cellulose Nanofibers Prepared by TEMPO-Mediated Oxidation of Native Cellulose. *Biomacromolecules* **2007**, *8*, 2485–2491. [CrossRef]
31. do Nascimento, E.S.; Pereira, A.L.S.; de Oliveira Barros, M.; de Aaraújo Barroso, M.K.; Lima, H.L.S.; de Fatima Borges, M.; de Andrade Feitosa, J.P.; de Azeredo, H.M.C.; de Freitas Rosa, M. TEMPO Oxidation and High-Speed Blending as a Combined Approach to Disassemble Bacterial Cellulose. *Cellulose* **2019**, *26*, 2291–2302. [CrossRef]
32. Santos, T.M.; Men de Sá Filho, M.S.; Caceres, C.A.; Rosa, M.F.; Morais, J.P.S.; Pinto, A.M.B.; Azeredo, H.M.C. Fish Gelatin Films as Affected by Cellulose Whiskers and Sonication. *Food Hydrocoll.* **2014**, *41*, 113–118. [CrossRef]
33. *ASTM D 5725-99*; Standard Test Method for Surface Wettability and Adsorbency of Sheeted Materials Using an Automated Contact Angle Tester. ASTM: West Conshohocken, PA, USA, 2008.
34. Pena-Serna, C.; Lopes-Filho, J.F. Influence of Ethanol and Glycerol Concentration over Functional and Structural Properties of Zein–Oleic Acid Films. *Mater. Chem. Phys.* **2013**, *142*, 580–585. [CrossRef]
35. *ASTM D882-12*; Standard Test Method for Tensile Properties of Thin Plastic Sheeting. ASTM International: West Conshohocken, PA, USA, 2018.
36. Wan, S.; Liu, Q.; Yang, D.; Guo, P.; Gao, Y.; Mo, R.; Zhang, Y. Characterization of High Amylose Corn Starch-Cinnamaldehyde Inclusion Films for Food Packaging. *Food Chem.* **2022**, *403*, 134219. [CrossRef]
37. Guo, S.; Fu, Z.; Yao, X.; Wang, X.; Wu, M. Impact of High Pressure Homogenization on the Properties of Potato Flour Film-Forming Dispersions and the Resulting Films. *J. Food Eng.* **2023**, *339*, 111281. [CrossRef]

38. Fu, B.; Mei, S.; Su, X.; Chen, H.; Zhu, J.; Zheng, Z.; Lin, H.; Dai, C.; Luque, R.; Yang, D.-P. Integrating Waste Fish Scale-Derived Gelatin and Chitosan into Edible Nanocomposite Film for Perishable Fruits. *Int. J. Biol. Macromol.* **2021**, *191*, 1164–1174. [CrossRef]
39. Cubas, A.L.V.; Bianchet, R.T.; de Oliveira, D.; Leonarski, E.; Cesca, K. Application of Non-Thermal Plasma as an Alternative for Purification of Bacterial Cellulose Membranes. *Sustain. Chem. Pharm.* **2022**, *29*, 100800. [CrossRef]
40. Wasim, M.; Shi, F.; Liu, J.; Zhang, H.; Zhu, K.; Tian, Z. Synthesis and Characterization of Curcumin/MMT-Clay-Treated Bacterial Cellulose as an Antistatic and Ultraviolet-Resistive Bioscaffold. *J. Polym. Res.* **2022**, *29*, 423. [CrossRef]
41. Mir, S.A.; Dar, B.N.; Wani, A.A.; Shah, M.A. Effect of Plant Extracts on the Techno-Functional Properties of Biodegradable Packaging Films. *Trends Food Sci. Technol.* **2018**, *80*, 141–154. [CrossRef]
42. Pająk, P.; Przetaczek-Rożnowska, I.; Juszczak, L. Development and Physicochemical, Thermal and Mechanical Properties of Edible Films Based on Pumpkin, Lentil and Quinoa Starches. *Int. J. Biol. Macromol.* **2019**, *138*, 441–449. [CrossRef]
43. Chhatariya, H.F.; Srinivasan, S.; Choudhary, P.M.; Begum, S.S. Corn Starch Biofilm Reinforced with Orange Peel Powder: Characterization of Physicochemical and Mechanical Properties. *Mater. Today Proc.* **2022**, *59*, 884–892. [CrossRef]
44. Alqahtani, N.; Alnemr, T.; Ali, S. Development of Low-Cost Biodegradable Films from Corn Starch and Date Palm Pits (Phoenix Dactylifera). *Food Biosci.* **2021**, *42*, 101199. [CrossRef]
45. Efthymiou, M.-N.; Tsouko, E.; Papagiannopoulos, A.; Athanasoulia, I.-G.; Georgiadou, M.; Pispas, S.; Briassoulis, D.; Tsironi, T.; Koutinas, A. Development of Biodegradable Films Using Sunflower Protein Isolates and Bacterial Nanocellulose as Innovative Food Packaging Materials for Fresh Fruit Preservation. *Sci. Rep.* **2022**, *12*, 6935. [CrossRef] [PubMed]
46. Papadaki, A.; Manikas, A.C.; Papazoglou, E.; Kachrimanidou, V.; Lappa, I.; Galiotis, C.; Mandala, I.; Kopsahelis, N. Whey Protein Films Reinforced with Bacterial Cellulose Nanowhiskers: Improving Edible Film Properties via a Circular Economy Approach. *Food Chem.* **2022**, *385*, 132604. [CrossRef] [PubMed]
47. Fernandes, G.D.J.C.; Campelo, P.H.; de Abreu Figueiredo, J.; Barbosa de Souza, H.J.; Peixoto Joele, M.R.S.; Yoshida, M.I.; Henriques Lourenço, L.D.F. Effect of Polyvinyl Alcohol and Carboxymethylcellulose on the Technological Properties of Fish Gelatin Films. *Sci. Rep.* **2022**, *12*, 10497. [CrossRef]
48. Baggio, E.; Scopel, B.S.; Rosseto, M.; Rigueto, C.V.T.; Dettmer, A.; Baldasso, C. Transglutaminase Effect on the Gelatin-Films Properties. *Polym. Bull.* **2022**, *79*, 7347–7361. [CrossRef]
49. Hernandez-Perez, P.; Flores-Silva, P.C.; Velazquez, G.; Morales-Sanchez, E.; Rodríguez-Fernández, O.; Hernández-Hernández, E.; Mendez-Montealvo, G.; Sifuentes-Nieves, I. Rheological Performance of Film-Forming Solutions Made from Plasma-Modified Starches with Different Amylose/Amylopectin Content. *Carbohydr. Polym.* **2021**, *255*, 117349. [CrossRef]
50. Kowalczyk, D.; Szymanowska, U.; Skrzypek, T.; Basiura-Cembala, M.; Materska, M.; Łupina, K. Corn Starch and Methylcellulose Edible Films Incorporated with Fireweed (*Chamaenerion Angustifolium* L.) Extract: Comparison of Physicochemical and Antioxidant Properties. *Int. J. Biol. Macromol.* **2021**, *190*, 969–977. [CrossRef]
51. Prachayawarakorn, J.; Kansanthia, P. Characterization and Properties of Singly and Dually Modified Hydrogen Peroxide Oxidized and Glutaraldehyde Crosslinked Biodegradable Starch Films. *Int. J. Biol. Macromol.* **2022**, *194*, 331–337. [CrossRef]
52. Martins, P.C.; Latorres, J.M.; Martins, V.G. Impact of Starch Nanocrystals on the Physicochemical, Thermal and Structural Characteristics of Starch-Based Films. *LWT* **2022**, *156*, 113041. [CrossRef]
53. Nurdiani, R.; Ma'rifah, R.D.A.; Busyro, I.K.; Jaziri, A.A.; Prihanto, A.A.; Firdaus, M.; Talib, R.A.; Huda, N. Physical and Functional Properties of Fish Gelatin-Based Film Incorporated with Mangrove Extracts. *PeerJ* **2022**, *10*, e13062. [CrossRef] [PubMed]

polymers

MDPI

Article

Novel Dextran Coated Cerium Doped Hydroxyapatite Thin Films

Carmen Steluta Ciobanu [1], Ionela Cristina Nica [2,3], Anca Dinischiotu [2], Simona Liliana Iconaru [1], Patrick Chapon [4], Bogdan Bita [5], Roxana Trusca [6,7], Andreea Groza [5,*] and Daniela Predoi [1,*]

1 National Institute of Materials Physics, Atomistilor Street, No. 405A, P.O. Box MG 07, 077125 Magurele, Romania; ciobanucs@gmail.com (C.S.C.); simonaiconaru@gmail.com (S.L.I.)
2 Department of Biochemistry and Molecular Biology, Faculty of Biology, University of Bucharest, 91-95 Splaiul Independentei, 050095 Bucharest, Romania; cristina.nica@drd.unibuc.ro (I.C.N.); anca.dinischiotu@bio.unibuc.ro (A.D.)
3 Research Institute of the University of Bucharest–ICUB, University of Bucharest, 050657 Bucharest, Romania
4 HORIBA Jobin Yvon S.A.S., 6-18, Rue du Canal, CEDEX, 91165 Longjumeau, France; patrick.chapon@horiba.com
5 National Institute for Laser, Plasma and Radiation Physics, 409 Atomistilor Street, P.O. Box MG 36, 077125 Bucharest, Romania; bogdan.bita@inflpr.ro
6 Department of Science and Engineering of Oxide, Faculty of Applied Chemistry and Materials Science, Materials and Nanomaterials, University "Politehnica" of Bucharest, 060042 Bucharest, Romania; truscaroxana@yahoo.com
7 Centre for Micro and Nanomaterials, University "Politehnica" of Bucharest, 060042 Bucharest, Romania
* Correspondence: andreea.groza@inflpr.ro (A.G.); dpredoi@gmail.com (D.P.)

Abstract: Dextran coated cerium doped hydroxyapatite ($Ca_{10-x}Ce_x(PO_4)_6(OH)_2$), with x = 0.05 (5CeHAp-D) and x = 0.1 (10CeHAp-D) were deposited on Si substrates by radio frequency magnetron sputtering technique for the first time. The morphology, composition, and structure of the resulting coatings were examined by scanning electron microscopy (SEM), energy-dispersive x-ray spectroscopy (EDX), atomic force microscopy (AFM), metallographic microscopy (MM), Fourier transform infrared spectroscopy (FTIR), and glow discharge optical emission spectroscopy (GDOES), respectively. The obtained information on the surface morphologies, composition and structure was discussed. The surface morphologies of the CeHAp-D composite thin films are smooth with no granular structures. The constituent elements of the CeHAp-D target were identified. The results of the FTIR measurements highlighted the presence of peaks related to the presence of ν_1, ν_3, and ν_4 vibration modes of (PO_4^{3-}) groups from the hydroxyapatite (HAp) structure, together with those specific to the dextran structure. The biocompatibility assessment of 5CeHAp-D and 10CeHAp-D composite coatings was also discussed. The human cells maintained their specific elongated morphology after 24 h of incubation, which confirmed that the behavior of gingival fibroblasts and their proliferative capacity were not disturbed in the presence of 5CeHAp-D and 10CeHAp-D composite coatings. The 5CeHAp-D and 10CeHAp-D coatings' surfaces were harmless to the human gingival fibroblasts, proving good biocompatibility.

Keywords: dextran coated cerium doped hydroxyapatite; composite coatings; surface morphology; chemical composition; biocompatibility; human gingival fibroblast cells

Citation: Ciobanu, C.S.; Nica, I.C.; Dinischiotu, A.; Iconaru, S.L.; Chapon, P.; Bita, B.; Trusca, R.; Groza, A.; Predoi, D. Novel Dextran Coated Cerium Doped Hydroxyapatite Thin Films. *Polymers* **2022**, *14*, 1826. https://doi.org/10.3390/polym14091826

Academic Editor: Cornelia Vasile

Received: 28 March 2022
Accepted: 27 April 2022
Published: 29 April 2022

1. Introduction

The production and characterization of hydroxyapatite materials in various chemical structures is a promising research topic, due to the medical and biological applications of these compounds. Natural hydroxyapatite (HAp) is the major mineral constituent extensively present in bones and teeth (enamel and dentine) and is largely used in reconstruction engineering. It can contain cations such as Mg^{2+} and Na^{2+}, or anions such as CO_3^{2-} and OH^- [1,2]. Nevertheless, synthetic hydroxyapatite $Ca_{10}(PO_4)_6(OH)_2$ is widely used for the repair and regeneration of bone hard tissue, considering its biocompatibility, non-toxicity and osteoconductive properties [1–4]. Hydroxyapatite can also encourage the growth and adherence of osteoblast and osteoclast cells, as well as their proliferation [3,4].

The crystalline lattice of HAp contains PO_4^{3-} tetrahedron groups linked to Ca^{2+} and OH^- groups [1,2]. The distance between PO_4^{3-} groups is large, and therefore, different cations or anions can be accommodated in the crystalline structure [1,5,6]. Anion substitutions of OH^- with F^- or Cl^- ions can occur in the apatite structure. Moreover, the cation substitution of Ca^{2+} can be performed with monovalent (Ag^+, Na^+, K^+), bivalent (Mg^{2+}, Sr^{2+}, Zn^{2+}, Ba^{2+}, Cu^{2+}) or trivalent (Ce^{3+}, Sm^{3+}, Eu^{3+}) ions [5–8].

The cation substitution of Ca^{2+} ions in the apatite lattice provides novel and improved properties to the doped hydroxyapatite materials, such as biocompatibility, bioactivity, non-toxicity, reduced bone fragility, and cell proliferation or growth [1,5–8]. In addition, as coatings, doped hydroxyapatite can promote interfacial bonding with implants, prostheses, or surrounding tissue. Na-doped HAp, or Sr-doped HAp can promote osteoblast and stem cell proliferation while the Zn-doped HAp and Mg-doped HAp can supplementarily increase the viability, adhesion, or spread of cells [1,2].

The substitution of Ca^{2+} ions with trivalent ions of rare-earth elements, such as Ce^{3+}, Sm^{3+}, and Eu^{3+}, in the HAp crystalline structure is very suitable, as their radii are very similar [1,7], even if the cytotoxicity of the synthetized compounds is still an open issue [1,9]. For 5% atomic concentrations, the doping of HAp powder by the co-precipitation method at lower doses of 100 micrograms/mL showed no significant cytotoxicity against mouse L929 fibroblast cells [10]. In addition, cerium can substitute calcium, increase the metabolism of bones, antimicrobial activity, as well as bone regeneration [1,9–11]. However, several studies indicated that the cytotoxicity, biocompatibility, and antimicrobial activity of Ce doped HAp are dependent on the synthesis method and dopant concentration [1,9–11].

When applied as coatings to cover medical implants or orthopedic prostheses, doped hydroxyapatite layers can improve the medical devices' properties by promoting strong interfacial bonds, improved biocompatibility, and antimicrobial activity. Several phyisco-chemical techniques are now used for the production of such layers: electrochemical deposition [12], micro-arc oxidation [13], sol-gel method [14], and biomimetic deposition [15]. In addition, plasma spray, magnetron sputtering discharges, pulsed laser [16,17] and electron deposition are methods used for the fabrication of doped hydroxyapatite layers.

Moreover, in order to improve the physicochemical and biological properties of such layers, biocompatible and nontoxic polysaccharides, such as dextran ($H(C_6H_{10}O_5)_xOH$), are often used [18]. Among various applications of dextran, we mention wastewater treatment [19], drug delivery, antithrombotic agent, tissue engineering, [20], blood supplement [18], etc. In the study conducted by Hussain, M.A. et al. [21] it has been shown that dextran can be used as a stabilizing agent for silver nanoparticles. In the paper elaborated by Predoi, D. and collaborators [22], it has been highlighted that the presence of dextran on zinc doped hydroxyapatite nanoparticles' surface does not alter their biological properties.

Furthermore, in our previous paper [7], we showed that Ce-doped HAp powders (produced by a co-precipitation method) can be used as a sputtering target for generation in radio frequency magnetron sputtering discharges of Ce-doped HAp layers with antimicrobial properties against gram-positive *Staphylococcus aureus* ATCC 25923, gram-negative *Escherichia coli* ATCC 25922, and fungal strain *Candida albicans ATCC 9002*. However, we generated, by radio frequency magnetron sputtering technique, a high diversity of HAp based coatings on metallic or polymeric substrates [7,8,23,24].

In the present work, we report, for the first time, our results regarding the morphology, composition, structure, and biocompatibility of dextran coated cerium doped hydroxyapatite ($Ca_{10-x}Ce_x(PO_4)_6(OH)_2$) thin films, with x = 0.05 (5CeHAp-D) and x = 0.1 (10CeHAp-D), obtained by radio frequency magnetron sputtering technique. This study demonstrates, for the first time, that the human gingival fibroblast cells retain their morphology in the presence of CeHAp-D composite coatings. Furthermore, this research showed, for the first time, that the behavior of gingival fibroblasts and their proliferative capacity were not disturbed in the presence of CeHAp-D composite coatings.

2. Materials and Methods

2.1. Materials

The powders of dextran coated cerium doped hydroxyapatite $Ca_{10-x}Ce_x(PO_4)_6(OH)_2$, with x = 0.05 (5CeHAp-D) and x = 0.1 (10CeHAp-D), were obtained using calcium nitrate tetrahydrate $Ca(NO_3)_2{\cdot}4H_2O$ (Sigma Aldrich, St. Louis, MO, USA, ≥99.0%), cerium nitrate hexahydrate $Ce(NO_3)_2{\cdot}6H_2O$ (Alfa Aesar ThermoFisher GmbH Company, Kandel, Germany, 99.97% purity), $(NH_4)_2HPO_4$ (Sigma Aldrich, St. Louis, MO, USA, ≥99.0%), ammonium hydroxide NH_4OH [Sigma Aldrich, St. Louis, MO, USA, 25% NH_3 in H_2O (T)], dextran $(H(C_6H_{10}O_5)_xOH$, (MW ~ 40,000), Merck, Kenilworth, NJ, USA), ethanol absolute, and double distilled water.

2.2. Synthesis of Dextran Coated Cerium Doped Hydroxyapatite (CeHAp-D) Powders

The powders of dextran coated cerium doped hydroxyapatite $Ca_{10-x}Ce_x(PO_4)_6(OH)_2$, with x = 0.05 (5CeHAp-D) and x = 0.1 (10CeHAp-D), were obtained in agreement with our previous studies [22] where the (Ca+Ce)/P molar ratio was 1.67. The powders were obtained through an adapted sol-gel method in ambient conditions. For this purpose, two solutions were obtained by dissolving the initial precursors in ethanol absolute. Therefore, a solution was obtained by dissolving the appropriate amount of $(NH_4)_2HPO_4$ in ethanol absolute. Secondly, $Ca(NO_3)_2{\cdot}4H_2O$ and $Zn(NO_3)_6{\cdot}6H_2O$ were dissolved in distilled water under vigorous agitation by a magnetic stirrer. The obtained solutions were dropped in a dextran solution. The resulting solution was stirred at 100 °C in order to obtain a gel. Then, the gel was washed several times with ethanol and double-distilled water [22]. The procedure is described in detail elsewhere [22].

2.3. Preparation of Dextran Coated Cerium Doped Hydroxyapatite Layers by Magnetron Sputtering Technique

The dextran coated cerium doped hydroxyapatite composite layers were deposited on silicon substrates by radio frequency (RF) magnetron sputtering technique. Firstly, sputtering targets of 5CeHAp-D and 10CeHAp-D in cylindrical shapes were obtained after the cold pressing of powders for a few minutes in air at atmospheric pressure. The geometrical sizes of the sputtering targets are the following: 2 mm thick, and 50 mm in diameter. Then, in turn, each sputtering target was placed inside the magnetron source for deposition of 5CeHAp-D and 10CeHAp-D layers, respectively. The target erosion zone had a diameter of about 4 cm.

The experimental setup is presented in Figure 1, and the deposition procedure was previously presented in detail in [23,24].

The magnetron source (acquired from K.J. Lesker Company, Hastings, UK) was coupled to a 13.56 MHz RF source through a matching box (Dressler, Cesar 136, Advanced Energy Industries GmbH & AEI Power GmbH, Metzingen, Germany). The substrate holder was at floating potential without being heated or cooled. Both types of layers were generated in the following experimental conditions: 50 W RF power, 180 min deposition time, 4×10^{-3} mbar Ar gas pressure (base pressure ~10^{-5} mbar), and 6 mln/min gas flow. For the 50.24 cm^2 plasma exposure area of the sputtering target, the power density was about 3.9 W/cm^2.

By applying RF electrical power to the magnetron source, the target was is struck and sputtered mainly by the Ar^+ ions produced in the plasma. Further, the atoms and ions sputtered from the target are deposited as layers on Si surfaces positioned on the sample holder centrally to the magnetron source head. The surfaces of the silicon substrates (5 mm × 5 mm × 1 mm) were optically polished. By maintaining the substrate holder at floating potential during the entire deposition process, a significant sputtering of the film, due to ion bombardment, was avoided.

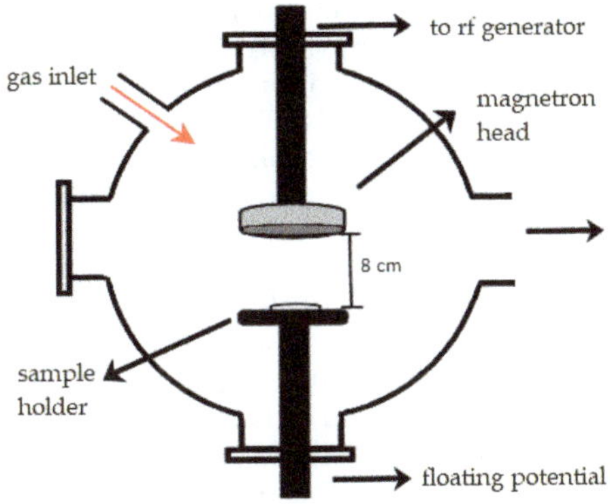

Figure 1. Experimental set-up.

The thicknesses of the layers were calculated by measuring the deposition rates of 5CeHAp-D and 10CeHAp-D in the RF magnetron discharge. The quartz microbalance (acquired from INFICON Company, Overland Park, KS, USA) positioned on the substrate holder centrally to the magnetron source measured a deposition rate of about ~0.1 Å/s (with a deviation of ~0.01 Å/s). The calculated thickness of the layers was about ~110 nm. The doping of hydroxyapatite with different concentrations of Ce ions did not significantly influence the values of the deposition rates for 5CeHAp-D and 10CeHAp-D layers, respectively. The temperature of the substrate holder was measured during the plasma depositions using a thermocouple probe. Thus, 10 min after plasma ignition, the temperature at of the surfaces of Si substrates was ~100, and after 3 h reached 135 °C.

2.4. *Physico-Chemical Characterisations*

The microstructure of 5CeHAp-D and 10CeHAp-D composite layers was characterized by scanning electron microscopy (SEM) using an FEI Inspect S scanning electron microscope (Hillsboro, OR, USA), in both high- and low-vacuum modes. The EDAX Inc. SiLi detector attached to the microscope allowed precise analysis of the elemental composition of the films by energy dispersive X-ray spectroscopy (EDS). The dispersive energy X-ray (EDS) spectra of 5CeHAp-D and 10CeHAp-D composite layers as well as the mapping of their elemental distribution have been recorded at 10 kV applied on the field emission gun. The Studies on the element distributions of 5CeHAp-D and 10CeHAp-D composite thin films were performed using glow discharge optical emission spectroscopy (GDOES) (Horiba Company, Longjumeau, France) [25]. The identification of functional groups, characteristic of 5CeHAp-D and 10CeHAp-D composite thin film structures, was performed by Fourier transform infrared spectroscopy (FTIR) studies. The FTIR measurements were performed with the help of a Perkin Elmer SP-100 spectrometer (Waltham, MA, USA). For these studies, the equipment was used in ATR mode. Moreover, all the FTIR spectra were collected in the spectral region of 550–4000 cm^{-1},with a scan resolution of 4 cm^{-1}. In addition, in agreement with the procedure described in detail in [26], the second derivative spectra of CeHAp-D composite layers were obtained.

Supplementary information regarding the morphology and roughness (root mean square roughness (R_{RMS}) parameter) of the 5CeHAp-D and 10CeHAp-D composite thin films were obtained by atomic force microscopy (AFM) analysis. An NT-MDT NTEGRA probe nano laboratory instrument (NT-MDT, Moscow, Russia), operated in semi-contact mode (using a silicon NT-MDT NSG01 cantilever (NT-MDT, Moscow, Russia) coated with a 35 nm gold layer), was used in order to obtain information about the surface topography of the samples. The atomic force microscopy images were obtained with a on surface areas of 5×5 μm^2. The collected data were analyzed with Gwyddion 2.59 software (Department of Nanometrology, Czech Metrology Institute, Brno, Czech Republic) [27].

The surface morphology of the obtained thin films was also studied with the help of the metallographic microscopy (MM) technique. Therefore, the MM studies were performed with the help of an inversed trinocular metallographic microscope OX.2153-PLM, (Euromex, Arnhem, The Netherlands). The MM images were obtained with a CMEX digital camera using the 10X objective of the microscope and ImageFocus Alpha software. The 3D representations of the MM images were obtained with the aid of ImageJ software [28].

2.5. Biological Evaluations

2.5.1. Culture of HGF-1 Fibroblasts on CeHAp-D Composite Coatings

The uncoated Si substrate and 5CeHAp-D and 10CeHAp-D composite coatings were sterilized by UV irradiation for 30 min and then placed in tissue culture plates. The human gingival fibroblasts (HGF-1 cell line purchased from American Type Culture Collection (ATCC), Cat. No. CRL-2014, Rockville, MD, USA) were seeded in 24-well culture plates (with or without 5CeHAp-D and 10CeHAp-D composite coatings) at a density of 5×10^4 cells per well in Dulbecco's Modified Eagle Medium (DMEM; Gibco/Invitrogen, Carlsbad, CA, USA), supplemented with 10% fetal bovine serum (FBS; Gibco/Invitrogen, Carlsbad, CA, USA), at 37 °C in a humidified atmosphere with 5% CO_2 (Figure 2). After 24 h of cell exposure to HAp-based samples, several biocompatibility tests were performed.

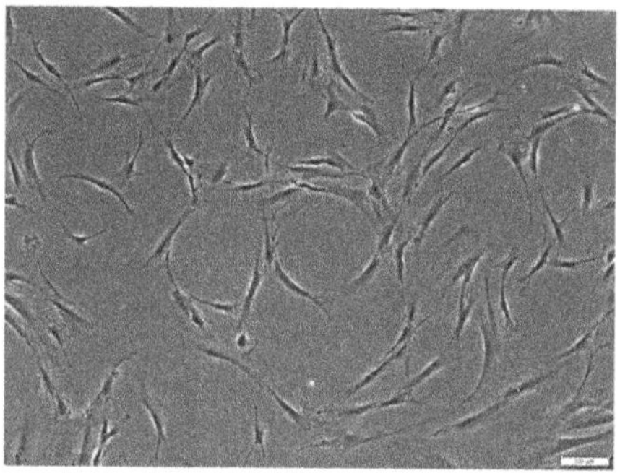

Figure 2. Representative image of HGF-1 cells cultured in DMEM medium supplemented with 10% fetal bovine serum before the exposure to HAp-based coatings. Scale bar: 100 μm.

2.5.2. Biocompatibility Assessment of CeHAp-D Composite Coatings

To test viable cell proliferation 3-(4,5-dimethylthiazol-2-yl)-2,5-diphenyltetrazolium bromide (MTT) assay was used. Authors chose lactate dehydrogenase (LDH) and MTT as good basic indicators to test cell viability and integrity, and chose nitric oxide (NO) as "a marker for inflammation".

The cell viability was measured using the MTT assay, which is based on the quantification of NAD(P)H-dependent cellular oxidoreductase enzyme activity in the viable cells. After 24 h, the medium was aspirated, and the cultured fibroblasts were incubated with a 1 mg/mL MTT solution at 37 °C for 2 h. The purple formazan crystals formed in the metabolic active cells were dissolved with 2-propanol and the absorbance was measured at 595 nm using a GENiosTecan microplate reader (GENiosTecan, Salzburg, Austria).

The LDH amount released into the culture medium was determined as a measure of cell membrane integrity and cell viability using a commercial kit (Cytotoxicity Detection Kit-LDH, Roche, Basel, Switzerland) by reading the absorbance at 490 nm using a microplate reader (GENiosTecan, Salzburg, Austria).

The level of NO released in the culture medium as an indicator of inflammation was assessed using the Griess reagent (a stoichiometric solution (v/v) of 0.1% naphthylethylendiamine dihydrochloride and 1% sulphanilamide in 5% H_3PO_4), after reading the absorbance at 550 nm.

In addition, the cell cytoskeleton morphology was visualized via fluorescence imaging, using cells fixed with 4% paraformaldehyde for 20 min and permeabilized with 0.1% Triton X-100—2% bovine serum albumin for 1 h. Filamentous actin (F-actin) was labeled with 20 μg/mL of phalloidin conjugated with fluorescein isothiocyanate (FITC) (Sigma-Aldrich, Munich, Germany) and the nuclei were counterstained with 2 μg/mL 4′,6-diamidino-2-phenylindole (DAPI) (Molecular Probes, Life Technologies, Carlsbad, CA, USA). Images were captured using an Olympus IX71 fluorescence microscope (Olympus, Tokyo, Japan).

2.5.3. Statistical Analysis

The GraphPad Prism software (Version 9, GraphPad, San Diego, CA, USA), based on one-way ANOVA with Tukey's multiple comparisons test, was used in order to conduct the statistical analysis of the obtained results in the revised manuscript. All data were expressed as mean value ± standard deviation (SD) of three independent experiments and a value of $p < 0.05$ was considered statistically significant.

3. Results and Discussion

The morphological characteristics of 5CeHAp-D and 10CeHAp-D targets are presented in Figures 3 and 4. Furthermore, information regarding chemical composition of the targets were obtained through EDS studies. The results of the EDS studies are also revealed in Figures 3 and 4. Therefore, it can be noticed that the main chemical elements present in the chemical composition of the targets are Ca, P, O, Ce and C. All these elements belong to the 5CeHAp-D and 10CeHAp-D chemical compositions. Both in the EDS spectra and in the obtained cartographies presented in Figures 3 and 4, the presence of impurities is not observed. Additionally, our results indicate the uniform and homogenous distribution of the main chemical constituents in the targets. The morphological features of the 5CeHAp-D and 10CeHAp-D composite coatings deposited on optically polished Si substrates were revealed after the scanning of their surfaces for various magnifications using SEM. Both films present granular structures on their surfaces (see Figures 5a and 6a), while the layers are not cracked or exfoliated. In Figures 5a and 6a the SEM images of the 5CeHAp-D and 10CeHAp-D composite thin films, with 5000× magnification, can be observed. Looking carefully at the SEM images from Figures 5a and 6a, it can be observed that the granular structure formed on the surface of the 5HApCe-D composite layer is denser than in the case of the 10HApCe-D composite layer. We suppose that the embedding of cerium into the hydroxyapatite structure of the sputtering target and their simultaneous co-deposition by magnetron sputtering technique is responsible for these granular structures. More than that, the increase of the cerium concentration into the 10HApCe-D composite sputtering target, in comparison with the 5HApCe-D sputtering target, conduce to the generation into the plasma of a double number of Ce ions, which bombard the substrate during the deposition process. This conduces to a more compact granular structure of the 10HApCe-D composite layer than the 5HApCe-D composite layer. The size of the grain decreases from ~600nm (in 5HApCe-D composite layer) to ~300 nm (in 10HApCe-D composite layer).

Our targets preserve the HAp structure. EDS studies of thin films of CeHap-D have shown that the (Ca+Ce)/P ratio is slightly changed to 1.66, instead of 1.67 (for the targets). As a result of EDS evaluation of the thin films, we could say that the coatings have the structure of hydroxyapatite doped with cerium, coated with dextran, and slightly deficient in Ca.

The formation of these grain-like structures could reflect the columnar grain growth of the films. Crystalline HAp films with grain morphologies were obtained by Lopez et al. [29] in RF magnetron sputtering plasma in a substrate holder floating configuration.

In our previous paper [7], when we deposited 5HAp-Ce layers on substrates placed on a grounded substrate holder in RF magnetron sputtering discharge, the grains were fine and hardly visible. Moreover, several studies [30,31] reported on the growth of columnar grains of calcium phosphate-based coatings, perpendicular to the grounded substrate, in RF magnetron sputtering discharges. The grain size depends on the plasma working conditions, such as gas pressure, deposition time, RF power, substrate holder–magnetron source distance and substrate bias. When the substrate is negative biased, the grain sizes of the coating are reduced due to the bombardment with ions of higher energy, thus favoring renucleation [30,31].

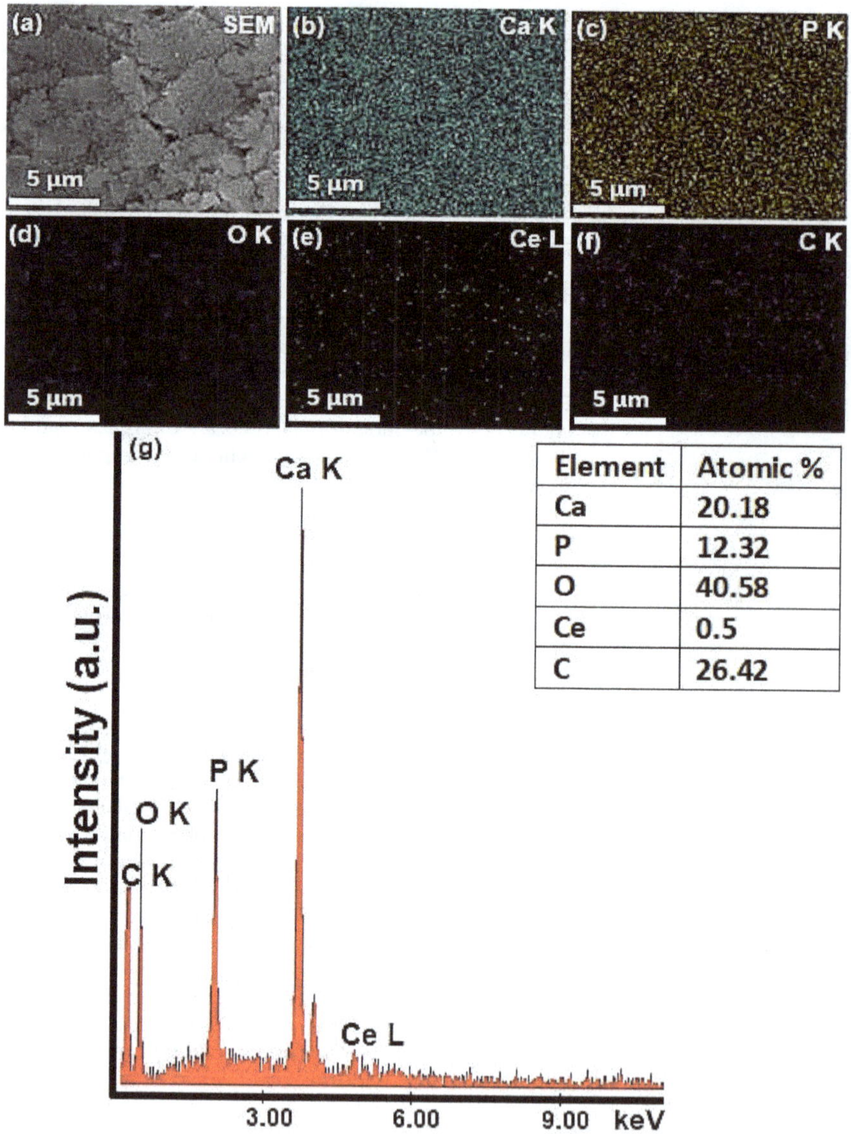

Element	Atomic %
Ca	20.18
P	12.32
O	40.58
Ce	0.5
C	26.42

Figure 3. SEM image (**a**), EDS elemental mapping of the constituent elements of 5CeHAp-D composite target (**b–f**) and EDS spectrum of 5CeHAp-D composite target (**g**).

Another explanation of grain formation could be given by the simultaneous co-deposition of dextran and Ce doped hydroxyapatite. In our recent study [23], we have shown that coagulation of hydroxyapatite–chitosan macromolecules on the substrate surface conduce to nm grain-like structure formation. Formation of polymer nanoparticles is governed by the negative charges [23,32] accumulated on the substrate surface during the depositions. Firstly, the polymer vapors nucleate in nm size particles (on the substrate surface) that further coagulate into larger particles.

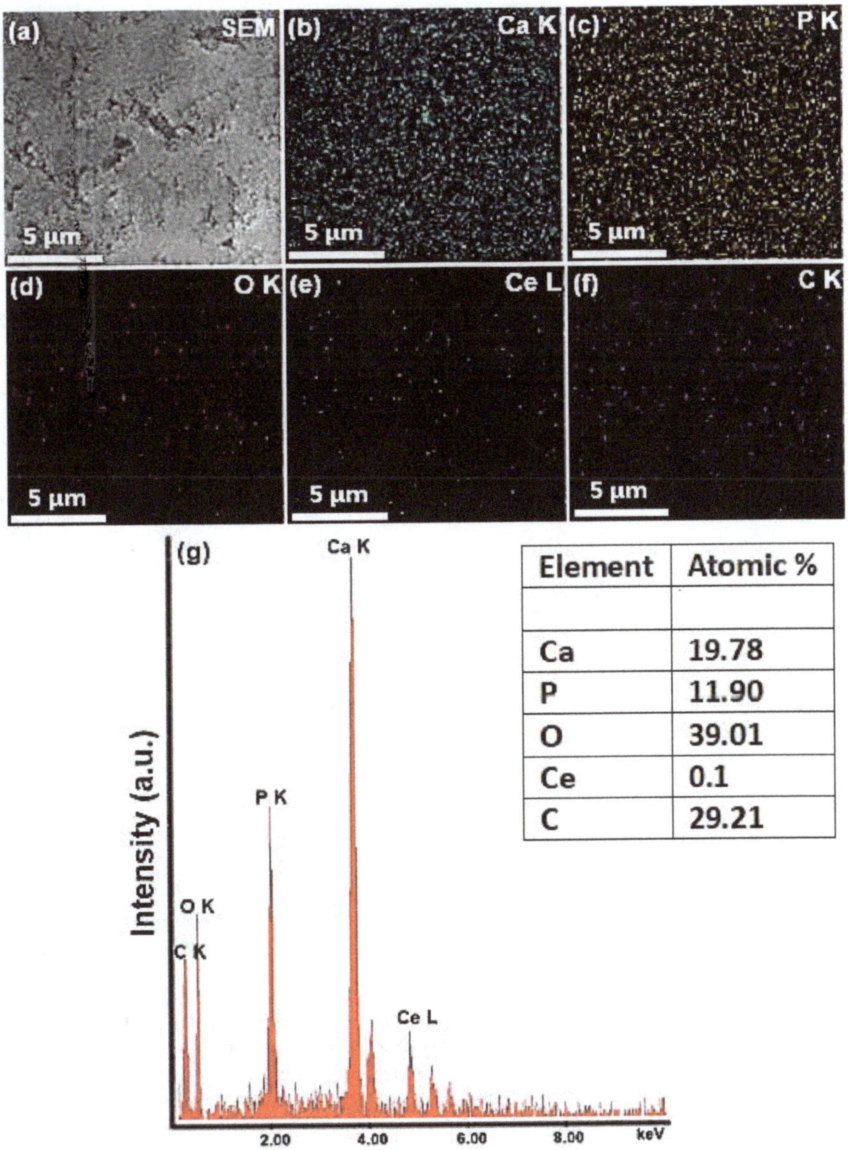

Element	Atomic %
Ca	19.78
P	11.90
O	39.01
Ce	0.1
C	29.21

Figure 4. SEM image (**a**), EDS elemental mapping of the constituent elements of 10CeHAp-D composite target (**b–f**) and EDS spectrum of 10CeHAp-D composite target (**g**).

The SEM–EDS elemental mapping of the 5CeHAp-D and 10CeHAp-D composite coatings, that corresponds to the images from Figures 5a and 6a, are presented in Figures 5b–f and 6b–f. The Ca, P, O, Ce and C elements are homogenously distributed over the entire analyzed areas. The EDS spectra (see Figures 5g and 6g) of 5CeHAp-D and 10CeHAp-D composite coatings indicated that the Ce atomic percentage in the 10HApCe layer is higher than in the 5HApCe-

D composite layer. The Ca K, P K, O K, Ce L and C K atomic percentages in both kinds of coating are shown in the tables from Figures 5g and 6g.

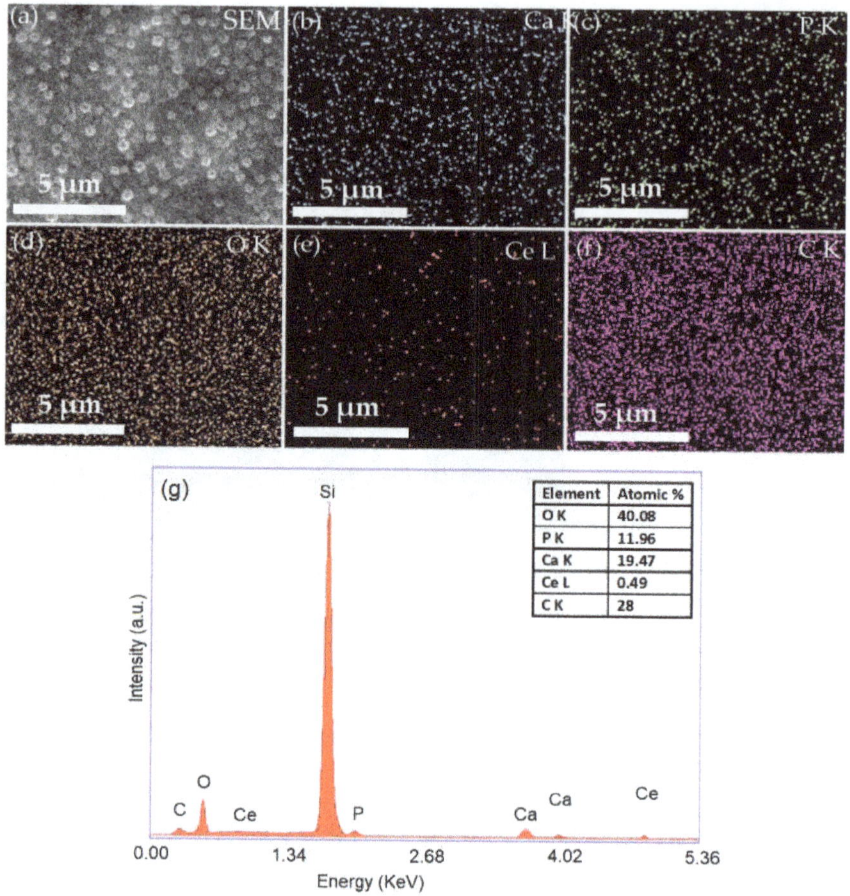

Figure 5. SEM image (**a**), EDS elemental mapping of the constituent elements of 5CeHAp-D composite coating (**b–f**) and EDS spectrum of 5CeHAp-D composite coatings (**g**).

The Ca/P ratio of calcium phosphate-based coatings depend on the RF magnetron plasma discharge working conditions [31,33] and substrate holder bias voltage. In RF magnetron discharge, the bombardment of the grounded substrates with negatively charged oxygen ions cause resputtering and removing of P, and PO_4^{3-} groups from the deposited layers. Therefore, a decrease in the number of negative charged ions that reach the substrates could limit the sputtering of phosphate ions from the deposited layers. We suppose that the simultaneous co-deposition of Ce and HAp on floating substrates could explain the measured (Ca+Ce)/P ratio of about ~1.6. In our previous studies on RF magnetron plasma depositions of carbon phosphate coatings on grounded substrates, we determined a Ca/P ratio of about ~1.4 [24].

Atomic force microscopy was also used to obtain information about the surface morphology of the 5CeHAp-D and 10CeHAp-D composite thin films. The results depicting the AFM 2D surface topographies as well as their 3D representation are presented in Figure 7a–d. The 2D AFM micrographs and the 3D representation of the surfaces of the 5CeHAp-D and 10CeHAp-D composite thin films suggest the continuous and uniformly deposition of the layers. Moreover, the AFM topographies of the 5CeHAp-D and 10CeHAp-

D composite thin films' surfaces highlighted that the layers do not present any unevenness, fissures, or any other type of discontinuity. The root means square roughness (R_{RMS}) parameters determined from the AFM topographies were R_{Rms} = 37.98 nm for 5CeHAp-D and R_{Rms} = 15.07 for 10CeHAp-D. The values obtained for the root mean square roughness imply that the surface topography of both samples is homogenous and does not present a significant roughness. On the other hand, the 2D AFM topographies, as well as their 3D representation, highlighted that the surface of the thin film is composed of nanostructured conglomerates, whose size decreased with the increase of the cerium ion concentration. These results are also in agreement with the SEM findings.

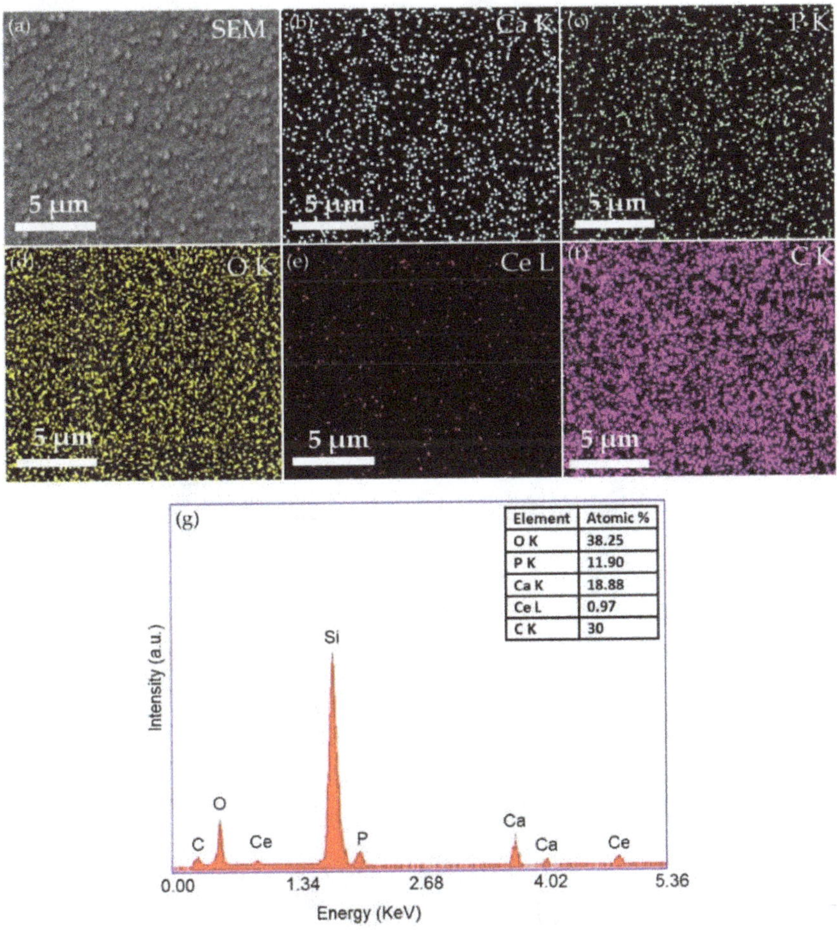

Element	Atomic %
O K	38.25
P K	11.90
Ca K	18.88
Ce L	0.97
C K	30

Figure 6. SEM image (**a**), EDS elemental mapping of the constituent elements of 10HApCe-D composite layer coating (**b–f**) and EDS spectrum of 10HApCe-D composite layer (**g**).

Complementary information about the surface morphology of the thin films was obtained using metallographic microscopy. The 2D metallographic images of the 5CeHAp-D and 10CeHAp-D composite thin films, as well as their 3D representation, are depicted in Figure 8a–d. The 2D images were recorded using the 10× objective of the metallographic microscope and their 3D representation was achieved using ImageJ software [28]. The results of the metallographic microscopy investigations revealed that the surfaces of both the 5CeHAp-D and 10CeHAp-D composite thin films are homogenous and do not present any fissures, cracks, or discontinuities, having the general aspect of a continuous and uniformly deposited layer. Furthermore, the 2D and 3D images also suggested that the surfaces of the studied thin films are nanostructured. More than that, the results highlighted that the grain size of the structures decreased with the increase of the cerium concentration.

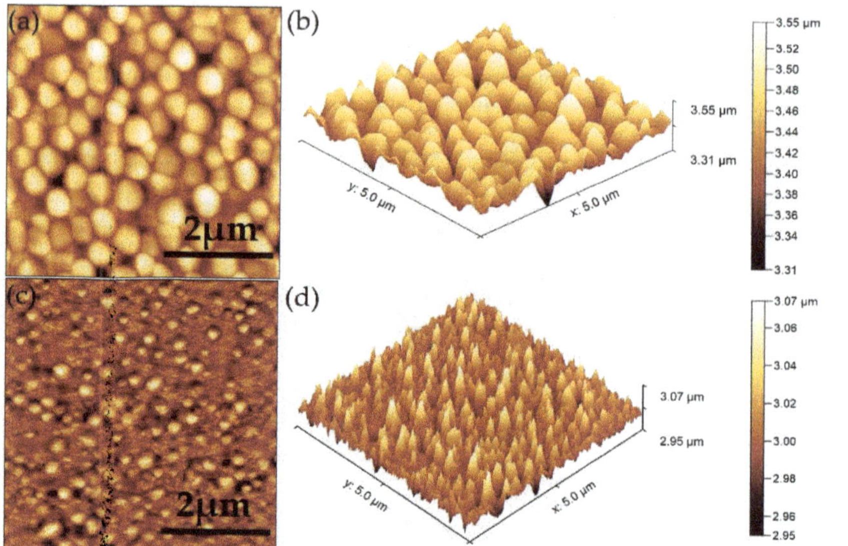

Figure 7. Typical 2D AFM images of surface topography obtained on 5CeHAp-D (**a**) and 10CeHAp-D (**c**) composite thin films and their 3D representation (**b**,**d**).

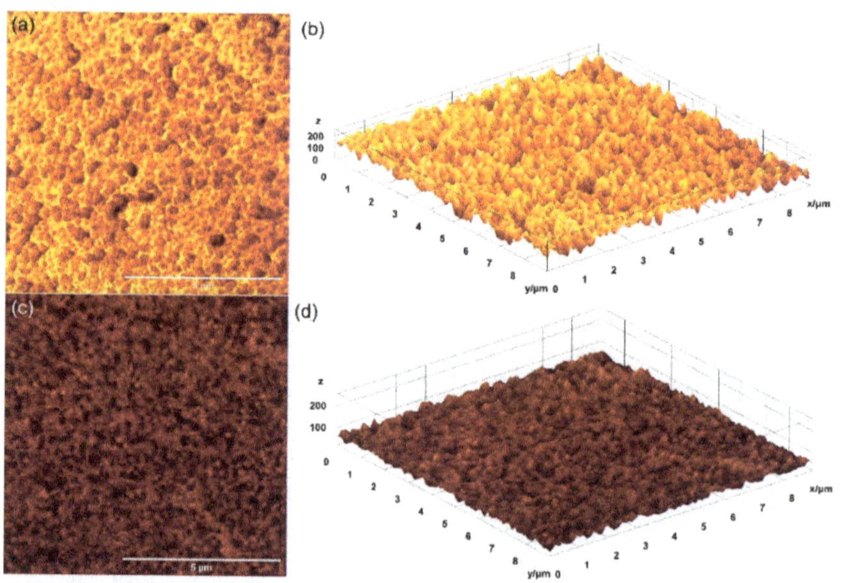

Figure 8. Metallographic microscopy image of the 5CeHAp-D (**a**) and 10CeHAp-D (**c**) composite thin films and their 3D representation (**b**,**d**).

Figure 9a,b present the depth profiles of the 5CeHAp-D and 10CeHAp-D composite thin films obtained by GDOES. By this technique, a large circular area of the layer

(4 mm in diameter) is sputtered from its surface to the substrate in an RF discharge. The atoms contained in the sample are excited by inelastic collisions in the RF plasma and the emission line signals are recorded and processed to provide the elemental composition of the analyzed sample. Thus, the distribution of elements contained in the 5CeHAp-D and 10CeHAp-D layers, namely: Ca, P, O, C, H and Ce, were revealed, starting from the layer surface (t = 0 s) to the substrate interface as a function of the sputtering time. Due to the changes of sputtering rate during the GDOES depth profiling analysis, mainly at layer/substrate interfaces, the conversion of sputtering time into sputtered depth is not accurate and not recommended [34]. In Figure 9a,b, the layer/substrate interfaces are marked by the simultaneous decreasing of the intensities of the depth profile curves of the elements contained in the layer (Ca, P, C, O, H, Ce) and increasing of the intensities of the depth profile curves of the element characteristic to the substrate (Si).

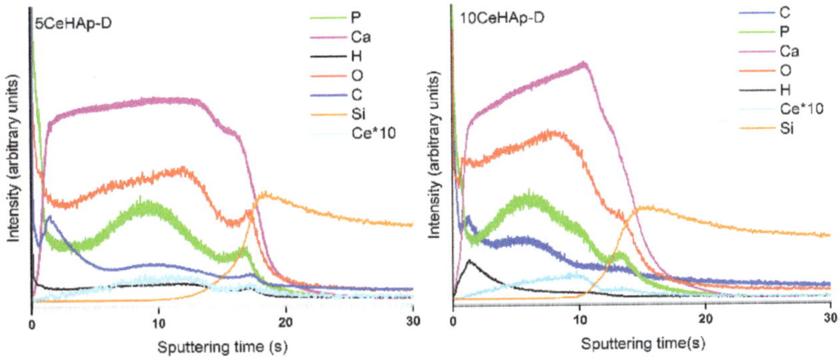

Figure 9. Compositional depth profiles of the 5CeHAp-D (**a**) and 10CeHAp-D (**b**) composite coatings obtained by GDOES.

The temporal evolution of Ca, P and Ce depth profile curves are similar (see Figure 9a,b), even at layer/substrate interface where some humps/peaks can be observed. These suggest the diffusion/implantation of Ca, P and Ce elements into the Si substrates as well as their linkages. Additionally, the sputtering time for the 10CeHAp-D sample is shorter than in the case of the 5CeHap-D sample, indicating a thinner 10CeHAp-D layer. These data are in agreement with SEM, AFM and metallographic microscopy analysis that shows a more compact structure of the 10CeHAp-D layer than of the 5CeHap-D layer.

The depth profile curves of C, O and H, which represent the chemical elements characteristic to dextran, are also observed in the graphs from Figure 9a,b. Their intensities are higher at shorter sputtering times (0–5 s) indicating the presence of a polymer, mainly at sample surfaces. These findings could sustain our assumption that the grains evidenced by the SEM images of the analyzed samples (see Figures 5a and 6a) could also be attributed to the nano-sized structuring of dextran on layer surfaces.

In Figure 10 the FTIR spectra obtained for both studied thin films are revealed. Mainly, the presence of the peaks related to the presence of ν_1, ν_3, and ν_4 vibration modes of (PO_4^{3-}) groups from the hydroxyapatite (HAp) structure and to the specific vibration bands of the dextran structure could be noticed. Therefore, the main peaks that are noticed (in the FTIR spectra of 5CeHAp-D composite thin films) at around 564 cm^{-1}, 604 cm^{-1}, 961 cm^{-1}, 1033 cm^{-1} and 1090 cm^{-1} are related to characteristic tetrahedral PO_4^{3-} groups from HAp [26,35,36]. In the FTIR spectra of the 10CeHAp-D composite thin films, the main maxima are found at about 566 cm^{-1}, 607 cm^{-1}, 963 cm^{-1}, 1035 cm^{-1} and 1092 cm^{-1} [26,35,36]. In the following section, we will only discuss the maxima obtained for the 5CeHAp-D composite thin films, with the maxima obtained for the 10CeHAp-D composite thin films being similar to these. The peak at around 564 cm^{-1} and 605 cm^{-1} corresponds to ν_4 vibration mode of (PO_4^{3-}) groups [26,35,36]. Moreover, the vibrational bands observed at about 1033 cm^{-1} and 1090 cm^{-1} belongs to ν_3 vibration mode of (PO_4^{3-})

groups from hydroxyapatite. In addition, the presence of the ν_1 vibration mode of (PO_4^{3-}) groups at 961 cm^{-1} was observed, as in Figure 10 [26,35,36]. The main maxima that are characteristic to dextran structure could be noticed in the following spectral domains: 950–750 cm^{-1}, 1200–950 cm^{-1} (due to C-O stretching), 1500–1200 cm^{-1} (due to symmetrical deformation of CH_2 and C-OH deformations, respectively) [37,38]. Therefore, the maxima found at about 772, 1100, 1430 and 1377 cm^{-1} in the FTIR spectra are characteristic of polysaccharide (dextran) structures [37,38]. The main peak that is observed in the spectral region 3100–3500 cm^{-1} belongs to adsorbed water. All these peaks belong to the HAp and dextran structures without noting the presence of additional maxima specific to the presence of impurities in the analyzed samples. In the FTIR spectra of the 5CeHAp-D and 10CeHAp-D composite thin films (Figure 10), it could be noticed that by increasing the cerium concentration in the samples the obtained spectra are slightly displaced, at the same time a slight decrease in the intensity of the peaks was observed (with the increase of the cerium concentration in the samples). The results obtained in this research are thus in agreement with those previously reported in the literature [26,35,36].

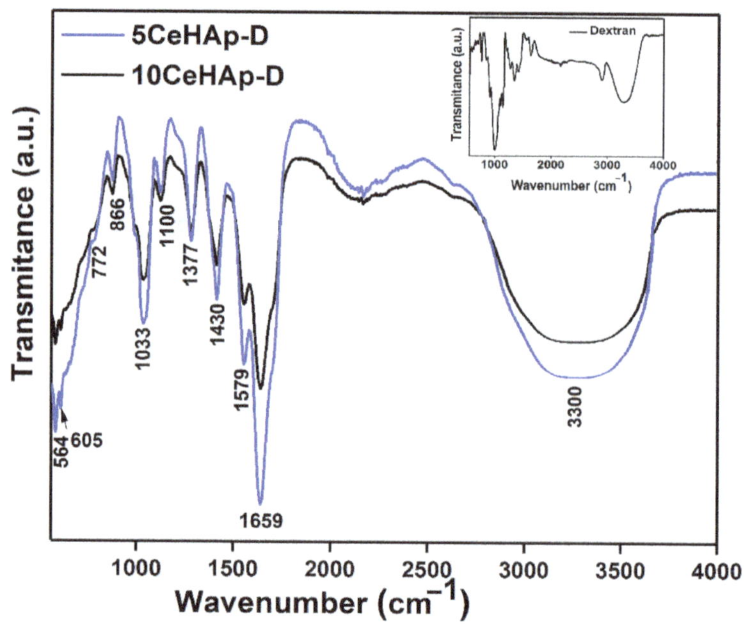

Figure 10. FTIR spectra of 5CeHAp−D and 10CeHAp−D composite thin films. The FTIR spectra of dextran is presented in the inset.

For a comprehensive understanding of the experimental results obtained by FTIR studies, a second derivative analysis was performed. The results of the second derivative study performed in the 550–1800 cm^{-1} spectral domain are revealed in Figure 11. Our results clearly highlight the presence of maxima associated to ν_1, ν_3, and ν_4 molecular vibration of (PO_4^{3-}) groups characteristic to HAp structure [26,39,40]. In addition, in the studied spectral domains the maxima associated with the characteristic vibration of dextran structures are revealed. In the spectral regions between 560–620 cm^{-1} and 900–1100 cm^{-1}, respectively, are found the maxima that appear due to the presence of ν_1, ν_3, and ν_4 vibration of (PO_4^{3-}) groups from the HAp structure [24,31,32]. Moreover, in Figure 11, the presence of the vibration bands specific to polysaccharides in the 950–750 cm^{-1}, 1200–950 cm^{-1}, 1500–1200 cm^{-1} spectral domains can be observed [37,38,41]. The maxima found around 1430 cm^{-1} and 1377 cm^{-1} belong to δ(C-H) and ν(C-H), meanwhile, the maxima found at 772 cm^{-1} belong to α-glucopyranose ring deformation [41].

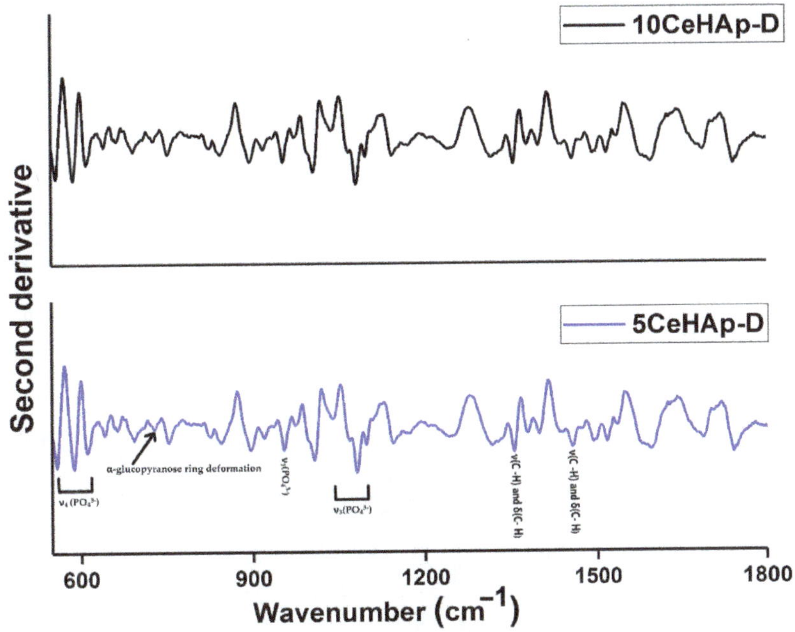

Figure 11. Second derivative spectra of 5CeHAp−D and 10CeHAp−D composite thin films.

In vitro cell response to uncoated Si substrates and different coatings containing dextran coated cerium doped hydroxyapatite were tested on normal human gingival fibroblasts. For this purpose, cellular viability, membrane integrity and their potential to generate an inflammatory response were evaluated and the results are presented in Figure 12.

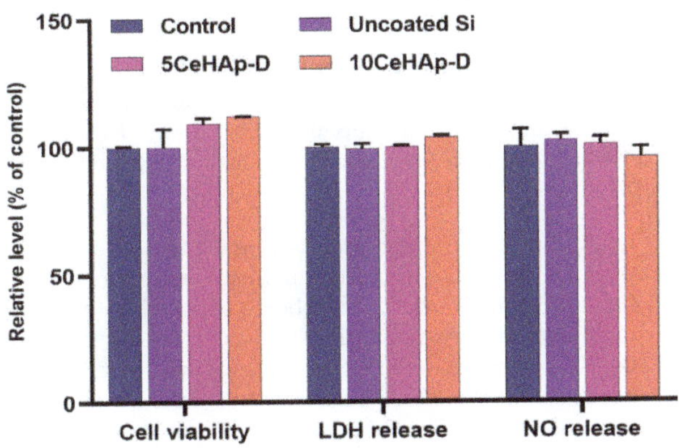

Figure 12. Biocompatibility of uncoated Si substrates and 5CeHAp-D and 10CeHAp-D composite coatings as shown by cell viability, lactate dehydrogenase (LDH), and nitric oxide (NO) release assays after 24 h exposure on normal gingival fibroblasts. Results are expressed as the mean ± standard deviation (SD) ($n = 3$) and represented relative to the untreated cells (control).

After 24 h of exposure to these bioactive coatings, the viability of gingival fibroblasts did not decrease but the small differences between the uncoated and the treated surfaces indicated that the best biocompatibility was exerted by the coatings containing dextran coated cerium doped hydroxyapatite. Moreover, the leakage of LDH into the cell culture medium was not recorded, demonstrating that none of the tested surfaces induced cell membrane permeabilization. These results were also confirmed by the NO release measurement that showed no significant changes between the controlled, uncoated Si surface and dextran coated cerium doped hydroxyapatite.

For a more in-depth characterization of the biological response induced by these modified Si surfaces in gingival fibroblasts, the morphology and actin cytoskeleton dynamics were also evidenced by fluorescence microscopy. The most representative images, as illustrated in Figure 13, were consistent with the results of the biocompatibility tests presented in Figure 12. Thus, it was shown that the human cells maintained their specific elongated morphology and established numerous focal adhesions after 24 h of incubation, which confirmed that the behavior of gingival fibroblasts and their proliferative capacity were not disturbed in the presence of the CeHAp-D composite coatings. These bioactive surfaces were harmless to the human gingival fibroblasts, proving good biocompatibility.

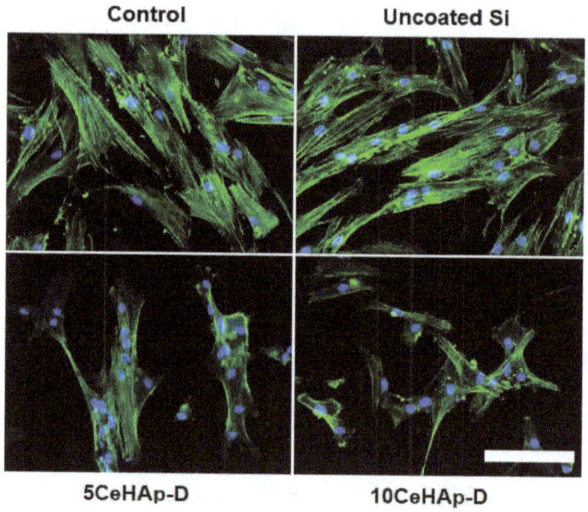

Figure 13. Actin cytoskeleton organization of gingival fibroblasts after 24 h of incubation with uncoated Si substrates and different dextran coated cerium doped hydroxyapatite coatings (5CeHAp-D and 10CeHAp-D). F-actin (green) was labeled with phalloidin-phalloidin-fluorescein isothiocyanate (FITC) and nuclei (blue) were counterstained with 4′,6-diamidino-2-phenylindole dihydrochloride (DAPI). Scale bar: 20 μm.

Additional information regarding the adherence and biocompatibility of the 5CeHAp-D and 10CeHAp-D composite thin films was obtained using AFM. The AFM topographies of the thin films' surfaces after being exposed with normal gingival fibroblasts were acquired at room temperature in normal atmospheric conditions on a surface of 20×20 μm^2. The AFM topographies of the normal gingival fibroblasts' adherence onto the surfaces of the uncoated Si disc and the 5CeHAp-D and 10CeHAp-D composite thin films are presented in Figure 14a–f. The 2D AFM topographies emphasized that on the surface of the investigated samples, typical patterns of the cellular morphology of normal gingival fibroblast cells could be seen, having typical flattened and elongated shapes [42–45]. Furthermore, both the 2D AFM topographies and their 3D representations highlighted that after an exposure of 24 h, the normal gingival fibroblasts exhibited a good adherence to the 5CeHAp-D and 10CeHAp-D composite thin films' surfaces, and there is also clear evidence of their spread all over the thin films' surfaces. On the other hand, the cells also adhered to the uncoated Si discs, but the AFM topographies revealed that they did not spread on the entirety of the Si discs' surface. The AFM results demonstrated that normal gingival fibroblast cells spread equally and formed a monolayer that exhibited characteristic elongated fibroblastic morphology on the surface of the 5CeHAp-D and 10CeHAp-D composite layers. The AFM results are in agreement with the viability assays conducted and suggest that the CeHAp-D composite thin films (with x = 0.05 and 0.1) do not exhibit any cytotoxic effect against the normal gingival fibroblasts after 24 h of exposure, making them suitable for the future development of biomedical devices.

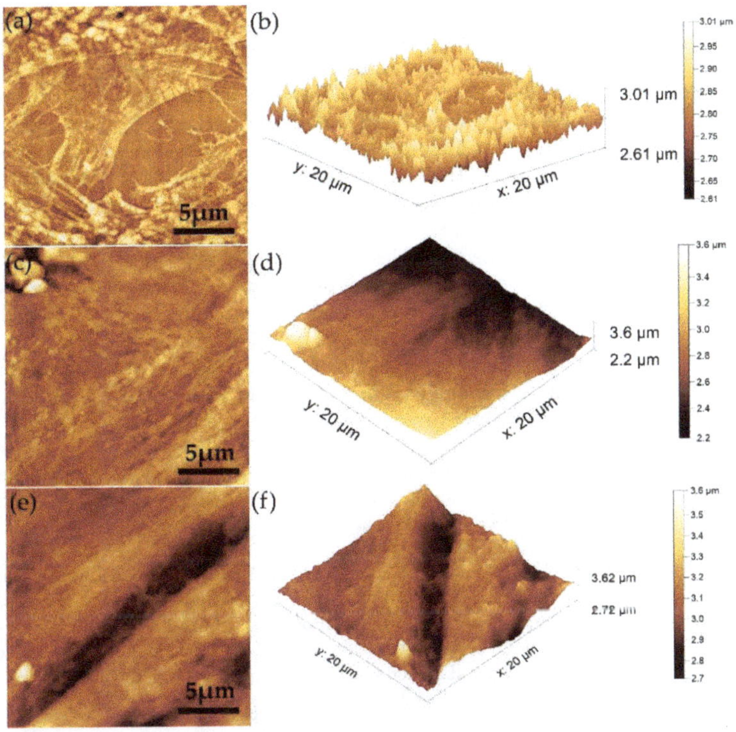

Figure 14. 2D AFM topography of gingival fibroblasts after 24 h of incubation with uncoated Si substrates (**a**), 5CeHAp-D composite thin film (**c**) and 10CeHAp-D composite thin film (**e**) and their 3D representation (**b**,**d**,**f**).

The present study sought to provide additional information on the behavior of dextran coated cerium doped hydroxyapatite thin layers, given that there is no other information in the existing literature. The composite material could have the properties of cerium (that stimulates the metabolic activity of organisms) [5], dextran (with a major role in reducing erythrocyte aggregation and platelet adhesion) and hydroxyapatite. The resulting composite material, thin layers of CeHAp-D, is a biocompatible and osteoconductive material with antibacterial activity [7]. The development of biocompatible CeHAp-D composite layers allowed us to obtain homogeneous coatings with compact granular structure and uniform distribution of constituent elements (Figures 5 and 6). It was thus highlighted that the surfaces of the composite coatings do not show unevenness, cracks, or any other type of discontinuities (Figures 7 and 8). Because the biological effects of cerium ions are not well known, gingival fibroblasts have been used in this research. This study proved that CeHAp-D biocomposite coatings can increase the viability of gingival fibroblasts. Furthermore, it has been established that the CeHAp-D biocomposite coatings did not induce cell membrane permeability. On the other hand, the measurement of NO release did not show significant changes between the control, the uncovered Si surface and the biocomposite coatings of CeHAp-D. This preliminary research also revealed that human cells maintained their specific elongated morphology and established numerous focal adhesions after 24 h of incubation (Figures 13 and 14). It was demonstrated that CeHAp-D biocomposite coatings are harmless to human gingival fibroblasts, proving good biocompatibility. Thus, conducting complex biological studies to fully understand the influence of different parameters involved, such as homogeneity, thickness or roughness of layers, is justified in conducting future studies.

4. Conclusions

The radio frequency magnetron sputtering technique had the ability to produce CeHAp-D composite coatings (x = 0.05 and 0.1) on silicon substrate with homogenous surface structure and were harmless to the human gingival fibroblasts. The uniformity of CeHAp-D composite coatings was proved by SEM, AFM and MM investigations. The EDS and GDOES analysis confirmed the presence of constituent element of CeHAp-D into the coatings. The obtained CeHAp-D composite coatings were shown to be homogeneous. The presence of main vibrational peaks characteristic to HAp and dextran structure in the studied samples was highlighted by the results of FTIR studies. This study shows that the CeHAp-D composite coatings, obtained by radio frequency magnetron sputtering technique, have very good biocompatibility against human gingival fibroblasts. The fact that the 5CeHap-D and 10CeHAp-D composite coatings were harmless to human gingival fibroblasts shows that these films could be used successfully in medical applications. This study, through its results, proves to be essential in the effort to develop biocomposites with superior performance in medical applications.

Author Contributions: Conceptualization, D.P. and A.G.; methodology, C.S.C., S.L.I., D.P., A.D., R.T., I.C.N., P.C. and A.G.; software, A.G. and S.L.I.; validation, D.P., A.G., A.D. and P.C.; formal analysis, C.S.C., I.C.N., B.B., S.L.I., D.P., A.G., A.D., R.T. and P.C.; investigation, C.S.C., I.C.N., B.B., S.L.I., D.P., A.G., A.D. and P.C.; resources, C.S.C., I.C.N., B.B., S.L.I., D.P., A.G., A.D., R.T. and P.C.; data curation, C.S.C., I.C.N., B.B., S.L.I., D.P., A.G., A.D. and P.C.; writing—original draft preparation, C.S.C., I.C.N., B.B., S.L.I., D.P., A.G., A.D. and P.C.; writing—review and editing, C.S.C., I.C.N., B.B., S.L.I., D.P., A.G., A.D. and P.C.; visualization, C.S.C., I.C.N., B.B., S.L.I., D.P., A.G., A.D. and P.C.; supervision, D.P. and A.G.; project administration, C.S.C.; S.L.I., D.P. and A.G.; funding acquisition, A.G. All authors have read and agreed to the published version of the manuscript.

Funding: This work was supported by the Romanian Ministry of Research and Innovation through the NUCLEU project PN 19 15/2019 and project PN-III-P2-2.1-PED-2019-0868 contract number 467PED/2020.

Institutional Review Board Statement: Not applicable.

Informed Consent Statement: Not applicable.

Data Availability Statement: Not applicable.

Conflicts of Interest: The authors declare no conflict of interest. The funders had no role in the design of the study; the collection, analyses, and interpretation of data; the writing of the manuscript; or the decision to publish the results.

References

1. Tite, T.; Popa, A.-C.; Balescu, L.M.; Bogdan, I.M.; Pasuk, I.; Ferreira, J.M.F.; Stan, G.E. Cationic Substitutions in Hydroxyapatite: Current Status of the Derived Biofunctional Effects and Their In Vitro Interrogation Methods. *Materials* **2018**, *11*, 2081. [CrossRef]
2. Eliaz, N.; Metoki, N. Calcium Phosphate Bioceramics: A Review of Their History, Structure, Properties, Coating Technologies and Biomedical Applications. *Materials* **2017**, *10*, 334. [CrossRef]
3. Sima, L.E.; Stan, G.E.; Morosanu, C.O.; Melinescu, A.; Ianculescu, A.; Melinte, R.; Neamtu, J.; Petrescu, S.M. Differentiation of mesenchymal stem cells onto highly adherent radio frequency-sputtered carbonated hydroxylapatite thin films. *J. Biomed. Mater. Res. Part A* **2010**, *95*, 1203–1214. [CrossRef]
4. Wang, P.; Zhao, L.; Liu, J.; Weir, M.D.; Zhou, X.; Xu, H.H.K. Bone tissue engineering via nanostructured calcium phosphate biomaterials and stem cells. *Bone Res.* **2014**, *2*, 14017. [CrossRef]
5. Sahu, T.; Bisht, S.S.; Ranjan, D.K.; Kerkar, S. Nanoceria: Synthesis and Biomedical Applications. *Curr. Nanosci.* **2013**, *9*, 588–593. [CrossRef]

6. Ouyang, Y.; Xie, Y.; Tan, S.; Shi, Q.; Chen, Y. Structure and antibacterial activity of Ce^{3+} exchanged montmorillonites. *J. Rare Earths* **2009**, *27*, 858–863. [CrossRef]
7. Predoi, D.; Iconaru, S.L.; Predoi, M.V.; Groza, A.; Gaiaschi, S.; Rokosz, K.; Raaen, S.; Negrila, C.C.; Prodan, A.-M.; Costescu, A.; et al. Development of Cerium-Doped Hydroxyapatite Coatings with Antimicrobial Properties for Biomedical Applications. *Coatings* **2020**, *10*, 516. [CrossRef]
8. Iconaru, S.L.; Groza, A.; Gaiaschi, S.; Rokosz, K.; Raaen, S.; Ciobanu, S.C.; Chapon, P.; Predoi, D. Antimicrobial Properties of Samarium Doped Hydroxyapatite Suspensions and Coatings. *Coatings* **2020**, *10*, 1124. [CrossRef]
9. Sundarabharathi, L.; Chinnaswamy, M.; Ponnamma, D.; Parangusan, H.; Al-Maadeed, S. Investigation of antimicrobial properties and in vitro bioactivity of Ce^{3+}-Sr^{2+} dual-substituted nano hydroxyapatites. *J. Am. Ceram. Soc.* **2018**, *102*, 144–157. [CrossRef]
10. Huang, W.; Mao, Z.; Chen, L.; Chi, Y.; Jiang, H.; Zimba, B.L.; Xiong, G.; Wu, Q. Synthesis and characterization of fluorescent and biocompatible hydroxyapatite nanoparticles with cerium doping. *Micro Nano Lett.* **2018**, *13*, 699–703. [CrossRef]
11. Rajeswari, D.; Gopi, D.; Ramya, S.; Kavitha, L. Investigation of anticorrosive, antibacterial and in vitro biological properties of a sulphonated poly(etheretherketone)/strontium, cerium co-substituted hydroxyapatite composite coating developed on surface treated surgical grade stainless steel for orthopedic applications. *RSC Adv.* **2014**, *4*, 61525–61536. [CrossRef]
12. Chakraborty, R.; Saha, P. A comparative study on surface morphology and electrochemical behaviour of hydroxyapatite-calcium hydrogen phosphate composite coating synthesized in-situ through electro chemical process under various deposition conditions. *Surf. Interfaces* **2018**, *12*, 160–167. [CrossRef]
13. Aktug, S.L.; Kutbay, I.; Usta, M. Characterization and formation of bioactive hydroxyapatite coating on commercially pure zirconium by micro arc oxidation. *J. Alloys Compd.* **2017**, *695*, 998–1004. [CrossRef]
14. Domínguez-Trujillo, C.; Peón, E.; Chicardi, E.; Pérez, H.; Rodríguez-Ortiz, J.A.; Pavón, J.; García-Couce, J.; Galvan, J.C.; García-Moreno, F.; Torres, Y. Sol-gel deposition of hydroxyapatite coatings on porous titanium for biomedical applications. *Surf. Coat. Technol.* **2018**, *333*, 158–162. [CrossRef]
15. Iijima, K.; Sakai, A.; Komori, A.; Sakamoto, Y.; Matsuno, H.; Serizawa, T.; Hashizume, M. Control of biomimetic hydroxyapatite deposition on polymer substrates using different protein adsorption abilities. *Colloids Surf. B Biointerfaces* **2015**, *130*, 77–83. [CrossRef] [PubMed]
16. Graziani, G.; Bianchi, M.; Sassoni, E.; Russo, A.; Marcacci, M. Ion-substituted calcium phosphate coatings deposited by plasma-assisted techniques: A review. *Mater. Sci. Eng. C* **2017**, *74*, 219–229. [CrossRef] [PubMed]
17. Duta, L.; Oktar, F.; Stan, G.; Popescu-Pelin, G.; Serban, N.; Luculescu, C.; Mihailescu, I. Novel doped hydroxyapatite thin films obtained by pulsed laser deposition. *Appl. Surf. Sci.* **2013**, *265*, 41–49. [CrossRef]
18. Zarrintaj, P.; Saeb, M.R.; Jafari, S.H.; Mozafari, M. Application of compatibilized polymer blends in biomedical fields. In *Compatibilization of Polymer Blends*; Elsevier: Amsterdam, The Netherlands, 2020; pp. 511–537.
19. Ahmed, S. Advanced green materials: An overview. In *Advanced Green Materials*; Ahmed, S., Ed.; Woodhead Publishing: Sawston, UK, 2021; pp. 1–13.
20. Varghese, S.A.; Rangappa, S.M.; Siengchin, S.; Parameswaranpillai, J. Natural polymers and the hydrogels prepared from them. In *Hydrogels Based on Natural Polymers*; Yu, C., Ed.; Elsevier: Amsterdam, The Netherlands, 2020; pp. 17–47. [CrossRef]
21. Hussain, M.A.; Shah, A.; Jantan, I.; Tahir, M.N.; Shah, M.R.; Ahmed, R.; Bukhari, S.N.A. One pot light assisted green synthesis, storage and antimicrobial activity of dextran stabilized silver nanoparticles. *J. Nanobiotechnol.* **2014**, *12*, 53. [CrossRef]
22. Predoi, D.; Iconaru, S.L.; Predoi, M.V. Dextran-Coated Zinc-Doped Hydroxyapatite for Biomedical Applications. *Polymers* **2019**, *11*, 886. [CrossRef]
23. Dreghici, D.; Butoi, B.; Predoi, D.; Iconaru, S.; Stoican, O.; Groza, A. Chitosan-Hydroxyapatite Composite Layers Generated in Radio Frequency Magnetron Sputtering Discharge: From Plasma to Structural and Morphological Analysis of Layers. *Polymers* **2020**, *12*, 3065. [CrossRef]
24. Groza, A.; Dreghici, D.B.; Ganciu, M. Calcium Phosphate Layers Deposited on Thermal Sensitive Polymer Substrates in Radio Frequency Magnetron Plasma Discharge. *Coatings* **2019**, *9*, 709. [CrossRef]
25. Sample Measurement with GDOES. Available online: https://www.horiba.com/int/scientific/technologies/glow-discharge-optical-emission-spectroscopy/sample-measurement-with-gdoes/ (accessed on 30 January 2022).
26. Iconaru, S.-L.; Motelica-Heino, M.; Predoi, D. Study on Europium-Doped Hydroxyapatite Nanoparticles by Fourier Transform Infrared Spectroscopy and Their Antimicrobial Properties. *J. Spectrosc.* **2013**, *2013*, 284285. [CrossRef]
27. Gwyddion. Available online: http://gwyddion.net/ (accessed on 30 January 2022).
28. ImageJ Website. Available online: http://imagej.nih.gov/ij (accessed on 29 January 2022).
29. López, E.O.; Mello, A.; Sendão, H.; Costa, L.T.; Rossi, A.L.; Ospina, R.O.; Borghi, F.F.; Filho, J.G.S.; Rossi, A.M. Growth of Crystalline Hydroxyapatite Thin Films at Room Temperature by Tuning the Energy of the RF-Magnetron Sputtering Plasma. *ACS Appl. Mater. Interfaces* **2013**, *5*, 9435–9445. [CrossRef] [PubMed]
30. Surmenev, R.; Vladescu, A.; Surmeneva, M.; Ivanova, A.; Braic, M.; Grubova, I.; Cotrut, C.M. Radio Frequency Magnetron Sputter Deposition as a Tool for Surface Modification of Medical Implants. In *Modern Technologies for Creating the Thin-Film Systems and Coatings*; Nikitenkov, N., Ed.; InTech: London, UK, 2017; pp. 1–36. [CrossRef]
31. Surmenev, R.A.; Ivanova, A.A.; Epple, M.; Pichugin, V.F.; Surmeneva, M.A. Physical principles of radio-frequency magnetron sputter deposition of calcium-phosphate-based coating with tailored properties. *Surf. Coat. Technol.* **2021**, *413*, 127098. [CrossRef]

32. Choukourov, A.; Pleskunov, P.; Nikitin, D.; Titov, V.; Shelemin, A.; Vaidulych, M.; Kuzminova, A.; Solař, P.; Hanuš, J.; Kousal, J.; et al. Advances and challenges in the field of plasma polymer nanoparticles. *Beilstein J. Nanotechnol.* **2017**, *8*, 2002–2014. [CrossRef] [PubMed]
33. Feddes, B.; Vredenberg, A.; Wolke, J.; Jansen, J. Bulk composition of rf magnetron sputter deposited calcium phosphate coatings on different substrates (polyethylene, polytetrafluoroethylene, silicon). *Surf. Coat. Technol.* **2004**, *185*, 346–355. [CrossRef]
34. Groza, A.; Surmeian, A.; Diplasu, C.; Luculescu, C.; Chapon, P.; Tempez, A.; Ganciu, M. Physico-chemical processes occurring during polymerization of liquid polydimethylsiloxane films on metal substrates under atmospheric pressure air corona discharges. *Surf. Coat. Technol.* **2012**, *212*, 145–151. [CrossRef]
35. Predoi, D.; Iconaru, S.L.; Predoi, M.V.; Motelica-Heino, M.; Guegan, R.; Buton, N. Evaluation of Antibacterial Activity of Zinc-Doped Hydroxyapatite Colloids and Dispersion Stability Using Ultrasounds. *Nanomaterials* **2019**, *9*, 515. [CrossRef]
36. Iconaru, S.; Predoi, M.; Chapon, P.; Gaiaschi, S.; Rokosz, K.; Raaen, S.; Motelica-Heino, M.; Predoi, D. Investigation of Spin Coating Cerium-Doped Hydroxyapatite Thin Films with Antifungal Properties. *Coatings* **2021**, *11*, 464. [CrossRef]
37. Mathlouthi, M.; Koenig, J.L. Vibrational Spectra of Carbohydrates. *Adv. Carbohydr. Chem. Biochem.* **1986**, *44*, 7–89. [CrossRef]
38. Ciobanu, C.S.; Iconaru, S.L.; Gyorgy, E.; Radu, M.; Costache, M.; Dinischiotu, A.; Le Coustumer, P.; Lafdi, K.; Predoi, D. Biomedical properties and preparation of iron oxide-dextran nanostructures by MAPLE technique. *Chem. Cent. J.* **2012**, *6*, 17. [CrossRef] [PubMed]
39. Manoj, M.; Mangalaraj, D.; Ponpandian, N.; Viswanathan, C. Core-shell hydroxyapatite/Mg nanostructures: Surfactant free facile synthesis, characterization and their in vitro cell viability studies against leukaemia cancer cells (K562). *RSC Adv.* **2015**, *5*, 48705–48711. [CrossRef]
40. Xie, J.; Riley, C.; Kumar, M.; Chittur, K. FTIR/ATR study of protein adsorption and brushite transformation to hydroxyapatite. *Biomaterials* **2002**, *23*, 3609–3616. [CrossRef]
41. Easo, S.L.; Mohanan, P. Dextran stabilized iron oxide nanoparticles: Synthesis, characterization and in vitro studies. *Carbohydr. Polym.* **2013**, *92*, 726–732. [CrossRef]
42. Predoi, D.; Groza, A.; Iconaru, S.L.; Predoi, G.; Barbuceanu, F.; Guegan, R.; Motelica-Heino, M.S.; Cimpeanu, C. Properties of Basil and Lavender Essential Oils Adsorbed on the Surface of Hydroxyapatite. *Materials* **2018**, *11*, 652. [CrossRef]
43. Warowicka, A.; Maciejewska, B.M.; Litowczenko, J.; Kościński, M.; Baranowska-Korczyc, A.; Jasiurkowska-Delaporte, M.; Koziol, K.; Jurga, S. MWCNT based matrices as a platform for adhesion and growth of cells. *Compos. Sci. Technol.* **2016**, *136*, 29–38. [CrossRef]
44. Al-Sabek, F.; Shostad, S.; Kirkwood, K. Preferential Attachment of Human Gingival Fibroblasts to the Resin Ionomer Geristore. *J. Endod.* **2005**, *31*, 205–208. [CrossRef]
45. Mukaddam, K.; Astasov-Frauenhoffer, M.; Fasler-Kan, E.; Marot, L.; Kisiel, M.; Meyer, E.; Köser, J.; Waser, M.; Bornstein, M.M.; Kühl, S. Effect of a Nanostructured Titanium Surface on Gingival Cell Adhesion, Viability and Properties against *P. gingivalis*. *Materials* **2021**, *14*, 7686. [CrossRef]

MDPI
St. Alban-Anlage 66
4052 Basel
Switzerland
www.mdpi.com

Polymers Editorial Office
E-mail: polymers@mdpi.com
www.mdpi.com/journal/polymers

www.ingramcontent.com/pod-product-compliance
Lightning Source LLC
Chambersburg PA
CBHW080938240926
51849CB00004BA/484